CSR, Sustainability, Ethics & Governance

Series Editors

Samuel O. Idowu, London Metropolitan University,
Calcutta House, London, United Kingdom

René Schmidpeter, Ingolstadt, Germany

For further volumes:
http://www.springer.com/series/11565

Christina Weidinger • Franz Fischler •
René Schmidpeter

Editors

Sustainable Entrepreneurship

Business Success through Sustainability

Springer

Editors
Christina Weidinger
Sustainable Entrepreneurship Award (SEA)
Vienna
Austria

Franz Fischler
European Forum Alpbach
Vienna
Austria

René Schmidpeter
Ingolstadt
Germany

ISBN 978-3-642-38752-4   ISBN 978-3-642-38753-1 (eBook)
DOI 10.1007/978-3-642-38753-1
Springer Heidelberg New York Dordrecht London

Library of Congress Control Number: 2013945182

Printed on acid-free paper

Springer is part of Springer Science+Business Media (www.springer.com)

Sustainability is in our hands!

Foreword

During the course of the last decade we have become increasingly aware of the fact that corporate social responsibility (CSR) is an integral part of a world in which competitiveness and sustainable development go hand in hand. In fact, CSR stands for a new kind of economic system that is based on value creation both for companies and for society, and in which the role that companies play is redefined. More and more European companies are integrating principles of social responsibility into their strategies and workflows. An ambitious CSR policy can have a positive influence on sustainable growth and job creation, while at the same time taking into account the interests of the company itself, its employees and stakeholders. As a result of the economic crisis, CSR is now even more relevant than before and is vital to the credibility of Europe's social market economy.

In this respect, companies should begin a process in which they identify, avoid and minimise the potentially negative effects of their activity on the environment. At the same time, they should maximise the value creation for the company owner or shareholders, other stakeholders and society in general. Therefore, companies should carry out a detailed risk-oriented evaluation (due diligence) that corresponds to the size and type of their activity.

A long-term strategic approach towards CSR, and the exploration of possibilities that innovative products, services and business models offer, can contribute to society's welfare and lead to higher-quality and more productive jobs. As corporate social responsibility requires a great degree of interaction with internal and external stakeholders, companies are able to foresee societal changes and to benefit from that.

CSR means going beyond legal regulations and implementing the social model of sustainable development. In today's world, an active commitment to social responsibility is equivalent to embedding social and environmental issues in business workflows. Some European companies are exceptionally advanced in their understanding of the risks and opportunities that are connected to various social problems, and have developed innovative business models through which they are able to offer solutions for some of these problems.

Corporate social responsibility and sustainability have become cross-divisional topics in an ever-increasing number of companies. The Commission recognises such good practices and strives to continue its support of these companies in their efforts to introduce social and environment-related innovations.

José Manuel Barroso
President of the European Commission

Foreword

During the last two decades three major developments have marked strongly our economic and social life: first, the fall of the iron curtain has laid the ground for the speeding up of globalised markets with a shift of production towards new markets as well as consumer behaviour. This opened enormous market potentials and new possibilities for entrepreneurs, but also uncertainties as it created sometimes greater instability in our working world.

Secondly, market growth and the related exploration of natural resources, the use of new chemicals and other substances together with a growing population have threatened our environment. Today, we see the consequences of climate change. These environmental impacts have consequences not only on our health, but imply as well a realistic threat to some regions in the world and their populations.

Finally, the financial and economic crises with the harsh implications that we are facing since 2008 have put into question our economic model. Speculation and focussing exclusively on financial profit has resulted in a disastrous impact not only on the financial and business sectors, but as a consequence, also on public budgets and on our social security system.

These major developments set the framework where we stand today. There is a huge debate on how to readjust our rules to put an end to the hollowing out of the real economy; the European Parliament has since then put forward important legislation which include measures on a stronger regulation of the financial and banking sector and strategies on the promotion of SMEs.

In the Treaty of Lisbon, the Member States of the European Union committed themselves to the principles of sustainable development and a highly competitive social market, aiming at full employment and social progress, and a high level of protection of the environment.

Politics can only set the framework for sustainable development. It is important that all actors of society contribute in a responsible and appropriate way. Business is the core of our society. Today, the European Union accounts for around 29 % of global GDP. According to forecasts, by 2050 the relative size of the EU in the world economy could be halved. These global developments and the upheavals of the last years should give us all the incentives necessary to build on sustainable entrepreneurship. Sustainable enterprises are solidly rooted in society; they have far-sighted business management which is characterised by long-term investment in training, infrastructure and innovation. They combine profit with social, cultural and environmental goals. In the future, sustainable entrepreneurship will be also decisive to how we can get more young people to work, but at the same time adjust the working reality to the demographic developments and give elder employees a place in the working world.

I welcome the initiative to award a prize for sustainable entrepreneurship as it gives just publicity for exemplary engagements; in this spirit this edition gives us – businessmen and politicians – an important impetus for a fruitful mutual debate.

Martin Schulz
President of the European Parliament

Foreword

Sustainable Entrepreneurship is another lens through which to view some of the questions that the European Commission addresses in its policy on corporate social responsibility (CSR).

The Commission defines corporate social responsibility as the responsibility of enterprises for their impacts on society.

Those impacts are usually positive, but can sometimes be negative. The Commission believes that both sides of that equation are important: maximising the positive impacts of enterprises on society, and identifying, preventing and mitigating possible negative impacts.

It is in the medium and long term that the excellence of European companies in the field of CSR will pay off. It will contribute to the creation of an environment in which our enterprises can grow, compete and innovate – to their own benefit and to the benefit of the countries and societies in which they operate.

A strategic approach to CSR is increasingly important to the competitiveness of enterprises. It can bring benefits in terms of risk management, cost reduction, access to capital, customer relationships, human resource management and innovation capacity.

For example, a small IT company competing for skilled graduates to join its workforce is likely to be more successful in attracting candidates if it can show that it has progressive policies in terms of employee welfare and professional development. Or, consider an oil and gas company, which is more likely to avoid costly public opposition to new operations if it works carefully to ensure that the human rights of affected communities are fully respected.

Because CSR requires engagement with internal and external stakeholders, it enables enterprises to better anticipate and take advantage of fast-changing societal expectations and operating conditions. It can therefore drive the development of new markets and create opportunities for growth.

In some consumer goods companies, for example, CSR becomes integrated with innovation. Through their cooperation with civil society organisations, such

companies can become more aware of the circumstances and health needs of poorer people, and can then help to develop new, commercially viable, hygiene or food products to help meet those needs.

By addressing their social responsibility, enterprises can build long-term employee, consumer and citizen trust as a basis for sustainable business models. Higher levels of trust in turn help to create an environment in which enterprises can innovate and grow.

I welcome the publication of this new book on sustainable entrepreneurship, which should help to further deepen our discussions on this important issue.

Antonio Tajani
Vice-President of the European Commission,
EU-Commissioner for Enterprise and Industry

Foreword

When our hunter-gatherer ancestors began to grow food crops and learned how to store it during times of plenty they gradually put more and more land under cultivation. Since that far off time humans have been destroying the natural world at an ever-increasing rate. The industrial revolution caused this destruction to accelerate. Meanwhile, the human population was constantly increasing, so that today, with more than 7 billion people on the planet, we face an ecological crisis. Habitats and animal and plant species are vanishing every-where. Mother Nature is resilient, but the time is fast approaching when she will be battered beyond her ability to restore herself. We are faced with a choice – to take action to protect the resources of our planet or carry on as usual, bringing more children into a diminishing world.

From one direction come the voices of those who put economic gain ahead of the interests of future generations, who believe that unlimited economic growth is an imperative for every country. There are millions of uninformed people who agree with them. And although there are also countless people who realize that endless economic growth on a planet with finite resources is, in the long run, unsustainable, they say and do nothing either because they refuse to change their comfortable lifestyles or because they feel helpless.

"We shall require a substantially new manner of thinking if mankind is to survive," Albert Einstein wrote. He understood what Mahatma Gandhi meant when he said "The planet can provide enough for every man's need but not enough for every man's greed." Only if we listen to those voices of wisdom can we turn things around: I still have hope that we shall.

A growing number of people now understand the need to protect natural resources, realizing that as we destroy animals and ecosystems our own future will also be affected. And this includes more and more corporate leaders who have realized that the materials they need from the developing world for their businesses are running out. This knowledge and understanding are so important for, as I often say, "only if we understand can we care: only if we care will we help: only if we help shall all be saved." Indeed, for today ethical values are moving into business, more people are speaking out for the poor, and the concept of fair trade has emerged. And, yes, fortunately nature is indeed amazingly resilient when we give it a chance.

I work with young people around the world. They are breaking down the barriers we have built between cultures, religions, nations and, above all, those between ourselves and the natural world. They are joining together, finding a voice, determined to make this a better world. What is needed is a critical mass of young people – the next parents, teachers, lawyers, politicians, and so on – who understand that while we need money to live, we should not live for money.

Finally, it is so important that we recognize that each one of us makes a difference – every day. If each one of us spends a few moments thinking about the consequences of the choices we make – what we buy, eat, wear, what we use for daily life – the cumulative impact on the planet will be huge. Knowledge and Understanding, Hard Work and Persistence, Love and Compassion: With these tools, linking head and heart we can, together, heal the world.

That's why I appreciate and truly support initiatives like the SEA that is raising awareness of these issues and hoping to move forward into an international arena. I wish Christina Weidinger all the best for this great project. May this idea change our minds, activate more people to take responsibility and, as a collective force, make our world a better place for our children and grandchildren and all future generations.

Jane Goodall
UN Messenger of Peace

Foreword

Fifty years ago nobody talked about 'Sustainable Entrepreneurs' or 'Sustainable Entrepreneurship' but if some scholars or practitioners did, nobody took notice of them or took them seriously. But today, our knowledge has gone through a reorientation process; which now means that our world yearns for sustainable entrepreneurs and the opportunities which flow from social entrepreneurship or eco-entrepreneurship as it is sometimes known.

What do sustainable entrepreneurs do? This is perhaps a question which naturally comes to the mind of anyone who is new to the field of 'Sustainable Entrepreneurship'. Let me borrow words from Schaltegger (2013) who argues that they are the twenty-first-century innovators 'who are opportunity oriented and aim to generate new products, services, production processes, techniques and organisational modes which substantially reduce social and environment impacts whilst simultaneously improving drastically man's quality of life'. Sustainable Entrepreneurs, Schaltegger notes further, destroy existing conventional, unsustainable production methods, products, market structures and consumption patterns through their convincing, superior and more sustainable offers (Schaltegger 2013). They are sustainable innovators.

A series of unexpected social and environmental disasters around the world prior to 1987 made the constitution of the Brundtland Commission on Sustainable Development inevitable. We are all too aware of the consequential effects of climate change, global warming and the scarcity and price increases of some of the natural resources in the world market. What about the recent global economic meltdown which continues to weaken the global economy and threatens the survival of many nation states? Corporate Social Responsibility (CSR) has now become a mainstream approach of recognising the social, environmental and economic impacts of business on society, and CSR continues to remind us what

we all need to do and ensure we do persistently in order to deal with all these social, environmental and economic issues of our time. It has become apparent that our world needs sustainable development–conscious managers and entrepreneurs who would take our world and its resources forward sustainably for this generation and our future generations.

The Chapters in this book by eminent scholars and world class business practitioners have competently attempted to provide us with the skills, knowledge and temperament required by sustainable entrepreneurs, sustainable managers and sustainable consumers of the twenty-first century. All these chapters have addressed issues of importance to all citizens of the world. I recommend the book unreservedly to all sustainable citizens of the world.

Finally, I congratulate Weidinger, Fischler and Schmidpeter for this fantastic contribution to the literature on sustainable development and for putting together these state-of-the-art contributions to sustainable entrepreneurship.

Samuel O Idowu
Editor-in-Chief, Encyclopedia of Corporate Social Responsibility
London Metropolitan Business School, UK

Reference

Schaltegger, S (2013) Sustainable Entrepreneurship (2013) in Idowu, S O, Capaldi, N, Zu, L and Das Gupta, A (Eds.), Encyclopedia of Corporate Social Responsibility, Springer, Berlin.

Preface

Introduction

The hype surrounding the buzzword 'sustainability' is currently assuming gigantic proportions. The fact that a Google search comes up with about 19.7 million results (as per 16th of July 2013) proves the overarching discussion about a sustainable future as well as sustainable business. Sustainability is on every agenda of corporate executives, and most managers are convinced that he or she is acting sustainably. The simple word of sustainability is right about to turn into a whole new industry in which a large number of CSR experts, PR and wording agencies, lawyers and management consultants build new business models, value chains and markets. One might raise the question whether these activities are really all sustainable or whether the whole discussion is only a facade – a bubble as many financial busts have proven to be.

Sustainable Entrepreneurship as a Business Strategy

In fact, sustainability as a concept seems already very much defined by the wider public. Sustainability describes the use of a renewable system in a way that ensures that this system is maintained in its basic characteristics and can renew itself naturally (Wikipedia, March 2013). The term originated in forestry but became known to the general public in the 1980s, in particular through the UN's Brundtland Report, called 'Our Common Future'. The topic was shifted even more to the centre of attention at the UN Conference on Environment and Development in Rio de Janeiro in 1992. Today we see sustainability primarily as a social, economic and ecological movement that meets the needs of the present without compromising the ability of future generations to meet their own needs (Brundtland Report).

Nevertheless, this perspective falls short when it comes to solving today's problems: (1) The aim of *not compromising* future generations has to be shifted to *safeguarding* them in the long term! And only those who think and act correspondingly are contributing towards the future. (2) A modern interpretation of sustainability needs also to bridge business and society in a constructive way. Only if business success and sustainability are considered to be two sides of the same coin can they foster each other. In order to become relevant for business, the creation of shared value needs to get into the focus of the sustainability discussion. Therefore, Sustainable Entrepreneurship is very much linked to the overall strategy of the company. It has to be integrated into the DNA of the company and not just as an add-on, as many CSR concepts have taken to be. Sustainable Entrepreneurship influences the whole company by widening its business scope from mere profit-driven goals to the creation of joint benefits and shared value. This is only possible if the innovation procedure of the whole company is reshaped towards an open as well as society-oriented search process that aims to find efficient sustainable solutions for the most pressing problems of our world. It becomes pretty clear that the current business models have to be expanded in order to foster product, service, process and management innovations more rapidly. The emergence of the debate on sustainable entrepreneurship has the power to transform not only our business organizations but also our societies as such. It increases the bottom-line results of corporations and creates societal and ecological value for people and nature at the same time. In this way business becomes part of the solution, rather than being considered part of the problem! Despite this positive view we are still at the starting point of the journey – in the status nascendi. So far sustainable entrepreneurship is just a change of paradigm and a new way of looking at businesses. Of course there are certain cornerstones like innovation, entrepreneurship and ethics, but how to link them together is still an ongoing discussion. Thus this publication can only provide the ingredients for the creation of new business models, and will not provide a recipe that fits to all. We are still at the beginning of a certainly steep learning curve providing the knowledge and know-how we need to create value for business and society at the same time. Trial and error will certainly be important as well as communicating already existing best practice. In the end it will be our collective mindset and experiences that will help our societies to overcome the most challenging problems through innovative business approaches. Currently we definitely need more rather than less entrepreneurship and business thinking!

About this Book

Sustainable Entrepreneurship might offer a new perspective on the relationship between business and society. Leading thinkers from the business world, academia as well as politics and civil society give their view on this new paradigm of sustainability. The current thought on sustainability, innovation and entrepreneurship is outlined by 35 pioneering experts. Linked together, these concepts build the basis for developing and exploring a new business paradigm that is able to foster economic, social and ecological values simultaneously. For the implementation of sustainable entrepreneurship, one needs not only the right perspective but also state-of-the-art instruments and inside knowledge. Both are provided and enriched by examples of application and in-depth statements by the authors of this book. The diversity of contributions offers a starting point in order to change our way of doing business. This book is for the new leaders of sustainability who are on the forefront in designing new business strategies, for anyone who wants to join the authors on their thought-provoking journey to a world that will be more sustainable and business-oriented at the same time.

Acknowledgements

We are grateful to all authors who have contributed to this volume. We know that all experts are very much engaged in a wide range of initiatives and projects in order to implement the ideas of sustainability, innovation and entrepreneurship. We really appreciate their time and expertise invested in this publication! Special acknowledgement goes to all of them as well as to the political and business supporters of the emerging concept of Sustainable Entrepreneurship. We would also like to thank all the promoters, partners, jury members and colleagues within the SEA initiative, who have very much contributed to this publication with numerous discussions and by sharing their personal insights into the current developments of sustainability. Our heartfelt thanks also go to Barbara Birkenmeyer, Stefanie Diem and Harald Hornacek, who supported the publication with their expertise, as well as Christian Rauscher and his colleagues from Springer, who gave not only their professional but also personal advice to this project.

Due to the continuous support of our business partners and supporters of the SEA, we dare to hope that it is only the beginning of the global spread and success of Sustainable Entrepreneurship. Our goal certainly will be to overcome today's unfruitful gap between business and society and make business enterprises an integral driver of sustainability. This is the least we can do for the next generation as well as for our own sake. Everyone who accepts our offer to join us on this way is welcome and we wish everyone the same fulfilment as we already had on the first miles of our journey towards a new and promising future. The mutual goal will be to develop an international network of high-level intellectual exchange. Together we can build the basis for a new business world in which enterprises are the key driver of sustainable solutions to the world's most pressing challenges. We hope that you will be part of it and wish you a brilliant start to your own personal journey into the world of Sustainable Entrepreneurship!

Christina Weidinger
Franz Fischler
René Schmidpeter

Contents

Part V   Looking Ahead

Linking Business and Society: An Overview

René Schmidpeter and Christina Weidinger

1 Introduction and Definitions

In recent years the approach of corporate social responsibility has been very much discussed. It started as a mere defensive/reactive approach (compliance oriented) and is now developing towards an innovative/proactive management concept (Sustainable Entrepreneurship). The term "Sustainable Entrepreneurship" recently emerged in the business world to describe this latest very entrepreneurial and business-driven view on business and society. Current definitions for Sustainable Entrepreneurship focus on new solutions or sustainable innovations that aim at the mass market and provide value to society. Entrepreneurs or individuals or companies that are sustainability driven within their core business and contribute towards a sustainable development can be called sustainable entrepreneurs, according to Schaltegger and Wagner (2011). Others argue that sustainable entrepreneurship stands for a unique concept of sustainable business strategies that focuses on increasing social as well as business value – shared value (Porter and Kramer 2011) – at the same time.

Although many argue currently that Sustainable Entrepreneurship has the potential to become the most recognised strategic management approach in our times, a lot of open questions do remain. This volume aims to provide underlying concepts of entrepreneurship, innovation and ethics in order to provide the pillars to further develop the concept of Sustainable Entrepreneurship. It certainly will not provide a single valid definition, but rather a framework of orientation for where the journey might go. However, the underlying assumption always is: those organisations that

R. Schmidpeter (✉)
Centre for Humane Market Economy, Salzburg, Austria
e-mail: rene.schmidpeter@gmx.de

C. Weidinger
SEA – Sustainable Entrepreneurship Award, Karlsplatz 1/17, Vienna 1010, Austria
e-mail: christina.weidinger@diabla.at

C. Weidinger et al. (eds.), *Sustainable Entrepreneurship*, CSR, Sustainability, Ethics & Governance, DOI 10.1007/978-3-642-38753-1_1,
© Springer-Verlag Berlin Heidelberg 2014

Fig. 1 Sustainable
Entrepreneurship – an
emerging business concept

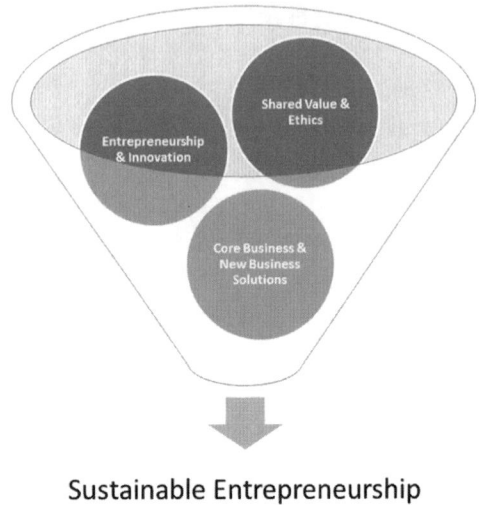

Sustainable Entrepreneurship

are able to develop business solutions to the most urgent social and ecological challenges will be the leading companies of tomorrow. Thus, Sustainable Entrepreneurship will not only be a key driver for our common sustainable future but also for business success. Sustainable Entrepreneurship is a progressive management approach to generate new products and services, management systems, markets and organisational processes that increase the social as well as the environmental value of business activities.

The main question thereby is how to increase competitiveness and economic value by integrating sustainability in the core business. Innovations will be core to achieving this alignment between business success and sustainability. These innovations do not come about automatically, but are rather initiated by entrepreneurs and managers who are leading their business in a new direction. We need new management approaches and processes that have a new normative paradigm: shared value instead of mere shareholder value. This ethical perspective provides a new way of shaping the role of business in society. This thinking provides a solid basis for a new capitalism where business is the main driver for social and ecological innovation. With this vision in mind sustainable entrepreneurs are remodelling the markets as well as the societies of the future (Fig. 1).

In order to further analyse the strengths and weaknesses of this upcoming management concept we need to look at theories and practices that already exist. Certain elements and foundations are already present in the field of innovation, entrepreneurship, ethics and sustainability. The aim is to get sustainability to the core of businesses by developing new business models. By linking these different fields of research and ideas we will be able to develop a new understanding of business and society. The following chapters certainly provide the necessary input and new thinking, but it will be up to the reader to draw the right conclusions.

Fig. 2 Content of the publication

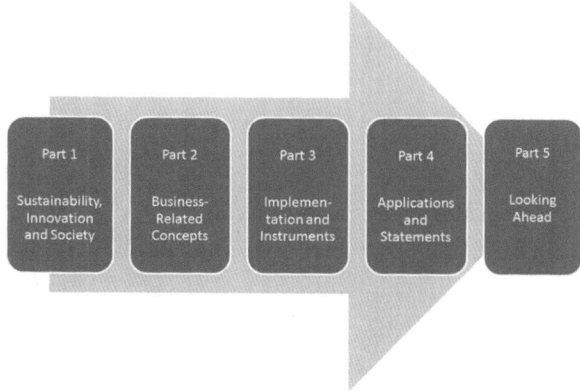

There is no one size that fits all – this is certain. The emergence of sustainable entrepreneurship is an innovation process in itself where the final outcome and impact lies in the future. Nevertheless, the best way to know more about the future is to shape it.

2 Overview of the Book

This publication is set up in five parts that give insights into the underlying concepts of sustainability and entrepreneurship as well as practical implementation tools and statements by recognised leaders from business, politics and civil society (Fig. 2).

The first part of the book describes the fundamental concepts of sustainability and innovation as well as their relation to the development of modern society. The second part explains the link between business and society and introduces different business concepts that integrate sustainability issues in strategic management thinking. The next part outlines practical approaches to fostering sustainable innovation as well as building corporate capacities to embed sustainable entrepreneurship in organisations. How Sustainable Entrepreneurship is applied in the modern business world and what youth, academia and civil society think about these new concepts can be learned from the statements of various leaders in their particular field. Last but not least, the cornerstones and future of a sustainable society as well as sustainable entrepreneurship are addressed in the final part. The fundamental questions of how egoism can be embedded in an innovative form of socio-economic thinking and how sustainability can be a driver for business success are elaborated.

Fig. 3 Sustainability,
innovation and society

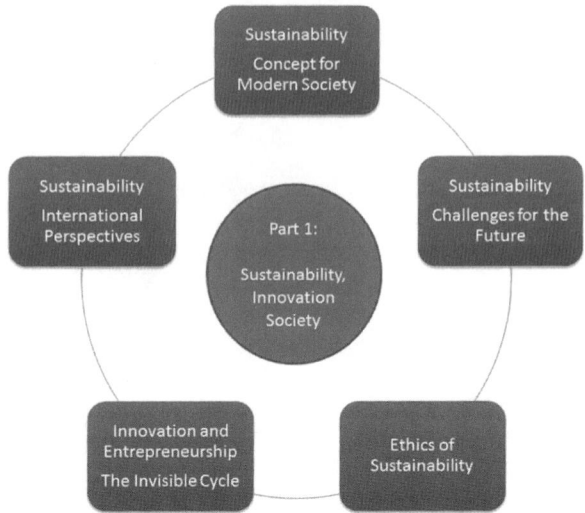

2.1   Sustainability, Innovation and Society

The first part gives a broad overview of theoretical thinking how sustainability can be integrated in society.[1] It introduces societal, sustainable, ethical, innovative and international perspectives on Sustainable Entrepreneurship (Fig. 3). Thereby it becomes clear that Europe has a pioneering role in the field of sustainability. It has always been trying to integrate the ecological and social factor into its economic model and has a long history of successful organizational models. In order to bring sustainability to a global scale we need intelligent solutions for the transformation from a non-sustainable society into a sustainable society. New approaches which do not lead to any losses in wealth, otherwise we will not remain politically capable of acting and maintaining social consensus. Management thereby has to be considered as a process that aims to the mutual advantage of all parties concerned. There is an invisible circle of innovation and entrepreneurship which determines whether business enterprises are successful or not. Only by understanding and leveraging individual differences as well as by fostering a trusting environment and soft values organisations are able to maximise their innovative potential. The task is to build new strategic business models that are lasting and sustainable.

Social and environmental issues have become imperative for businesses as well as governments on an international level. Significant achievements in the field of sustainability thus can only be achieved through collaboration between business, governments and NGOs. Business and its stakeholders need to work together in order to develop innovative business solutions to the most pressing problems.

---

[1] Contributors to the first Part: Franz Fischler, Estelle L.A. Herlyn and Franz Josef Radermacher, Robert B. Rosenfeld, Clemens Sedmak, Liangrong Zu.

Part 2: Business-Related Concepts

Sustainable Entrepreneurship — Driver for Social Innovation

Entrepreneurship — the Role of Business in Society

Sustainable Entrepreneurship — the Evolution of CSR

Sustainable Entrepreneurship — Next Stage of Responsible Business

Opening the Door to Opportunities — Real Sustainability Management

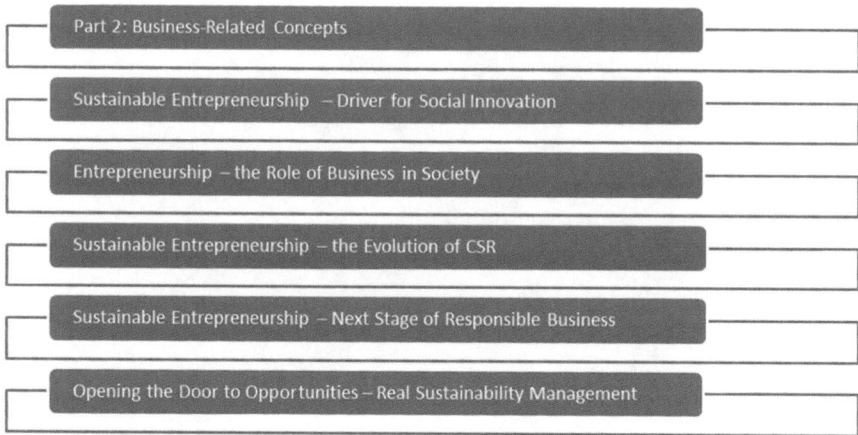

Fig. 4 Business-related concepts

2.2 Business Related Concepts

The second part of the publication describes different perspectives on how business models are able to integrate sustainability (Fig. 4).[2] It shows how different concepts have developed and how a new paradigm of Sustainable Entrepreneurship emerges in the business world. At the moment concepts of sustainable entrepreneurship and social innovation are becoming increasingly relevant to business, governments and NGOs worldwide. Social innovation becomes part of the regular innovation process within corporations. True leadership, open innovation, entrepreneurial spirit, change agents are important to build up an innovation for long-lasting success.

Especially in today's times of high uncertainty and insecurity, global trends are driving the change towards sustainability. The core business models are affected and new perspectives on value creation emerge. Innovation is the key for business success and sustainable value creation. Thus, the concept of corporate social responsibility has been transformed dramatically over the last couple of years. From a pure philanthropic perspective (sponsoring and donations) as well as a mere legal compliance approach it has been developed to an explicit responsible management issue. Now with the concept of sustainable entrepreneurship a dramatically new stage has arrived. It comprises the general question of the overall contribution of enterprises to urgent social challenges. This new strategic positioning of businesses in society aims at increasing social and business added value at the same time (shared value). Sustainable Entrepreneurship will be the next stage of responsible business. It becomes clear that companies are required to take a more active role than in the past when it comes to solving social and ecological problems.

---

[2] Contributors to the second Part: Thomas Osburg, Bradley Googins and Manuel Escudero, René Schmidpeter, Mara Del Baldo, Michael Fürst.

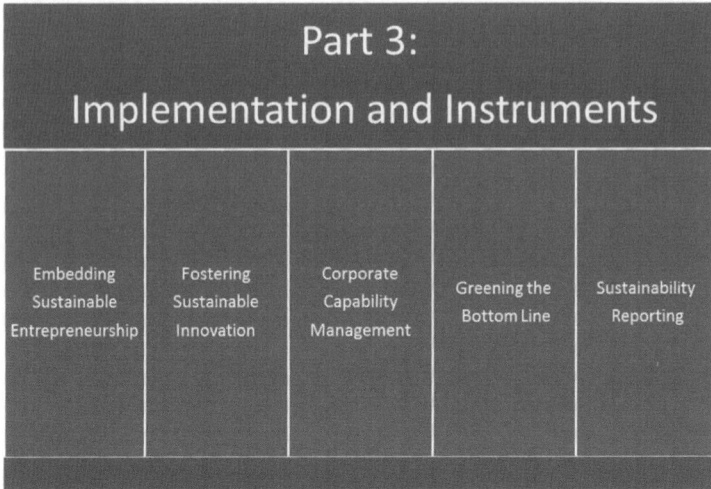

Fig. 5 Implementation and instruments

The concept of entrepreneurship applied to the question of responsibility and sustainability of business helps to reframe societal challenges into opportunities.

2.3   Implementation and Instruments

The third part of the book deals with approaches to implementing Sustainable Entrepreneurship. Questions answered are how to embed Sustainable Entrepreneurship, foster Sustainable Innovation, build-up a corporate capability management, how to green the bottom line and how to report on sustainability issues (Fig. 5).[3]

Research has proven that the stronger the business case of sustainability projects and strategies, the better the internal alignment within the organisation. Understanding the business relevance of social and environmental impacts is crucial in this perspective. Different drivers and types of strategic innovation are needed to embed a sustainability strategy successfully.

In most large organisations building structures and a culture of innovation is a challenging task and is only possible by applying new conceptual thinking. Innovative system have to use the most valuable resources – human creativity and imagination – to the fullest. Organisational leaders and managers have to decide now whether they want to be part of defining the future or leave it to the fast-growing

---

[3] Contributors to the third Part: Aileen Ionescu-Somers, Peter Vogel and Ursula Fischler-Strasak, Daniel Verlásquez Norrman, Martin Riester and Wilfried Sinn, Marc R. Pacheco, Matthias Fifka.

community of entrepreneurs. Innovative capability management approaches, enable businesses to access a broader field of idea sources by providing a blueprint for designing, implementing and operating a process that continuously improves capabilities step by step. Sustainable entrepreneurship thus can also be applied to greening the bottom line. New energy-efficient products, clean technologies and green jobs can be created through environmentally friendly legislation as well as innovative business models. Thus, the government can establish incentives and the framework within which the free market can expand and develop sustainable growth. The positive effect of sound sustainability policies on the economy and that intelligent sustainability approaches lead to a win-win situation for both society and business. These positive impacts can be shown by sustainability reporting. This field becomes more and more important. Companies that introduce sustainability reporting thus can gain a competitive advantage against those who are unprepared. This will strengthen the business case of sustainability for business and society.

2.4 Statements and Looking Ahead

The fourth part shows how the novel thinking of Sustainable Entrepreneurship is already applied and provides further insights from leading thinkers from business, politics, academia and civil society.[4] The various contributions clearly show that sustainability is affecting all parts of society and that a mutual approach has to be developed. It is about partnering in order to solve the urgent problems of our society. Entrepreneurship will be key to foster social and sustainable innovations (Fig. 6).

The last part of the publication has not only the goal to wrap up the different views, concepts and examples of the contributions in the book, but also to provide some further ideas as well as to look ahead into the future of sustainable entrepreneurship.[5] What can a successful concept of sustainable entrepreneurship look like and what are the drivers for its realisation? How far have we already travelled on the path towards a sustainable future? What is the role of business in shaping the future of our society? Why does Sustainable Entrepreneurship provide business success? The cornerstones of a new management concept are described and how it will change the future of doing business. Sustainable Entrepreneurship can be the spearhead in providing entrepreneurial solutions to the most pressing ecological, economic and social challenges (Fig. 7).

---

[4] Contributors to the fourth Part: Felix Finkbeiner and friends from *Plant-for-the-Planet*, Walter Rothensteiner from *Raiffeisen Zentralbank Österreich AG*, Markus Beyrer from *Business Europe*, Almgren Gunilla from *UEAPME*, Stefan Crets from *CSR Europe*, Jakob von Uexkull from the *World Future Council*, Claudia Kemfert from the *Hertie School of Governance*, Katherina Reiche from the *Federal Ministry for the Environment, Nature Conservation and Nuclear Safety*.

[5] Contributors to the fifth Part: Ernst Ulrich von Weizsäcker und Christina Weidinger.

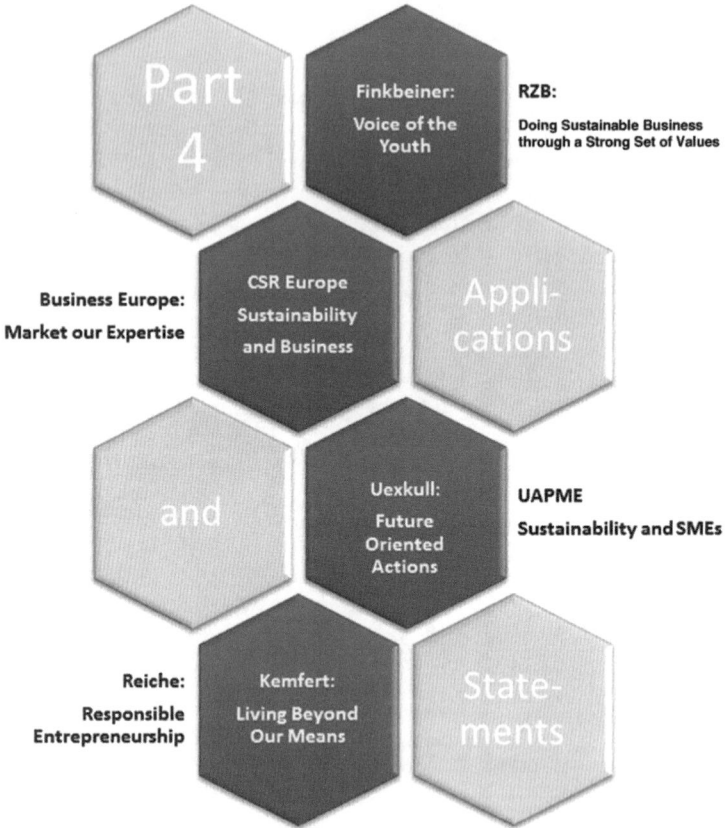

Fig. 6 Applications and statements

3 Conclusion

Although there are differing views on sustainability, one thing is clear: businesses will play a key role when it comes to making our societies more sustainable. Only through innovation, new business models and the creativity and imagination of entrepreneurs will we use all the capacities necessary to tackle world challenges. We are already on our way: (1) There has been an on-going discussion on corporate social responsibility – a discussion that contributes towards highlighting the positive role of business. Instead of defensive thinking, more and more businesses are applying sustainable business strategies that focus on increasing social and business value at the same time. (2) International standards and platforms of mutual engagement and exchange have been built up over the last decades. A mutual learning process and collaboration between businesses, politics and civil society with regard

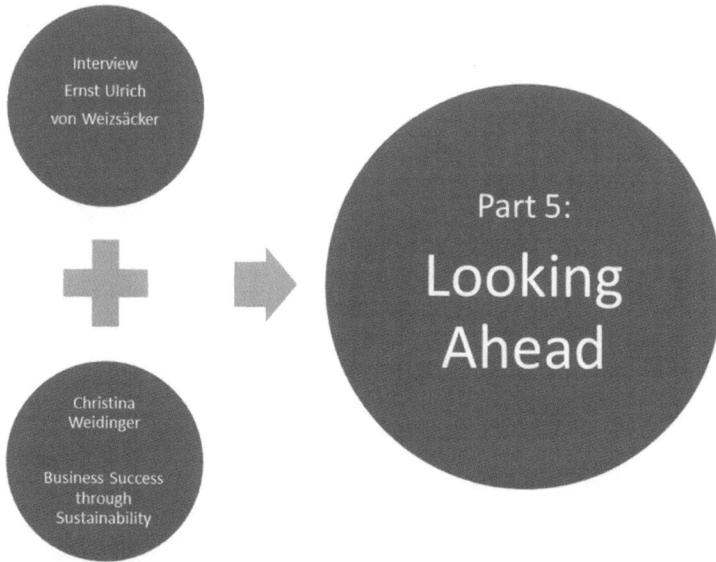

Fig. 7 Looking ahead

to addressing the challenges of today will be a key for developing an open and cooperative way towards sustainability. (3) In Europe the goal of the idea of a social market economy and eco-social market economy has always been to balance social, economic and ecological needs. Nevertheless, this thinking now has to prove itself in the context of globalisation and worldwide competition. It certainly needs to be further developed and adapted to the actual context. Especially the positive role of business in providing social end ecological solutions as well as being a co-creator of new international and regional frameworks has to be further considered. (4) Right now social innovation is emerging as a very powerful discussion when it comes to finding new solutions to our current problems. The EU Commission states the importance of innovations that are both social in their ends and in their means. We urgently need new solutions that address societal challenges in a new entrepreneurial way. To this end, Sustainable Entrepreneurship is seen as a key driver of innovative solutions for the development of a sustainable society (Fig. 8).

Sustainable Entrepreneurship should ultimately be described as the 'sweet spot' – to use an innovation management term – where business interests meet sustainability needs by developing new products, services and processes, new markets as well as new business and management models. A concept that is worth investing time and resources in, because it will not only help to safeguard the world of our children but will also lay down the base for the future business success of those who are part of this new movement.

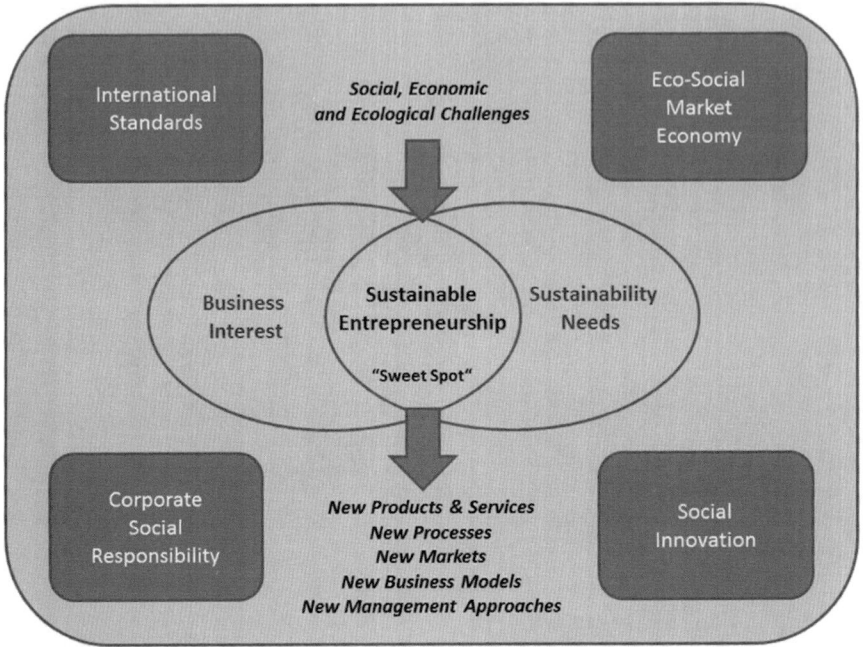

Fig. 8 Sustainable Entrepreneurship – an innovative management concept (Adapted from Eva Grieshuber in Schneider and Schmidpeter 2012)

Literature

Porter, M. E., & Kramer, M. R. (January–February 2011). Creating shared value. *Harvard Business Review* 3–17.
Schaltegger, S., & Wagner, M. (2011). Sustainable entrepreneurship and sustainability innovation. Categories and interactions. *Business Strategy and the Environment, 20*(4), 222–237.
Schneider, A., & Schmidpeter, R. (2012). *Corporate social responsibility.* Springer Gabler, Berlin Heidelberg.

Part I
Sustainability, Innovation and Society

Sustainability: The Concept for Modern Society

Franz Fischler

1 Introduction

When reading the daily newspaper and pursuing public debates, one may easily be inclined to declare the word 'sustainable' the negative buzzword of an entire generation. What isn't sustainable these days? From the forestry, agriculture and fishing sectors to the industry sector including the financial sector, tourism, airlines – just about everything must be sustainable or at least become sustainable. In the meantime, it is possible to make a sustainable impression with sustainability and sustainability has a sustainable effect on our thinking, sometimes also our actions.

Why then should this overused word be employed as the name for a future social concept, as indicated in the heading of this article? There are several reasons for this:

1. The current global problems, with which we have been grappling for decades now – one has just to think of globalisation, climate change, the population explosion in certain parts of the world and the senescence in others, the increasing scarcity of resources, the decline of biodiversity, among other things – make it increasingly clear that the existing economic, social and societal models have had their day and it is time to seek new solutions.
2. This new solution still lacks a name and the term 'sustainability' incorporates many elements that must play a role in the new model and be based on the concept of sustainability found in the Brundtland Report.
3. Whichever term is selected, it can only take shape through the description of the content.

F. Fischler (✉)
European Forum Alpbach, Franz-Josefs-Kai 13/10, Vienna 1010, Austria
e-mail: Franz.fischler@alpbach.org; christiane.schwaiger@alpbach.org

C. Weidinger et al. (eds.), *Sustainable Entrepreneurship*, CSR, Sustainability,
Ethics & Governance, DOI 10.1007/978-3-642-38753-1_2,
© Springer-Verlag Berlin Heidelberg 2014

2 Sustainability and Society

2.1 Origins of Sustainability

The term 'sustainability' has its origins in the forestry sector and was used for the first time by Hannß Carl von Carlowitz in 1713, exactly 300 years ago, in his book *Sylvicultura oeconomica* to explain the simple principle that you cannot harvest more wood from a forest than it can grow, if you wish to durably preserve the forest (Carlowitz 1713). Over the next 200 years, the word 'sustainability' remained a forestry term, and only since the Club of Rome and the publication of the book *Limits to Growth* and the resulting discussion has the widespread application of the word become common practice (Meadows et al. 1972). However, I well remember the initial difficulties in the use of the term 'sustainability'. As a young agricultural engineer, my request that the term 'sustainable agriculture' be included in the resolution to be passed at a congress in Copenhagen in the late 1970s was rejected on the grounds that the translators were unable to find an appropriate term for the French translation.

In 1987, the Brundtland Report brought the decisive breakthrough and a certain degree of clarity. It contains the following crucial sentence: "Sustainable development is development that meets the needs of the present without compromising the ability of future generations to meet their own needs." (Hauff 1987)

A second definition in the Brundtland Report is cited less often, however, as it is much more politically demanding. It reads: "In essence, sustainable development is a process of change in which the exploitation of resources, the direction of investments, the orientation of technological development, and institutional change are all in harmony and enhance both current and future potential to meet human needs and aspirations." This process of change is precisely what it must be all about if we want to move forward and make our societal, economic, and social systems fit for the future. To me, future viability is the most important qualifying feature for sustainability. In the last few decades, an alternative approach has often been used to describe future, not yet precisely recognisable social developments. Instead of giving the 'unborn baby' its own name, upcoming developments have simply been labelled with the name of the past period, prefixed with a 'post'. Thus, we speak of the post-industrial era and post-capitalism. More recent manifestations include post-democracy and the post-national era.

2.2 Sustainability in a Post-Industrial Society

In 1969 Alain Touraine used the term 'post-industrial society' for the first time. His core assessment was: "A new type of society is now being formed. These new societies can be labelled post-industrial to stress how different they are from the industrial societies that preceded them." (Touraine 1969) Daniel Bell then went one step further in 1973 and provided an initial description of the content, thus in fact ushering in the era of the knowledge-based society with his thoughts. He wrote: "The concept of a 'post-industrial society' emphasises the centrality of theoretical knowledge as the axis around which new technology, economic growth and the stratification of society will be organised." (Bell 1976) This is precisely what has long since happened in the former industrialised countries. Now, most of the hopes of growth in these countries are based on having an edge in knowledge and technology coupled with a prosperous development of the largest sector, namely the services sector. This development has been accelerated enormously by modern information technologies and, if the futurologists are to be believed, we are at the dawn of a new era.

Since the functioning of industrial society was directly linked to the functioning of capitalism, the question of which financial system should accompany the post-industrial era soon surfaced. Good old capitalism believed it would remain the only economic system for all time after the collapse of the Soviet Union and the demise of dialectical materialism and communist central planning, but very shortly after 1989 the weaknesses of capitalism became apparent, especially for post-industrial societies, with the liberalisation of financial and capital markets. Apart from that, the original function of banks, namely to serve as a service facility for the private and public real economy, has increasingly dwindled. It is not surprising then that even the protagonists of capitalism express doubts with regard to the continuation of the existing economic system. In his opening speech, C. Schwab, the architect of the World Economic Forum in Davos, raised the question: "What comes after capitalism?"

He wanted to indicate the need for an intense debate about post-capitalism. All in all, it is about the economic and social model of the future, which must definitely meet the following criteria:

- A sturdy balance between economy, ecology and social aspects must be achieved.
- Economic growth and the use of non-renewable resources must be decoupled.
- The economic system must remain stable even with very low or no GDP growth.
- A new understanding of growth must be developed, namely growth in quality, focusing particularly on increasing the quality of life.
- Innovations play a central role and become the main engine of growth. Therefore, knowledge-based economies rely on research and development and invest in people's minds.
- As a transfer site for goods and services, the market must continue to maintain this role, even win it back in certain cases; market participants must, however, also comply with state regulations, particularly in order to ensure a level playing field for stakeholders.

2.3 Quality of Life as Key Measure of Success

Precisely because quality of life is to be a key measure of success for the economic and social model in the future, it is no longer enough to focus solely on economic activities. Social life also includes the cultural dimension of our life together. In this context, culture is to be understood in very broad terms. It involves creating and maintaining our cultural goods, as well as everyday culture such as eating and drinking, cultivating relations with friends and neighbours, preserving the diversity of our landscapes, recreational activities, sports and much more. All in all, it is about home. Larissa Krainer, Rita Trattnigg and others are therefore calling for 'cultural sustainability'. (Krainer and Trattnig 2007)

In my opinion, it is quite justified to assign the name of 'sustainability' to this new model, because it contains all of the elements required by the Brundtland Report and has, to a certain extent, already proven its feasibility as well as the lack of alternatives.

Europe occupies a certain pioneer role in this area and already intensely discussed the European sustainability model several years ago in the convention on the issue of the Constitutional Treaty, codifying it in Article 3 of the target catalogue. The Article states (European Union 2008):

> The Union shall establish an internal market. It shall work for the sustainable development of Europe based on balanced economic growth and price stability, a highly competitive social market economy, aiming at full employment and social progress, and a high level of protection and improvement of the quality of the environment. It shall promote scientific and technological advance. It shall combat social exclusion and discrimination, and shall promote social justice and protection, equality between women and men, solidarity between generations and protection of the rights of the child. It shall promote economic, social and territorial cohesion, and solidarity among Member States. It shall respect its rich cultural and linguistic diversity, and shall ensure that Europe's cultural heritage is safeguarded and enhanced.

2.4 Sustainability and Europe

In 2010 these principles were cast into a concrete political strategy and are now implemented with the EU 2020 strategy. The core of this strategy is described as "smart, inclusive and green growth". (European Commission 2011) For the first time, verifiable quantitative objectives were specified in order to create a new balance between economic success based primarily on innovations, a high level of employment and a fair distribution of economic achievements, as well as a resource-saving environment and energy policy.

What has not yet been accomplished, or at best rudimentarily, in the context of regional policy is the greater integration of the culture factor into our economic model. This is probably linked to the fact that some attempts have been made to develop new parameters for measuring the qualitative logic of growth but these approaches have not yet found their way into economic analyses and growth forecasts. The main obstacle lies in our obsession with believing that everything that has to do with economics must be expressed in numbers. Thomas Sedlacek (2013) demands "that we learn to respect what cannot be expressed in numbers, the 'soft' aspects. That we learn to respect things like aesthetics or the value of a beautiful view. This, however, requires a huge shift away from the obsession with numbers and towards something that is not that easy to grasp."

It is to be hoped that we are not so tangled up in austerity versus prosperity debates that the current decade ends up being a lost decade, and that the viability of the new European economic and social model can be demonstrated.

But this will not solve all of our problems. Just as the nation state was developed as a suitable political model at the beginning of the industrial age and capitalism, a corresponding model must now be developed for the post-industrial and post-capitalist phase. It is therefore no coincidence that a debate on post-democracy and post-national Europe has arisen.

Colin Crouch (2004) coined the term post-democracy. He does not, however, use it to describe a new model of democracy, but to denounce the progressive democratic decadence of western societies. He points out the growing imbalance between the interests of society and those of businesses, and describes the frequent trend of politics increasingly becoming a matter of elites and lobbies while the general public's interest in politics diminishes further and further.

In the meantime, authors are entering the picture who not only bemoan the decline of democracies organised by nation states, but are also intensely focusing on post-national governance models. In Austria, the ideas of Robert Menasse (2012), culminating in the demand for the "invention a new post-national democracy", are being intensely debated.

In my opinion, such a post-national democracy should build on the European values established in the Lisbon Treaty and the European economic and social model, should provide a new dynamic division of labour between regions, nation states and the European Union and include democratic reforms. These reforms must strengthen civil rights and establish the primacy of politics over the economy without unnecessarily restricting its freedom. Furthermore, the principle of majority decision making should be established at all levels, the instruments of direct democracy expanded and efficient decision making and institutional structures created. This alone illustrates how great a task it would be to build a sustainable post-national Europe.

But back to sustainability in the sense of the new economic and social model, as described in the Lisbon Treaty and the EU 2020 strategy. This marks the right path towards the future and can secure Europe a top spot in a globalised world. The thing is, it is often underestimated how far we have to go from a programmatic model to economic reality.

The second part of this chapter therefore focuses on the key players in the game of supply and demand and the practical application of sustainability in enterprises.

3 Key Players in the Field of Sustainability

3.1 Legal Frameworks

There can be no doubt that both a clear legal framework as well as interesting incentives and the initiative of market operators are necessary in order to bring about the required sturdy balance between economy, ecology and society while including the culture factor. But what constitutes the right balance between the three? It cannot be determined in theory, but must find its ideal state over time. At present, the pendulum is swinging too much in the direction of state intervention; it should therefore be encouraged to swing back by strengthening the personal interest of market operators and making adequate use of private standard setting. It will certainly not be possible to get by in the future without competition rules, norms and minimum standards, neither for products and services, nor for labour and social standards or the environment. In doing so, these must usually be minimum standards that ensure the safety of people and the environment. Of course, these standards remain a 'moving target', depending on the progress in science and technology.

Other legal instruments can also be used to pilot towards sustainability. Here, we primarily need to think about taxes and social security contributions. Both serve to organise redistribution and secure social stability. This is not a new concept and was already introduced with the social market economy, but it will also be required in the future. How else should one organise the education and health system, safeguard against unemployment or accidents, or provide for old age?

In the wake of the growing need to curb resource consumption, the idea of using taxes to improve the sustainability of the planet has rapidly gained importance. However, the steering effect, e.g. for the consumption of fossil fuels, remains limited. This has to do with the fact that many states prefer to employ the so-called Floriani principle instead of setting binding objectives, and moreover gain unfair comparative advantages at the expense of countries that wish to achieve greater sustainability by implementing drastic measures. Occasionally, mistakes or abrupt changes are made in the design of measures targeted at reducing resource consumption, which then massively undermine confidence in such measures.

Incentives of various kinds are much more motivating, be they tax bonuses, investment incentives, dedicated concessionary loans or innovation funding. Incentives are now likely the largest area with which the public sector seeks to promote sustainability. In our country, we tend to overshoot to mark, resulting in a funding labyrinth in which hardly anyone can find their way.

All in all, it is a matter of establishing the frequently-used term 'level playing field', that fictional place where the balance of sustainability is to be born. This field must also contain a balance between the four dimensions of sustainability, as well as a balance between the relationship of state intervention and private entrepreneurship. Ignoring these principles not only threatens to stifle market forces, but also impedes the responsibility of economic operators, whether as consumers or producers, to live sustainably. Private freedom to act must therefore remain as large as possible and may be restricted only insofar as this is necessary for public services. The invisible hand of the market should be allowed as much freedom as possible and only be 'taken by the hand' if it starts to go astray.

This freedom is fundamental and also constitutes the foundation for giving sustainability a chance to be put into practice. Sustainability is not philanthropy or a kind of rigmarole used to prettify a company's economic activities. Sustainability is much more. It is a matter of ethical responsibility to make the economy sustainable. This benefits not only the entrepreneurs, but also their staff and everyone who purchases the companies' products and services. This means that sustainability must be incorporated into the supply and demand relationship as an additional criterion for determining the value of a product or service. Therefore, it makes sense to review sustainability performance via certifications and mark them in an appropriate manner in order to make the products comparable. In this way, sustainability performance can then bring about economic success. In order to be effective, sustainability must become an essential part of our economy. However, it is not possible to express this performance in a kind of 'sustainability currency', which makes its handling rather complex for the time being.

3.2 Sustainability and Corporate Social Responsibility

Thus far, the best way to put sustainability into practice in the free economy is by means of corporate social responsibility (CSR). The general societal objectives and the sustainability model of the Lisbon Treaty can thus be taken to the entrepreneurial level and integrated into markets. For this reason, various kinds of CSR activities have been carried out for years. These include voluntary social and environmental services, private sustainability standards, sustainable operations and sustainability-based business relationships culminating in new business models in which sustainability constitutes the core business.

The issue concerns nothing less than "the voluntary contribution of companies to sustainable development" (Veit Sorger 2012). I would expand this viewpoint further and also involve the customers and suppliers of the companies as well as consumers. The supplier-customer relationships and a growing awareness of sustainability on the part of consumers are providing the 'new economy' with great opportunities.

In 2001 the European Commission already defined CSR as "a concept whereby companies integrate social and environmental concerns in their business operations and in their interaction with their stakeholders on a voluntary basis." (EU Commission 2001)

In its communication of 2011, the Commission notes with satisfaction the progress made since the Green Paper of 2001.

- The number of enterprises committed to the rules of the United Nations' Global Compact grew from 600 in 2006 to 1,900 in 2011.
- The number of organisations registered in the community system for eco-management and auditing (EMAS) rose in the same period from 3,300 to 4,400.
- The number of European companies that publish sustainability reports in accordance with the guidelines of the Global Reporting Initiative climbed from 270 to over 850 within 5 years.

These developments demonstrate two things: firstly, the rapid rise in the number of companies committed to CSR is considerable; secondly, only a small number of companies participate in the official CSR programmes. There are many reasons for this: ranging from the sometimes low commitment of the member states, to imprecise definitions of CSR, a low level of awareness and a lack of clarity about the benefits of the CSR concept.

To mitigate these weaknesses and increase opportunities for CSR, the EU Commission (2011) proposes a number of additional considerations in its communication in 2011:

- A simpler definition, according to which CSR is "the responsibility of companies for their impact on society".
- Orientation according to international principles (OECD, Global Impact, ISO26000, etc.).
- The Commission draws attention to the multidimensional character of CSR.
- It calls for an enhancement of the role of public authorities and other stakeholders.
- Drawing attention to the great relevance of CSR for all companies.
- The need to step up the social dialogue about CSR.

Thirteen letters of intent from the Commission followed, detailing how it would like to achieve a widespread application of CSR. The EU expects considerable commitment from the member states and the adoption of national action plans. In Austria as well as in many other countries, such an action plan has not yet seen the light of day.

4 Sustainability as Core Business

The initiatives of the European Commission are laudable, but the crucial point remains the self-interest of stakeholders in CSR, particularly in pushing for its implementation in and by companies. This can involve many things, ranging from simple improvement measures to raise corporate social and environmental standards, increasing energy and hence cost effectiveness, integrating sustainability requirements in brands, quality labels and seals of approval, integrating CSR into the core business or even completely reorienting enterprises with the aim of creating sustainability as a core business. The protagonists of sustainability are therefore generally advised not to wait for political decisions but rather to be innovative, take the initiative and be that famed step ahead of the competition. Because this is precisely what will make sustainability pay off.

Literature

Bell, D. (1976). *The coming of post industrial society: A venture in social forecasting.* Kindle Edition.

Crouch, C. (2004). *Post democracy.* Oxford: Polity Press.

European Union (2008). Official Journal of the European Union, consolidated versions of the Treaty on the European Union and the Treaty on the functioning of the European Union, 2008/C 115/01 Communications and notices, Volume 51, 2008.

European Commission (2001). Green book: Promoting a European framework for corporate social responsibility (CSR), COM (2001) 366, Brussels 2001.

European Commission (2011). Communication to the European Parliament, the Council, the European Economic and Social Committee and the Committee of Regions, A renewed EU strategy (2011–14) for Corporate Social Responsibility (CSR), COM (2011) 681, Brussels 2011.

Hauff, V. (Ed.). (1987). *Unsere gemeinsame Zukunft, Der Brundtland Bericht der Weltkommission für Umwelt und Entwicklung.* Greven: EggenkampVerlag.

Köppl, P., & Engert, P. (Eds.). (2012). *Corporate Social Responsibility und Nachhaltigkeit, Vom Idealismus zur betrieblichen Realität.* Vienna: Linde.

Krainer, L., & Trattnigg, R. (Ed.). (2007). Kulturelle Nachhaltigkeit, Konzepte, Perspektiven, Positionen, Reike Wissenschaft Bücher, Alpen Adria University Klagenfurt.

Meadows, D., et al. (1972). *The limits to Growth: A report to the Club of Rome.* New York: Universe Books.

Menasse, R. (2012). *Der europäische Landbote.* Munich: Hanser Verlag.

Sedlacek, T., & Orrell, D. (2013). *Bescheidenheit – Für eine neue Ökonomie.* Munich: Hanser Verlag.

Sorger, V., quote from 2012, he was President of the Federation of Austrian Industries.

Touraine, A. (1969). La société postindustrielle, Paris Denoel.

von Carlowitz, H.C. (1713). Sylvicultura oeconomica.

Sustainability: Challenges for the Future

Estelle L.A. Herlyn and Franz Josef Radermacher

1 Introduction

Sustainability is today an overarching orientation line of world politics. However, debate and implementation differ heavily. The aim of this paper is to show that sustainable development is strongly coupled to implement a sustainable economy on a global scale. This challenge eventually means implementing the 35-year-old concept of a global eco-social market economy. A vital element in this concept is the market which creates competition under a given regulation and economic constraint system in order to provide goods and services. In addition, a second constraint system must meanwhile assure sustainability in its ecological and social dimension.

An eco-social market economy thus combines two constraint systems. The achieved economic performance continues to be measured via a GDP-like system. The goal of an appropriate global increase in living standards will be pursued in an eco-social market economy as well. However, all mentioned restrictions (such as the amount of allowed CO_2 emissions) will be taken into account. A formula known as fundamental identity derives from this: Market economy + sustainability = global eco-social market economy. Regarding the current debate on an "improved" definition of wealth and progress, we suggest to work with two rather than just one indicator.

This paper widely consists of a translation of the publication Herlyn, E.; Radermacher, F. J.: Ökosoziale Marktwirtschaft: Wirtschaften unter Constraints der Nachhaltigkeit, in: Rogall et al. (2012): Jahrbuch 2012 I 2013 Nachhaltige Ökonomie, Marburg.

E.L.A. Herlyn
FOM (Hochschule für Ökonomie und Management), Essen 45141, Germany
e-mail: estelle@herlyn.com

F.J. Radermacher (✉)
Universität Ulm, Ulm 89069, Germany
e-mail: Franz-josef.radermacher@uni-ulm.de

C. Weidinger et al. (eds.), *Sustainable Entrepreneurship*, CSR, Sustainability, Ethics & Governance, DOI 10.1007/978-3-642-38753-1_3,
© Springer-Verlag Berlin Heidelberg 2014

2 Market Economy: Regulated Competition

Experiences in the past have shown that the market is a central and unsurpassed factor in the creation of wealth. It is therefore an important ingredient of any sustainable future. In this context, the market is a "flexible" and "adjustable" concept which has developed many peculiarities over the centuries. An early form of the market was the barter trade which at times featured strict regulations. In historic Venice, for example, barter trade was only permitted through Venetian brokers while direct trade was prohibited. Ricardo's free-trade theory with its concept of a "free" barter trade economy for the benefit of all parties involved manifests a counter-reaction to this system (Samuelson 2004).

Over time, the market has developed from barter trade to a structured high-performance system which creates goods and services and allows for as well as implements innovation (Schumpeter 1912). The significance of money as a means of bartering and payment, as a means of value storage and value measure continually increased. Pure barter trades still exist today in the form of barter transactions. The global financial system gains an increasingly central role in today's modern world. Even the latest global financial crisis has not alleviated this phenomenon. The global financial system has a catalytic effect and greatly expands the production of goods and services as well as their worldwide exchange. Today's monetary and financial system enormously reduces transaction costs of economic trading. It allows for value transfer from today to the future, it provides what is known as rescheduling between short-term and long-term financing and allows for broad risk distribution and securing in real-economic processes. Regulating the financial system as a part of market economy is of vital importance due to the system's immense significance. It is monitored and greatly influenced by the nations. This sector features *supra-national regulations* to a vast extent.

Depending on the specific regulations, a great variety of market developments is possible. Markets may develop in the form of Manchester capitalism, social market economies or casino capitalism just as well as they may develop into mercantilism or state capitalism as it exists in China today. Market always means regulated competition. This bears analogies with athletic competitions: Competition always generates performance, i.e. efficiency – a good ratio of input and output, low costs, short times or large quantities. However, it is the rules which characterize the individual market with its specific peculiarities (and thus the efficiency) and the same applies for the manifestation of a sport.

The market-creating regulations set an *initial* market-structuring *constraint system*. They bear vital significance for a market to be able to perform. The four great freedoms (in their individual development) are an important part of the market-structuring regulations (Debroy et al. 2011):

1. Freedom of property
2. Freedom of contract
3. Freedom for innovation
4. Freedom to take out and/or grant loans

The creation of *innovations* is one of the vital contributions of markets in the long term as innovations have allowed and still allow us to broadly increase wealth. Nations nowadays subsidize innovation in competition amongst each other. They set technical standards, e.g. regarding emission standards for cars and thus greatly influence technological developments and the environmentally relevant parameters in automotive vehicles. They act as purchasing parties with vast purchasing volumes and thus as a demanding party. They push innovation in further sectors by funding military budgets.

The assertion of interests in markets is done as per certain legalities: Those with the greatest economic power and the greatest financial volumes have the best chances to assert their own interests. This is a principle contrary to democratic principles. In a democracy, each and every constituent has a vote irrespective of his economic possibilities. To believe that markets create democracy is illusionary. A market environment may create autocratic or plutocratic structures just as well. Societies under participative-democratic governance lean towards social market economies (Held 2007), towards a positioning of property for the public welfare and thus towards a regulatory policy and governance which meets the interests of the vast majority of people. There is a balance between the democratic principle oriented towards the interests of all people and the principles of the market oriented towards economic success. The necessity for such a compromise comprises the basis for good solutions in the form of social democracies and social market economies (Weizsäcker and Picht 1964).

3 Wealth and Economic Performance

In the context of a market economy, the terms of wealth and economic performance as well as their measurement play an important role. In a very common sense, wealth is a performance of civilizations (Kay 2004). It is based on *distribution of labor* and *cooperation* and builds upon the respective performance of previous generations. ("We are all dwarfs standing on the shoulders of giants.") Wealth comprises more than the goods and services produced in markets. The non-monetary forms of wealth which lay outside of markets, such as time and muse, an intact environment as well as functioning families with children, need to be added. It is difficult to quantify these complex dimensions. A GDP-like measurement of the economic performance can only quantify wealth by accounting for the goods and services produced in market-oriented processes in the more narrow sense. Currently, wealth measurement is exclusively based on the GDP. However, a number of national and international committees are currently working on alternative approaches.

The GDP is thus a central key figure for success in today's economic system. It is used to quantify the rendered economic performance. The criticism regarding this key figure in the context of the sustainability debate is, amongst others, based on the fact that the figure's almost unconditional maximization represents the primarily pursued goal without considering the ecological and social boundaries or constraints which result from sustainability objectives, planetary boundaries and various other human demands. However, due to the great significance of this key figure, the certainly justified criticism should not lead us to the wrong conclusion of simply abandoning the GDP as a key figure. In an eco-social market economy, a GDP-like key figure will continue to play a central role. The entire economy, however, needs to be framed by a superior system of restrictions and regulations which will enforce the ecological and the social aspects of sustainability. The creation of economic performance is thus a maximization task under social and ecological constraints in the sense of a mathematical optimization theory. The adherence with the constraints takes top priority. The constraints are to be adhered with at all costs albeit this may result in reductions concerning the economic performance measures such as GDP in some cases. This means in particular that we will distinguish between the measurement of the economic performance and the measurement of sustainability proximity. Integrating both into one key figure raises significant methodical problems of a very basic nature.

4 Growth: Change of the Economic Performance

The growth debate has become very emotional at the beginning of the twenty-first century. Do we want a post-growth economy? In the wealthy part of the world or worldwide? What are the chances, what are the risks? Or, to ask more basic questions: Does humankind need growth? Does the economic system need growth? Is economic growth the natural enemy of the concept of sustainability? What must a post-growth society look like? Do we need selective growth? What is to grow and what is to dwindle?

Objective analysis reveals that growth as per the traditional quantitative definition means change in the scope of economic performance with regard to the chosen type of measurement over a period of years. It is thus about the change of a figure quantifying a monetized economic growth, today this figure usually is the GDP or a derived GDP-like figure (such as the NNP instead of the GNP). Annual changes can be either positive or negative to the same extent, which means that a rotation of positive growth, consistency and negative growth is possible. All these options have occurred in market economies in the past.

The theory of markets does not necessarily require growth – as frequently claimed – for the market to function. However, it is a fact that "political business" and/or compromising among people or states with different objectives is much easier under conditions of growth. Also, in today's markets, growth would probably favor high employment levels, although this statement also implies a question mark.

Matters of distribution are usually easier to address when "the cake is growing" albeit the persisting claim of equal profiting during growth periods is to be seen critically and differentiated and has turned out to be incorrect in the end (Herlyn

2012). From the individual perspective, much more significance is assigned to the individual income than to the average GDP and/or the total economic income. Seen from a purely mathematical standpoint, the per-capitum GDP may even grow with declining population rates albeit the overall GDP is on the decline. In the past years, only the income of the most wealthy decile in Germany grew noticeably, despite moderate growth rates. Medium-level income remained unchanged for the most part while low-level income even declined (Heitmeyer 2011). In the USA, this development was even more dramatic.

The term growth thus defines the change in the appropriately quantified economic performance within the afore-mentioned constraint system type 1. For now there is no immediate factual connection to sustainability. The current challenge is to integrate sustainability into the existing system nevertheless as the current system is not sustainable despite all debates and activities. Important parameters, such as the global CO_2 emissions, rather point towards an ever-increasing deterioration of the status quo. The situation is equally bad regarding resources and energy, the global starvation issue, the "exploitation" of the real economy and the nations through an inadequately regulated financial system and the resulting debt crisis. The crucial positions of points must be set now to solve these issues. For reasons of comprehension and communication, the necessary incorporation of sustainability into the existing system of key figures should, under aspects of the present text, not be done via a radical change or even by abolishing the existing GDP definition but rather by means of integration of all economic activity into a further system of restrictions (type 2) which grants the compliance with ecological and social parameters. The reasons for this approach will be given subsequently.

5   Sustainability Expressed via a System of Constraints

Ideally, sustainability may be described as a system of constraints (e.g. with regard to acceptable CO_2 emission levels worldwide). Operationalization then requires a second constraint system for the economic, social and ecological sectors. (Note: Sustainability-oriented restriction and/or constraint systems may be disjunctive to the constraint systems for measuring economic performance.) Scientific literature as well as publications from the entrepreneurial and political sectors mention various approaches for the development of such constraint and/or indicative systems. Examples are the ecological footprint (Wackernagel and Beyers 2010) and the concept of planetary boundaries (Rockström 2010). A joint study by large companies displays the "non-sustainable global development" on the national level by considering the ecological footprint on the one hand and the Human Development Index (HDI) on the other hand (World Business Council for Sustainable Development 2010).

Questions as to consistency as well as global extendability and verifiability are always present. Albeit Germany is considered to be a global trailblazer in many aspects concerning sustainability, a sustainable Germany cannot exist in a non-sustainable world in the long term. This, however, only applies in relative consideration taking the economic performance into account. If all people lived like the people do in Germany, the eco-systems would immediately collapse. In terms of per-capita CO_2 emissions, Germany ranks significantly behind France, a consequence of the high proportion of nuclear power generation in France.

There are various forms of labor division when implementing sustainability. In order to achieve the stipulated restrictions, different actors have different means at their disposal. In this way, politics can help towards compliance with the stipulated regulations through regulatory instruments (product- and process-related legislation), market economic instruments (such as taxes, subsidies, certificates) and sideline instruments such as sanctions or cooperation with companies. On the company level, self-obligation plays an important role. The standards of the Global Compact, the Global Reporting Initiative or the ISO 26000 standards provide orientation. In the past, there have been cases where politics followed companies' lead and turned a previously voluntary obligation into a legal obligation. Even ethical ties from religion all the way to the concept of the "honest merchant", ideally up to the level of operational management, could have positive effects. As a consequence of the close watch kept by society and critical NGOs as well as consumers, especially large brand-name companies face considerable pressure and the obligation to turn to the issue of sustainability and provide transparent reports on their activities. They move towards sustainability due to the economic effects of this pressure – although, cautiously. Industry segment codes of conduct as for example implemented in the semi-conductor industry or as "Responsible Care" in the chemical industry are noteworthy as well.

We may expect the knowledge especially as to ecological parameters of sustainability and as to the urgency of compliance to expand over time. In an extremely dynamic world, the matter of sustainability carries the character of a *dynamic equilibrium*. New findings and necessities result in new demands as to sustainability which in turn must result in an appropriate expansion of the constraint system for sustainability. Furthermore, the principle of caution is to be taken into account.

On our way towards a sustainable world, we must expect the already precarious situation to call for risky measures. The principle of caution in this context means that the present problems cannot be tackled solely with the hope for technological progress (such as new energy sources or energy systems) as the probability for uncontrollable risk can in most cases not be kept at a sufficiently low minimum.

Sustainability may in general be operationalized if the described approaches are implemented, potentially at the price of considerable loss in wealth. Whether or not the global society can pull this off, is a totally different question. Furthermore, the operationalization process is everything but trivial for a number of reasons. Two urgent challenges, which both require an appropriate system of guiding rails and/or restrictions shall subsequently clarify this.

6 Worldwide Cap of CO$_2$ Emissions

There is a global consensus on the achievement of the 2°C goal. However, the international community of nations is unable to reach any mutual decisions in terms of climate protection which would be worth mentioning. Especially against the background of the WBGU's (*translator's remark: The German Advisory Council for Global Environmental Changes*) budget restriction it becomes obvious that any further delay in time in connection with the lowering of CO$_2$ emission levels will drastically aggravate the situation and thus present higher requirements in terms of a solution. The situation has meanwhile aggravated to such an extent that only a very elaborate contractually secured constraint system maintains the chance to still reach the 2°C target (Radermacher 2011). A "soft cap", agreed upon by the nations of the world as per the Cancún-Copenhagen-Compromise-Formula is an important factor and the main restriction in this regard. This cap is to be tied to a certificate system and a climate fund. This fund is to motivate the non-industrialized nations to participate as well. They will receive funding for economic, ecological and social development.

We may expect a "soft cap" to not be sufficiently low in order to meet the WBGU's budget restrictions and thus to reach the 2°C target and that a gap of 600–800 billion tons will remain to the allowable CO$_2$ emission levels. This gap could partly be closed without any loss in wealth in the north and with a wealth perspective in the south (negotiation gap). The governments of the world would have to decide annually on a dynamic cap which would induce a volume of CO$_2$ emission rights for suspension, which could be closed by interested parties. If the private sector, that is to say private persons, companies and organizations, funded such suspensions, the initial "soft cap" may turn into the most stringent cap still compatible with a growth perspective and yet suitable to find political acceptance. The stipulation of a second "more stringent" cap would have to be implemented dynamically in e.g. 1-year intervals, depending on the overall economic situation, the current efficiency level etc.

Moreover, the remaining open portion of the 600–800 billion tons of CO$_2$ emissions gap which amounts to 200–400 billion tons, depending on the method of calculation, must be withdrawn from the atmosphere (sequestration gap) (Radermacher 2011).

A *global reforestation and landscape restoration program* on 150 million hectares by 2020 and on 500 million hectares by 2050 would be especially suitable to close the sequestration gap. Such a program could also be funded by the private sector which has the opportunity (just as with the closure of the negotiation gap) to position itself climatically neutral. A comparable program for the maintenance of existing forest areas would have to be added. Delumbering currently creates an annual 6 billion tons of CO$_2$ emissions which must be avoided at all costs in the future. In order to allow for the described program, the definition of climate neutrality would have to be appropriately protected on the UN level and respective expenditures by companies would have to be deductable from the taxable income.

Fortunately, many private actors are now voluntarily, investing into such a program under CSR type consideration, e.g. Deutsche Bank, Deutsche Bahn, Paketdienst DPD, Bundesverband des Schornsteinfegerhandwerks etc., see also, for the private side Berliner Appell: Klimaneutral handeln (www.klimaneutral-handeln.de) and for the state of Hessen (Hölscher and Radermacher 2012) and for the branch (GdW 2012).

This example makes clear how complex the regulatory demands may become which can in the end serve to achieve sustainability. Simple solutions such as "the prices will regulate the matter" are generally not successful as they implement ecological matters at the cost of social balance. The poor are in the end kept from developing and from access to resources. In the context of the climate issue, a north–south-partnership must involve the environmental, social, taxation, financial and economic sectors in order to operationalize the reduction of the accumulated CO_2 emissions over 40 years below the WBGU's budget restriction through the described approach.

7 Balance of Income Distribution in a Global Perspective

Today it is common knowledge that the distribution of income plays a central role for the situation of a given society apart from the per-capita GDP key figure. At least in the realm of the OECD nations and their inequality level it is obvious that the prevailing inequality level of income is more important for the welfare of the society than the level of the per-capita GDP. Interestingly, a lower inequality level adds to the positive effects of a great number of other social parameters (Wilkinson and Pickett 2009). That is to say that the focus on one parameter is advisable to describe income distribution balance as an important aspect of social sustainability. This parameter in itself addresses an important aspect of sustainability and indirectly many other figures such as life expectancy, school performance or criminal statistics. The balance of income distribution not only bears great significance in the social sector but also in the economic sector. Interdependencies of income distribution and growth as well as income distribution and wealth are thus uncontradicted. It is obvious that the matter of distribution is characterized by diametrically different interests (Pestel and Radermacher 2003), (Herlyn 2012). The Indian law of full employment in the countryside (MNREGA) is of high interest in this context and must be honored (Jacobs 2012) (Eco-social Forum Germany, Advisory Council 2012).

8 Fundamental Identity

The term of sustainable market economy, which implies a combinability of the two great concepts of sustainability and market, inevitably raises the question as to whether the parallel implementation of both leitmotifs is possible on principal. Today's world is far from being sustainable. Among the representatives of companies and the civil society are those who doubt the potential parallelism of both concepts. Even more doubts exist as to whether there may in addition be growth (as per its current definition). If the uttered doubts were indeed justified, it would probably mean a disaster for humankind. In this case, the two goals of global environmental protection on the one hand and economic development of the non-industrialized nations of the world on the other hand, which were agreed upon during the UN conference in Rio in 1992, would have to be abolished. We would have to decide either for the goal of achieving a high wealth level for everybody, which would, however, result in the irrevocable destruction of the environment, or for a by far lower wealth level, which would be compatible with sustainability. The goal would have to be to aim at a lower wealth level as politically acceptable or as desirable or advantageous for other reasons (Miegel 2010).

A promising approach, which might make the combination of both concepts feasible, is the a little more than 35-year-old concept of an eco-social market economy. In order to succeed, it would have to be implemented without any loopholes worldwide. Correctly implemented, even (positive) growth which is compatible with sustainability may be possible in a context considering today's situation. The reasons for this will be given below.

Eco-social market economy is an operationalization of the idea of a sustainable market economy. It is a market economy which complies not only with the constraint system of type 1, which gives eco-social market economy its specific economic characteristics by producing goods and services, but also complies categorically and primarily with the constraint system of type 2 which not only grants sustainability but compels it. Today's wealth, enhanced by wealth growth in a developing world, can be maintained if we manage to at least maintain today's (monetized) production level for goods and services in the developing world despite the additional limitations enforced by the goal of sustainability, and if we manage to substantially increase the production level in the non-industrialized nations over the next decades. A GDP-like definition will continue to play an important role in asserting the development. Such a development is still possible from today's point of view and it is actually necessary if a sustainable world is to be successful in peaceful cooperation with a population of approximately 10 billion people as of 2050. Resource efficiency must improve greatly through technological and organizational progress, which corresponds to a decoupling of growth and resource consumption which has been a central concern of the Club of Rome for 35 years. For this purpose, the characteristics of wealth must clearly change from resource orientation towards service orientation and certain sufficiency demands must materialize via restrictions. Achieving goals, however, becomes increasingly difficult (see 2°C target). Each year, prospects of success diminish and the risk for inevitable loss in wealth increases the longer we delay an appropriately forced restructuring of society. We have clearly demonstrated this previously through the example of CO_2 emission levels.

Decisive factors to be considered are as follows:

1. The population level worldwide,
2. The per-capita GDP,
3. A resource efficiency parameter (which puts resource consumption in relation to a GDP unit),
4. The availability of resources at the current consumption rate,
5. The balancing of the income distribution.

A broadly acceptable distribution of the (global) wealth is thus a central guideline for the social aspects of sustainability. This aspect has meanwhile been adopted by a number of committees on the national and international level. The competition implemented by means of a constraint system of type 1 is the decisive driving factor for the generation of wealth. To which extent a society can or is willing to "submit" to such a driving force is a question of regulations within a constraint system of type 2. What is known as a boomerang or rebound effect must be prevented as well – increased resource consumption as a consequence of improved resource efficiency and the resulting decrease in price. This effect is easy to comprehend when considering the allegedly "paperless office" which is actually the location of the highest paper consumption in human history.

All considerations finally lead to what is called the fundamental identity:

Market Economy (or Wealth) + Sustainability = Eco-social Market Economy

The proof of the fundamental identity means proof of the compliance of market and sustainability (alternatively of wealth and sustainability) on the one hand and a global eco-social market economy on the other hand. It involves a huge list of arguments (Radermacher and Beyers 2011), (Pestel and Radermacher 2003) leading to an operationalization of the definition of sustainable market economy.

9   Green GDP: Green Economic Performance – Green Growth

Every market which meets the described constraint system of type 2 is a sustainable market. This is a fact, completely independent of the GDP definition which quantifies the production of goods and services while considering a constraint system of type 1. The constraint system of the market is additionally imposed with the constraint system for sustainability which enforces sustainability and takes priority in compliance. We can expect this to negatively affect the potential scope of produced goods (initially). Services, which are mainly characterized by dematerialization, are less affected. This also carries the possibility of a negative growth that is to say of a declining amount of all produced and monetized goods and services, especially at the point in time when the sustainability restrictions are strictly complied with. If we operate over night without an extended adaptation phase systemically within the framework of sustainability constraints, we should realistically expect a negative growth at first. The decision as to such an adaptation phase is up to society. Such a phase is probably not politically agreeable unless in the context of a disaster. For all other instances, we will choose a step-by-step approach.

Generating future positive growth despite compliance with sustainability restrictions would be great news and would require a massive increase in resource efficiency. We can expect such a positive growth rate to be smaller than it would be if we faced no restrictions. This is a consequence of today's "exploitation" at the cost of natural assets. The slower growth, however, would be compensated for by the long-term sustainability. Further "exploitation" will sooner or later lead humankind into neo-feudal structures or into ecological collapse (Radermacher and Beyers 2011).

If sustainability restrictions are complied with, any development of the GDP, whether positive or negative, will induce growth compatible with sustainability. This connection reflects the previously described fundamental identity. In contrast, the operational implementation of the sustainability restrictions is tremendously difficult. Politics and companies but also people have a hard time not only talking about their implementation but putting them into reality since they are aware of the negative consequences for the economic development. In addition, there are individual worries of no longer being able to compete if commencing to implement the changes required individually – a typical *situation of prisoner's dilemma.*

Especially if seen from the international standpoint, today's situation is unfortunately characterized by the players looking for loopholes in the form of lower environmental standards in developing countries, lack of climate protection in international sea and air traffic, child labor, low wages or much too rapid spending of remaining resources. The problems arising here are dealt with in literature under the heading of the "Trilemma of Globalization" (Rodrik 2012). The EU optically abates its CO_2 emissions by transferring CO_2-intense production to China. China has to take the "blame" afterwards. In such a constellation, the developing part of the world is unwilling to agree to the required international constraints, for example in terms of CO_2 emissions, especially if the wealthy part of the world shows no intentions for cross-funding, for example in the form of green technology transfer. The matter of transfer will – and there is no other way – assume a key role in the worldwide implementation of sustainability.

In the context of CO_2 emission levels, this affects financial mechanisms. A global reforestation and landscape restoration program, funded by the private sector of the wealthy parts of the world for the maintenance of the status of *climate neutrality*, for example, is a promising concept (Hölscher and Radermacher

2012). Another transfer matter is a better balance of income distribution, especially under global aspects. Funding of a global minimal daily allowance in this sector is overdue, for example funded by means of a global tax on the consumption of common resources.

10   Green Growth for Worldwide Wealth Is Still Possible

The previous explanations show that "green" growth is always possible, but can also be negative. A finite world implies that growth rates continue to decline. In positive cases, however, this does not exclude constant absolute growth and with a world population decreasing in numbers at some point in time maybe even lead to a relative annual growth.

So, why does this paper represent the statement that a perspective of wealth on the level of the industrialized nations is yet possible for 10 billion people by the year 2050? The statement is based on the EU-funded research projects Asis and Terra 2000, which date back approximately 10 years and were conducted in the context of the Information Society Forum of the EU (Mesarovic et al. 2003), (Radermacher et al. 2011). These projects aim at technological breakthroughs combined with greatly improved Global Governance and significant dematerialization. That is to say they aim at an increase in resource efficiency as addressed for years, amongst others, by the Club of Rome with the terms Factor 5 and Factor 10 (Schmidt-Bleek 1998), (von Weizsäcker et al. 2010).

One central topic is to avoid the boomerang or rebound effect (Neirynck 1994). We tap the asymmetric growth potential of developing countries in relation to developed countries (leapfrogging). For over 70 years, the combination of medium-level growth rates of approximately 1.5 % in the industrialized nations and approximately 7 % in non-industrialized nations will lead to an average global growth rate of approximately 4 % in a world then populated by approximately 10 billion people. The immense growth in population will have taken place in the non-industrialized nations. The resulting social balance will be compatible with sustainability and will approximately equal the social balance in the EU today. Poverty will have been overcome by then, the potential of whole humankind will have developed, women's and minorities' rights will have been broadly implemented. In this constellation, global population numbers will commence to decrease as of 2050. This will significantly improve the situation in terms of sustainability. Innovation processes will begin to slow down due to the achieved balance of wealth (re-discovery of slowness). The protection of the environment and of the resource basis is compatible with this perspective and is virtually fertilized by it. This applies as well for the reaching of the 2°C target including a reduction of the global CO_2 emission levels in the steady state to approximately 12 billion tons of CO_2 emissions as of 2050 (see the previously given notes for details).

The decisive matter, of course, is the matter of environment and resources. As, for example, repeatedly pointed out in the works of the member of the Club of Rome E.-U. von Weizsäcker, the annual increase of energy and resource efficiency by approximately 4 % is a prerequisite for this period. Such an increase seems feasible, especially considering the high level of eco-inefficiency of industrial processes in the non-industrialized nations. The lowering of climate gas emission levels worldwide to approximately one third of today's volume is also feasible in connection with further innovations in the energy sector. Approximately as of 2050 the global population level in this scenario will commence to decrease which also lowers the ecological pollution levels.

The relative costs for energy will remain acceptable to the people thanks to appropriate annual increases in price, which just equal the efficiency gains, the boomerang effect is avoided as well as the over-burdening of socially weaker people. Energy is a key resource in this context. In combination with the previously described global reforestation and landscape restoration program until 2050, the 2°C target may yet be attainable.

Fundamental innovations in the energy sector for green growth are a vital matter beyond today's existing solutions. We need to act much more bravely in this regard. The Desertec approach, promoted by the Club of Rome for years now, is a key technology which especially targets wealth improvements in North Africa. Solar chimney power plants could also be a key component. Intelligent grids, direct current high-voltage connections across great distances and potentially the conversion of power into methane for subsequent usage via gas distribution systems are especially important for improved distribution and transportation in the solar and wind sector. In addition, significant efforts should be made in the area of deep geothermics. Green growth for a wealthy world is possible, however, it calls for global empathy, a broadened view and broad-scope innovations in many technological sectors as well as in governance (Radermacher 2010), (Radermacher and Beyers 2011).

11   On Our Way to Sustainability

The situation to start from today is a non-sustainable society which nevertheless intends to become sustainable without having to face any losses in wealth. Politics must grant this if they want to maintain potential for consensus and thus to remain capable of acting. Finding intelligent solutions to this challenge is our task. The decisive figures to measure success are as follows:

1. Current system performance, today measured as GDP
2. Current distance to a status of sustainability

   1. Pursuing this aim in a sustainable world, we will try to close the gap to a status of sustainability year by year. For this purpose, we would have to monitor the degree of compliance with the constraint system of type 2 over

time. This process will last for several decades. For the important parameters of CO_2 emission levels and income distribution, this was indicated. The annual improvement factor will become a further restriction on our way to sustainability. As previously indicated, closing in on the status of sustainability will happen at the cost of system performance unless significant innovations allow for new dematerialized and low-energy value-added possibilities. Only when the status of sustainability has been achieved system performance can once again become a dominant key figure and efforts may be directed towards increasing this performance – if this still meets people's mentality.

2. The described path is interesting against the background of the current debate, for example in terms of transition to alternative energy in Germany. It is well possible for this transition to be feasible at extremely low increases in wealth and may be even result in losses in wealth. However, this option may be the only way in the medium-term perspective to once again reach a positive status while insisting on our current technology structure in order to avoid losses in wealth may well result in significant losses in wealth and crises in the long term. The matter of long-term planning is a key element of sustainability.

Literature

Debroy, B., Bhandari, L., Aiyar, S. (2011). Economic Freedom of the States of India, Washington; see: http://www.cato.org/economic-freedom-india/.

Eco-social Forum Germany, Advisory Council (2012). On the social dimension of sustainability. Balanced income distribution, adequate wages and vanquishing of hunger and poverty as key issues. Policy document.

GdW - Bundesverband deutscher Wohnungs- und Immobilienunternehmen e. V. (2012). Strategie der Wohnungswirtschaft zur Umsetzung der Energiewende. GdW Position, Berlin.

Heitmeyer, W. (2011). Die rohe Bürgerlichkeit, in: DIE ZEIT Nr. 39.

Held, D. (2007). *Soziale Demokratie im globalen Zeitalter*. Frankfurt: Suhrkamp.

Herlyn, E. (2012). Einkommensverteilungsbasierte Präferenz- und Koalitions-analysen auf der Basis selbstähnlicher Equity-Lorenzkurven – Ein Beitrag zur Quantifizierung sozialer Nachhaltigkeit. Buch zur Dissertation, Wiesbaden.

Herlyn, E., & Radermacher, F. J. (2010). Ökosoziale Marktwirtschaft – Ideen, Bezüge, Perspektiven, Internal FAW/n Report, Ulm.

Hölscher, L., & Radermacher, F. J. (Eds.). (2012). *Klimaneutralität – Hessen geht voran*. Wiesbaden: Springer Vieweg.

Jacobs, G. (2012). From freedom movement to prosperity. The role of human and social capital in India's development.

Kay, J. (2004). *The truth about markets. Why some nations are rich but most remain poor*. London: Penguin.

Mesarovic, M., Pestel, R., Radermacher, F. J. (2003). What future? Contribution to EU Project TERRA.

Miegel, M. (2010). *Wohlstand ohne Wachstum*. Berlin: List Taschenbuch.

Neirynck, J. (1994). *Der göttliche Ingenieur*. Renningen: Expert Verlag.

Pestel, R. & Radermacher, F. J. (2003). Equity Wealth and Growth – Why Market Fundamentalism makes Countries Poor, Manuskript zum EU-Projekt TERRA, Ulm.

Radermacher, F. J. (2010). Energie und Klima – Chancen und Risiken, Beitrag für die Jubiläumsausgabe der Fachzeitschrift SEV/VSE (Fachzeitschrift und Verbandsorgan des Schweizerischen Elektrotechnischen Vereins).

Radermacher, F. J. (2011). Wege zum Zwei-Grad-Ziel – Wälder als Joker, in: Politische Ökologie 127.

Radermacher, F. J., & Beyers, B. (2011). *Welt mit Zukunft*. Hamburg: Murmann.

Radermacher, F. J., Riegler, J., Weiger, H. (2011). Ökosoziale Marktwirtschaft – Historie, Programm und Perspektive eines zukunftsfähigen globalen Wirtschaftssystems, München.

Rockström, J. (2010). Planetary boundaries, in: Nature.

Rodrik, D. (2012). *The globalization paradoxon: Democracy and the future of the world economy*. New York: Norton.

Samuelson, P. A. (2004). Where Ricardo and Mills rebut and confirm arguments of mainstream economists supporting globalization. *The Journal of Economic Perspectives, 18*, 135–146.

Schmidt-Bleek, F. (1998). Das MIPS-Konzept. Weniger Naturverbrauch – mehr Lebensqualität durch Faktor 10, München.

Schumpeter, J. A. (1912). *Theorie der wirtschaftlichen Entwicklung*. Berlin: Duncker & Humblot.

Wackernagel, M., & Beyers, B. (2010). *Der Ecological Footprint. Die Welt neu vermessen*. Hamburg: Murmann.

Weizsäcker, C. F., & Picht, G. (1964). *Bedingungen des Friedens*. Göttingen: Vandenhoeck & Ruprecht.

Weizsäcker, E. U., Hargroves, K., & Smith, M. (2010). *Faktor Fünf: Die Formel für nachhaltiges Wachstum*. München: Droemer.

Wilkinson, R., & Pickett, K. (2009). *The spirit level – Why equality is better for everyone*. London: Penguin.

World Business Council for Sustainable Development (2010). Vision 2050 – The new agenda for business.

Innovation and Entrepreneurship: The Invisible Cycle

Robert B. Rosenfeld

1 Introduction

Why is it that some corporations thrive for decades, only to diminish into financial struggles, demoralizing layoffs, and often, bankruptcy? Perhaps there exists an invisible force or phenomenon that has eluded the leaders of yesterday and today, not unlike microscopic organisms, that were made visible by the advent of the microscope in 1590. If we could turn a "creative lens" on these invisible forces and phenomena, we may discover an opportunity to revitalize the organization and restore it to a time of achievement and success.

As an organization ages and grows, it becomes more difficult to sustain entrepreneurship. There could be an 'Invisible Cycle' that challenges sustainability. Recognizing the cycle could lend a beginning to the process for promoting sustainability. Exposing its microscopic, moving parts and the social element of individual preferences could provide the opportunity to encourage and sustain business growth.

2 Corporate Life Cycles

In his book, *Barbarians to Bureaucrats: Corporate Life Cycle Strategies*, Lawrence Miller provides insights into an invisible cycle. He explains the seven stages of the Corporate Life Cycle and the seven corresponding leadership archetypes that dominate each stage. "During growth, leaders respond creatively to challenges. During decline, they respond mechanically, relying on responses that have been successful in the past...Creative response is the essential function of leaders.

R.B. Rosenfeld (✉)
Innovating Consulting, 693 East Avenue, Rochester, NY 14607, USA
e-mail: gerti.neubauer@aon.at

C. Weidinger et al. (eds.), *Sustainable Entrepreneurship*, CSR, Sustainability,
Ethics & Governance, DOI 10.1007/978-3-642-38753-1_4,
© Springer-Verlag Berlin Heidelberg 2014

The moment leaders relax and rely on yesterday's successful response in the presence of today's challenge, the decline begins" (Miller 1989, p. 2).

Stage 1: Inspiration and Innovation
Stage 2: Crisis and Conquest
Stage 3: Specialization and Expansion
Stage 4: Systems, Structure, and Security
Stage 5: The Tight Grip of Control
Stage 6: Alienation and Revolution
Stage 7: The Synergist Prescription

What remains clear is that corporations tend to move through this invisible cycle, and while not annually, it is similar to the four seasons (spring to summer, summer to fall, and so on). Many corporations remain in decline because they are unable to determine their current cycle. To return to a state of growth, it is helpful to recognize the signs of impending decline.

One sign of decline can be defined as an overly "collaborative" style of decision-making. Collaboration is usually a good thing. In fact it fosters trust, which is a critical element to innovation. But when taken to an extreme, collaboration can devolve into bureaucracy and can even become an obstruction. Collaboration can imply that people, teams and/or departments are more or less equal partners in decision making and problem solving. Collaborative teams function "democratically", making decisions by seeking input from each member. As a result, this inclusive process tends to be very time consuming. The feedback can eventually evolve into 'rules,' 'by-laws' and 'analysis paralysis.' That is not to say that collaboration is inappropriate. It is merely a reflection of being in a late stage in the 'invisible' cycle.

Once these cycles become visible and recognizable, evolution commences and challenges abound. The most jarring challenge is to alter the directional flow of the cycle by trying to return the organization to earlier stages of the cycle, to its *entrepreneurial beginnings,* by *swimming upstream to spawn and returning to the shallows.*

Entrepreneurial decision making is more "monocratic" in nature, having a leader who makes decisions which team members scramble to fulfill. One important element of an entrepreneurial centric organization is the ability to produce more revolutionary (out-of-the-box/unique/disruptive) ideas that hopefully lead to real innovation.

Important note, coming up with a unique and creative idea is one thing, but executing and turning that idea into an actual innovation (something of quantifiable gain to the organization), is a much more complicated process.

In the beginning, the transition from the initial entrepreneurial style to a more collaborative style can be a challenging process, but is one that naturally occurs as organizations evolve. Once that natural transition has played out and an organization has arrived at the more collaborative style, reverting back to the entrepreneurial style is both complex and challenging. Creating an environment where BOTH styles coexist is important. BOTH styles are necessary for sustainability, but having both styles often create conditions for conflicts.

Collaborative-centric organizations react to entrepreneurial activities and values like oil reacts to water. Entrepreneurs are deterred and frustrated by the collaborative decision making style because this practice tends to take on less risk and moves at a deliberately slow pace. Entrepreneurs can see this collaborative behavior as a 'barrier-to-success' and 'too late'. Likewise, collaborators are often uncomfortable with entrepreneurial action-based, fleet-footed, shoot-from-the-hip decision making. Entrepreneurial style can appear somewhat reckless and inconsiderate of the inclusive behavior inherent in collaborative teams. Also, expansionary and revolutionary innovations, key elements for many entrepreneurs, are shunned by collaborators who prefer to maintain the status quo.

Assuming the organization needs to facilitate entrepreneurial behavior, expansionary or revolutionary innovation and fleet-footed decision making, how can these seemingly incompatible behaviors coexist? Businesses require a combination of entrepreneurial and collaborative behaviors in order to flourish. How does one unite two diverse styles and cultures to revitalize growth without damaging the organization? How do we enable entrepreneurial spirit to thrive in a historically collaborative environment?

One answer is to provide a catalyst for breakthrough success by utilizing and LEVERAGING the strengths and differences between the two cultures (and the people who enjoy each culture), not effacing them. The active deployment of tools that enable us to recognize these invisible forces and also to understand the positive application of differences allows the organization to learn the art and science of leveraging differences. This can provide the means for diverse teams to work together productively and promote revitalization.

3 Leveraging Differences

In 1934, Walt Disney assembled all of his artists in an empty sound stage and acted out his vision for a full-length animated film.[1] This became the script for the film that his brother, Roy, and his wife, Lillian, tried to talk him out of doing. Most of the entertainment world referred to his production as "Disney's Folly." But in December 1937, *Snow White and the Seven Dwarfs* was released. (Bennis and Biederman 1997) (Walt Disney: A Biography 2008) (Snow White and the Seven Dwarfs 2009).

Much has been written about the technical and business issues related to the production of *Snow White*. The film used new technologies, including rotoscope,

---

[1] This chapter is adapted from Rosenfeld et al. (2011). Leveraging Differences. *The Invisible Element: A Practical Guide for the Human Dynamics of Innovation* (pp. 53–73). Rochester: Innovatus Press.

to provide more realistic human animation and the multiplane camera to add depth. Walt initially estimated that the film could be produced for $250,000, or about 10 times the budget for producing a typical short film at that time. Actual production costs exceeded $1.7 million. Walt was betting the future of Disney Studios, and even his own house, on the success of the film. Over Walt's objections, his brother Roy showed a partially completed portion of the film to Bank of America. After viewing the film, a call was made by the banker: "Give Mr. Disney the money." The initial release of the film brought in about $4.2 million in revenue.

One arena that received far less publicity was the blending of unique talents utilized to create *Snow White*. While Walt provided the creative genius and Roy provided the business acumen, it also took hundreds of artists, sound people, photographers, etc., to turn the creative idea into a breakthrough innovation. Walt hired the best artists he could find from around the world. He allowed them to pursue their own passion. While they could have drawn most anything, Walt encouraged them to focus on what they were most passionate about: faces, people, animals, etc. His only requirement was that they do it extremely well.

What Walt Disney demonstrated was an understanding that people have unique skills and passions. He allowed people to pursue them in an environment that treated people as individuals rather than interchangeable parts. He allowed the potential for greatness to emerge.

Almost 70 years later, Hong and Page demonstrated mathematically that a diverse group of intelligent problem solvers will outperform a non-diverse group of the best problem solvers, thus supporting Disney's approach. (Hong and Page 2004). Embedded in this notion is the understanding that looking at how a person thinks and valuing diversity of thought processes is more important than merely measuring a person's IQ or examining educational credentials.

Only by understanding and leveraging individual differences, coupled with fostering a trusting environment and appropriate soft values, can innovative potential ever be truly maximized.

Leveraging differences is vital for success today. In a lot of respects, the world is becoming flat. But when looking at individual differences, the world is anything but flat. According to management guru Peter Drucker, "The most important contribution management needs to make in the twenty-first century is creating a 50-fold increase in the productivity of knowledge work and the **knowledge worker**." (Drucker 2008).

Our experience has shown us that many managers struggle with leveraging the knowledge worker. They do not recognize there is an invisible difference between the manual worker and the knowledge worker. Knowledge workers think, act, and behave differently. Their wants and needs are also different. All of these differences dictate the need to create specific environments, systems, and processes so knowledge workers can be most productive.

How do you leverage these invisible differences? By matching people to the appropriate task, problem, or job.

Individuals are not interchangeable components; they each have unique preferences, skills and capabilities. Understanding these is critical to having them in the most effective roles.

Fig. 1 Individual attributes fall along a continuum from visible to invisible. They are all important in understanding and leveraging the unique skills that each of us have. How well we leverage them can either support the creation of something new or kill it!

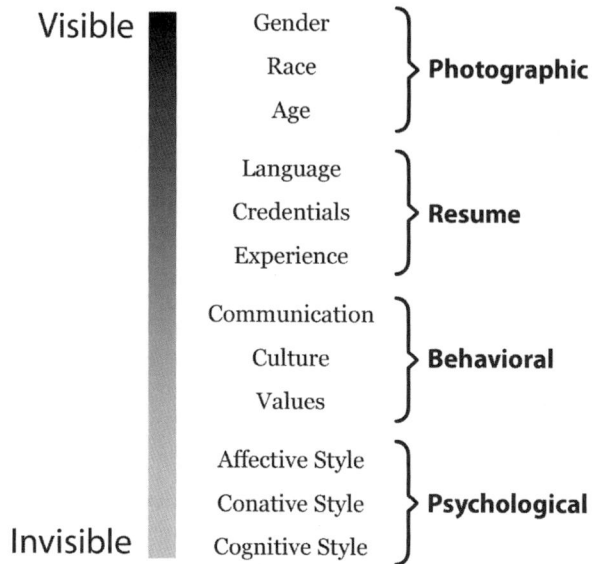

Visible

Gender	⎫
Race	⎬ **Photographic**
Age	⎭

Language	⎫
Credentials	⎬ **Resume**
Experience	⎭

Communication	⎫
Culture	⎬ **Behavioral**
Values	⎭

Affective Style	⎫
Conative Style	⎬ **Psychological**
Cognitive Style	⎭

Invisible

It doesn't make sense to have people who are "process-driven Six Sigma experts" trying to create out-of-the-box revolutionary concepts. It also doesn't make sense to assign an individual with a predisposition towards revolutionary innovation the task to streamline an existing process. Yet, we continually see managers trying to cut-and-paste people into roles for which they are not a natural fit. If you use the wrong people for the wrong thing, you're going to get the wrong results.

To leverage individual differences, each employee's visible attributes and skills must be analyzed and understood, including their invisible predispositions. The following figure illustrates the individual attributes that fall along a spectrum ranging from the visible to invisible. That is, there are attributes which can be readily seen as a "photograph" and are very visible. At the other end, there are psychological attributes which are largely invisible to most of us (Fig. 1).

When teams are selected today, most of the attention is given to the visible (or upper) end of the scale: Very little attention is devoted to the lower (or invisible) end of the spectrum: Do we have the right values mix or the most appropriate psychological mix? Do we have people most comfortable with revolutionary innovation focusing on Six Sigma? Do we have people most skilled at evolutionary innovation working on breakthrough ideas? Only by considering attributes at the invisible end of the spectrum can we select the best team members.

To begin understanding individuals, we need to have a basic understanding of their psychological makeup, as presented via three axes utilized in psychology (Huitt 2001):

- **Affect** – *To feel*: It is associated with a person's emotional state.
- **Conation** – *To act*: It is an aspect of a person's mental processes or behaviors directed toward action or change, including impulse, desire, volition, and striving.
- **Cognition** – *To know*: It is how people take in and process information, and how they put that information to work to make decisions and solve problems.

4   Tools and Instruments

There are numerous tools or instruments that may be used to differentiate between people through their affect, conation, and cognition. In our work, we have used many – such as the Myers- Briggs Type Indicator (MBTI), Kirton Adaption-Innovation Inventory (KAI), Fundamental Interpersonal Relations Orientation-Behavior (FIRO-B), Strength Development Inventory (SDI), Intercultural Development Inventory, and the Kolbe System, among others. While these are excellent tools to help address specific questions, until the ISPI™ was developed, there was not an effective way to integrate the information from the individual tools into a composite picture. Without this integrated view, leveraging the information was difficult for the innovation leader.

The Innovation Strengths Preference Indicator® (ISPI™)

We have utilized the theories that lie behind all of the previously listed tools with success. However, they were not developed with the innovator in mind. Therefore, we decided to create a tool for the innovation leader and individual innovators using similar principles but in different applications. The Innovation Strengths Preference Indicator® (ISPI™) is a tool that combines the three different psychological axes into a single indicator. It highlights an individual's predispositions toward a certain type of innovation, as well as how they prefer to interact with others. The results of the ISPI™ cover 12 different orientations. They are:

Innovation Orientation™ (iO™): How You Prefer to Innovate

1. Overall ISPI™ (your total for Ideation, Risk, and Process)
2. Ideation (your approach for generating new ideas)
3. Risk (your approach for taking risks)
4. Process (your approach for establishing and following processes)

The iO™ scale is explained on a scale from Extreme Builder to Extreme Pioneer.

A "builder" may be seen as linear, methodical and organized. They tend to focus on optimizing things and doing things better. They accept the problem definition and work within the problem definition. They can excel at (and enjoy) work requiring attention-to-detail over long stretches of time. Builders tend to enjoy cultures that are fixed, defined and less risky. They also may prefer innovation that is more evolutionary (structured/process-based/incremental).

A "pioneer" may be seen as unconventional and spontaneous. They tend to challenge the current paradigm and problem definitions. They like to create really new things and tend to do things very differently. Pioneers may have creative ideas that can be the kick-start to revolutionary innovations (but they are unlikely to be great at implementation and follow through).

Innovation Orientation Modifiers™ (iOM™): How You Prefer to Innovate with Other People

5. Control (your approach for taking charge or allowing others to do so)
6. Relationship (your approach for establishing personal relationships)
7. Networking (your approach for establishing and being part of networks)
8. Input (your approach for seeking information: concrete/visionary)
9. Flow (your approach to pursuing divergence or convergence)
10. Passion (your approach for taking action)
11. Output (your approach toward making decisions)
12. Energy (How you seek energy to solve problems)

Malcolm De Leo, an innovation leader who has used the ISPI™ extensively inside an organization describes the ISPI™ as showing people how they create, interact, learn, and execute.

The results of the ISPI™ can be used for individual awareness, team development and analysis, as well as for creation and analysis of organizational innovation systems.

Recall that cognition means "to know." It is used to refer to the human capacity for processing information, applying knowledge, and dealing with change. It's how people take in information, make meaning of information, and also how they put that information to work to make decisions and, ultimately, solve problems.

People generally possess the skills to be creative, generate novel thought, solve problems, and interact with each other synergistically. It is true that we each have a unique cognitive orientation, but very few of us know how to capitalize on this orientation or understand and leverage differences between individual orientations. The key is to understand and leverage the unique capabilities in each of us.

5 Six Legends

As an example of different people excelling in a similar field, we are going to describe six legends in the field of science and innovation (by perceived ISPI™):

Extreme Builder: Friedrich Wilhelm Herschel (1738–1822) was an astronomer who used a 40-foot telescope to methodically map double stars and moons. He discovered Uranus and two of its moons, as well as the 6th and 7th moons of

Saturn. He also designed and manufactured telescopes. (Plicht 2007) (Taylor and Saey 2006). In his field of scientific pursuit, Herschel went very narrow and deep.

Builder: Marie Curie (1867–1934) was a pioneer in the field of radioactivity and the first person to receive two Nobel Prizes. (Marie Curie 2008). Her field of interest was radioactivity (a term she coined). Included in her many accomplishments were the discovery of polonium and radium. She and her husband refrained from patenting the process for isolating radium so that the scientific community could use the process. Curie's work and interests were broader than that of Herschel; however, it was still somewhat focused.

Mid-Range Builder: George Washington Carver (1864–1943) was an agricultural scientist who revolutionized the economy of the southern United States. He invented over 300 products from peanuts, ranging from peanut butter to extractions of peanut oil. He also invented over 100 products from sweet potatoes. He created the concept of crop rotation and soil conservation. And he was the first African American faculty member at Iowa State University. (Fishbein 2008) (George Washington Carver 2008). Carver was a little broader than Curie; however, he focused in one general area and was very methodical in his approach.

Mid-Range Pioneer: Thomas Edison (1847–1931) was an inventor with 1,093 patents to his name. His areas of invention include the phonograph, electricity, the light bulb, film projectors, motion pictures, kinetophone, and kinetoscope. Edison's expertise was in testing and refinement. (Beals 1999). His goal was to make things people could use. "Never waste time inventing things people do not want to buy." Edison used a think tank philosophy, recruiting many ambitious inventors who became known as being part of Thomas Edison's Muckers. Edison was much broader than Herschel, Carver, or Curie. He pursued many different innovations; however, all had the common thread of practicality and all were developed through extensive refinement and experimentation.

Pioneer: Benjamin Franklin (1706–1790) was known for many things. As an entrepreneur, he was one of America's earliest innovators. He saw the value of the "Double Bottom Line" (which refers to creation of wealth and social capital). Franklin was also known as a printer, inventor, scientist, economist, philosopher, statesmen, and musician. His efforts contributed to establishing fire protection, libraries, and sanitation services. Some of Franklin's innovations included swim fins, the stove, bifocals, and the harmonica. He was also known as a peacemaker and revolutionary. (A Quick biography of Benjamin Franklin 2008) (Benjamin Franklin: Glimpses of the Man 2008). Franklin's actions are summarized well by his quote, "If you would not be forgotten, as soon as you are dead and rotten, either write things worth reading, or do things worth the writing." Unlike the previous four legends, Franklin's interests were much broader. He moved easily from one pursuit to another, incorporating both scientific and social innovations.

Extreme Pioneer: Leonardo da Vinci (1452–1519) was a "Renaissance Man." (Renaissance Man 2006) (Dickens 2005). As an unrepentant left-hander who sometimes wrote backwards, he spent a lot of time pondering universal truths. Da Vinci's work is known throughout the fields of art, architecture, mechanics, and medicine (for his understanding of the human body). For example, his work led to discoveries in the organs and artery system of a woman and an embryo in the uterus. He tinkered with the giant catapult, cannon, flying machine, and a tank-like vehicle. Da Vinci also brought to the world his paintings, such as The Last Supper, Mona Lisa, and Virgin and Child with St. Anne. Even with all of these accomplishments, da Vinci failed to finish much of what he started because his interests were too broad. Da Vinci's interests were the broadest of the six legends, so broad that he would be easily distracted from completing many of them to pursue a different one that interested him more at the time.

These six legends were all scientists, but you can identify legendary experts in any other field and find the same range of innovation preferences. All are experts in some regard, but they do so using different orientations or approaches to innovation or problem solving. Within your organization, depending on what you are trying to do, you would tap different experts for different types of innovation.

6 The Innovation Relay Race

In an ideal innovation world, you could start by gathering your "da Vincis" and letting them "play" by generating new ideas. At the same time, you would not want this group leading the implementation charge.

The Relay Race would start with your "da Vincis" who would then pass the ideas to your "Franklins", and they would pass to your "Edisons" and so forth. At the end, your "Herschels" would be optimizing the idea.

iOM™ – Innovation Orientation Modifiers At a high level, the iO™ is someone's preferred way to innovate. The iOM™ is the way a person prefers to interact with others. Each of these "experts" have unique ways of working with others. We must weigh both iO™ and iOM™ to optimize their ability to work together. Their awareness of these differences enables them to create a better opportunity for success.

6.1 ISPI™ Summary

There is no right or wrong answer for any information relative to the ISPI™. No one type is better or worse than another. Yet, there are people who are better suited to be working on certain tasks than others. It doesn't mean that others can't make themselves perform all tasks, but we feel it makes sense to match people's natural psychological preferences, which are a truly hidden benefit, to their work.

Advice for the Innovation Leader

Innovation leaders must be comfortable with understanding and identifying differences between people. You must be "bilingual" in your ability to communicate, listen to, and be heard by builders, pioneers, and differing iOM™s. You must appreciate and positively recognize the strengths of all people and realize that they are not interchangeable.

The following model shows a basic way for builders and pioneers to be used in relation to the timeline of a new product or process. (Rosenfeld and Servo 1984). It is comparable to a relay race. At the start, you want to go heavy on the pioneers. As the product matures, more mid-range pioneers and mid-range builders should be integrated. At the end, the builders are accountable for optimization, and the pioneers are working on developing a new S-curve. Note that an S-curve for a product or service can last 20 years, 20 months, or 20 weeks.

You can have ideal business practices, technological aids, and creativity tools, but if you have the wrong people in the wrong phase of the innovation process, it will be difficult to meet your innovation goals. For innovation to flourish, you need the right people with the right skills, innovation Orientation™ (iO™), and innovation Orientation Modifiers™ (iOM™) – as well as the right tools for working on the right tasks at the right times (Fig. 2).

As an innovation leader, you need to:

- Make it a practice to note the iO™ and iOM™ of the people you work with. Over time, your eyes and ears will help you identify personality orientations.
- Use iO™ and iOM™ as one of the variables in putting together your project teams.
- Use your knowledge of iO™ and iOM™ to guide your approach when you try to present to, influence, or manage others.
- Help others understand their invisible differences and how to leverage them.

As the leader, you link the human pieces of the innovation puzzle with the appropriate business and technical pieces. In addition, your job is to help others understand their differences and why the differences within the team are a benefit. The degree to which you value differences will determine the degree to which others will value them.

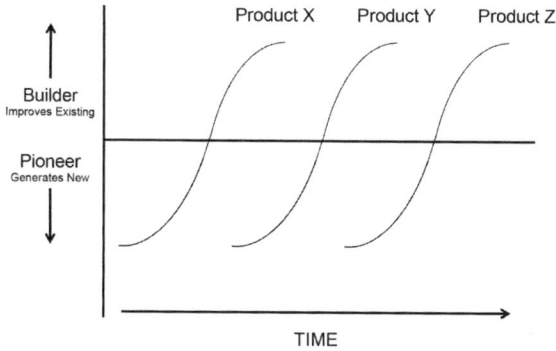

Adapted from Rosenfeld, and Servo

Fig. 2 As a product goes through its life cycle, the ratio of pioneers to builders shifts. At the beginning of the cycle, the mix is weighted toward the pioneers. As the product matures, the mix shifts to a combination of pioneers and builders and then to mostly builders. As this shift occurs, the pioneers begin development of the next new product to ensure organizational sustainability. This is critical during times of massive change

6.2 Remember

Differences are a gift! Look for the gifts, respect and leverage them!

7 Parting Thoughts

We can foster Sustainable Entrepreneurship by: recognizing the phase of the 'Invisible Cycle' in which our business exists, returning to the beginning of the 'Invisible Cycle', and utilizing the ISPI™ to identify key players, understand their preferences, as well as help team members understand, appreciate, and leverage each other's differences.

By embracing differences and leveraging them, we develop social capital and business value. This will help the organization return to the fertile, innovative ground of entrepreneurship, while still sustaining the necessary process base to maintain the business. By recognizing and honoring the inherent relationships between business, social capital and leveraging differences (for examples: between builder and pioneer) we can realize a quantifiable gain while creating new strategic business models that are lasting, renewable, and sustainable.

Literature

A Quick Biography of Benjamin Franklin. Retrieved 2008, from US History. org: http://www. ushistory.org/franklin/info/index.htm

Adapted from Rosenfeld, R. B., Wilhelmi, G. J., & Harrison, A. (2011). Leveraging differences. In *The invisible element: A practical guide for the human dynamics of innovation* (pp. 53–73). Rochester: Innovatus Press.

Beals, G. (1999). *The Biography of Thomas Edison.* Retrieved 2008, from Thomas Edison.com: http://www.thomasedison.com/biography.html

Benjamin Franklin: Glimpses of the Man. Retrieved 2008, from The Franklin Institute: http:// www.fi.edu/franklin/

Bennis, W., & Biederman, P. W. (1997). *Organizing genius.* Cambridge: Perseus Books.

Dickens, E. (2005). *The Da Vinci notebooks.* New York: Arcade.

Drucker, P. *Peter Drucker on Knowledge Worker Productivity.* Retrieved 2008, from Knowledge Worker Performance.com: http://www.knowledgeworkerperformance.com/Knowledge-Worker- Management/Drucker-Knowledge-Worker-Productivity/default.aspx

Fishbein, T. *The legacy of George Washington Carver.* Retrieved 2008, from Iowa State University: http://www.lib.iastate.edu/spcl/gwc/bio.html

Friedman, T. (2005). *The world is flat.* New York: Farrar, Straus, and Giroux.

George Washington Carver. Retrieved 2008, from Gale Cengage LearningTM: http://www.gale. cengage.com/free_resources/bhm/bio/carver_g.htm

Hong, Lu, & Page, Scott (2004). *Groups of diverse problem solvers can outperform groups of high-ability problem solvers.* Retrieved from The National Academy of Science: http://www. pnas.org/content/101/46/16385. full.pdf + html

Huitt, W. (2001). *The Mind.* Retrieved 2007, from Educational Psychology Interactive, Valdosta State University: http://chiron.valdosta.edu/whuitt/col/summary/mind.html

Knowledge work and knowledge worker were first used by Drucker. A knowledge worker is an individual valued for their ability to interpret information within a specific area.

Marie Curie. *The Nobel Prize in Physics 1903.* Retrieved 2008, from http://nobelprize.org/ nobel_prizes/physics/laureates/1903/marie-curie- bio.html

Marie Curie. Retrieved 2008, from Wikipedia: http://en.wikipedia.org/wiki/Marie_Curie

Miller, L. M. (1989). *Barbarians to bureaucrats: Corporate life cycle strategies* (p. 2). New York: Fawcett Columbine.

Plicht, C., & Herschel, F. W. Retrieved 2007, from http://www.plicht.de/chris/06hersch.htm

Renaissance Man. Retrieved 2006, from Museum of Science: http://www.mos.org/leonardo/bio. html

Rosenfeld, R., & Servo, J. (1984). Business and creativity: Making ideas connect. *The Futurist, XVII*(4), 21–26.

Snow White and The Seven Dwarfs. Retrieved 2009, from Film Reference: http://www. filmreference.com/Films-Se-Sno/Snow-White-and-the-Seven-Dwarfs.html

Taylor, P., & Saey, S. *Herschel Club – Friedrich Wilhelm Herschel.* Retrieved 2006, from The Astronomical League: http://www.astroleague.org/al/obsclubs/herschel/fwhershs.html

Walt Disney: A Biography. Retrieved 2008, from Disney Archives: http://disney.go.com/vault/ read/walt/index.html

Sustainability: Ethical Perspectives

Clemens Sedmak

1 What Exactly Do We Mean by "Economic Activities"?

According to Aristotle economics can be categorized either as Politics, Technology or Ethics. Regardless which philosophical approach one chooses, however, economics is basically about pursuing conceptions of good life, about managing life and life situations. "Economics" is linked to running an "oikos", a household – *household* economics can be considered a central reference point for any kind of management structure. Household economics demand that a household be set up, maintained and if necessary extended and should be a place providing the structure and framework needed for a 'good' life. The household itself is not the 'good' life, it provides the favourable conditions – environment – in which a good life can take root and develop. Central to good management and good development are a sense of care, permanence and persistence, moderation, regularity and a sense of neighbourliness. One vital aspect which should not be overlooked here is that the house and its household is not created by some 'invisible hand' but is the product of human making: a structure built on, via and enabling human existence.

Entrepreneurial activity of any kind involves the application of certain "action plans" to fulfil and satisfy human needs where "goods" available are in scarce supply. So what we have here are in fact three core concepts: "taking action", "having need" and "providing goods". Human "action" suggests a pattern of behaviour based on counsel and deliberation; it is an attitude comprising space and scope and is thus inextricably bound with the potential and possibility to plan. Human action is ethically important since it assumes human action is based on a selection process of alternatives, in other words: P acts, even if P could have acted differently. If we follow this notion through, we can see that the concept of action is anchored in the concept of freedom, which is commonly acknowledged as playing

C. Sedmak (✉)
Universität Salzburg, Zentrum für Ethik und Armutsforschung, Mönchsberg 2a,
Salzburg A-5020, Austria
e-mail: clemens.sedmak@sbg.ac.at

C. Weidinger et al. (eds.), *Sustainable Entrepreneurship*, CSR, Sustainability,
Ethics & Governance, DOI 10.1007/978-3-642-38753-1_5,
© Springer-Verlag Berlin Heidelberg 2014

an essential role in the history of economic thinking. In David Landes', "the wealth and poverty of nations" the term freedom is regarded as central in determining the economic success – or lack of it – in different nations and civilisation. (Landes 1999, pp. 232–234) Fundamentally, economics is founded on a system of exchange. For any exchange system to work there must be a degree of scope and space, and equally a need of certain goods and/or commodities and the will or desire to satisfy that need. Any successful exchange partnership requires long-term co-operation capabilities. The rise of the Rothschilds was essentially due to the capabilities of Mayer Amschel Rothschild (1744–1812) to think and plan ahead long term and to prove himself as a reliable business partner. (cf. Elon 1988: Chap. 3; Wilson 1988, pp. 38–44; Ferguson 1999) The economic and social rise of Mayer Amschel Rothschild, who was essentially the founder of the Rotschild dynasty (the success story of its day and beyond), was mainly propelled and made possible through his social connections with people in high places e.g. aristocracy and members of the royal families of Europe, and his willingness to co-operate with their demands e.g., as with Wilhelm IX, the wealthy Elector of Hesse-Kassel. Mayer Amschel proved himself to be a trustworthy and prudent business partner, with marked business acumen revealing itself in his willingness to trade at the expense of forgoing initial profits to the benefit of establishing long-term business relations; he was thus far-sighted in his capability to accept "delayed gratification".

Managing trade or an economy means co-operation. Co-operation is what triggers an economy, and competition taking place fairly in any trans-action is also an expression of co-operation, in which each business partner adheres to the rules of the game, rules founded on the well-being of all 'players' and which guarantee minimum standards in the process of co-operation and competition. Management is by its very nature a process which transforms, transforms resources or raw materials into a commodity or relationship (as Aristotle describes it) to the mutual advantage of all parties concerned. Apart from the root production or manufacture of commodities which plays an essential role in defining the entrepreneur behind the goods, any trading process –"trans-action" – depends on an exchange of those goods. Something is given in exchange for something else; or to put it another way, that something which is received symbolises that which has been given and visa versa. The minimum requirement in any transaction is an exchange of equivalents of some kind, be it merely symbolic: there has to be some notion of parity in the exchange. Philosophically speaking, we are dealing with the concept of justness in the exchange process, with commutative fairness. Since there is "similarity in the parity" of the goods traded with each other, the act of exchanging those goods demands a degree of comparability and a degree of trust, – faith inasmuch as the long-term overall development – manufacture and trade of goods cannot be pre-empted, foreseen or guaranteed. Trade depends on trust because there is always an element of risk inherent in any trading process. Trust is engendered of social necessity and suggests a readiness – willingness – on the part of the one giving – to entrust his valuable goods to a third party without having any guarantee that these goods will not be misused or damaged in the transfer process. If we now consider employers and employees as part of this process, then we will see that there is an exchange of payment for service in the same way as there is payment for goods received; trading partners will trade services with each other, buyers and sellers will exchange goods for money. Thus, clearly, trust is very much at the heart of any trading business: trying to win back a loss of trust ("trust repair"), is an arduous and costly task. (cf. Lewicki and Wiethoff 2000) A producer has to rely on her workers on the factory floor, distributors and customers. However, in all three cases we see the need of the one to trust the other. Customers have to trust in the producer to deliver according to their needs. Customer satisfaction is the prime goal.

"Needs" are necessities of life which demand to be fulfilled by their very nature and are the trigger of action and management. Realisation of primary needs does not however go far enough and according to Manfred Max-Neef we need to distinguish between "satisfiers", "pseudo-satisfiers" and "destroyers", i.e., between those goods which satisfy needs in real terms, those which only seemingly do so, and pseudo goods which feign capacity to satisfy and do in fact endanger the very nature of the need, (one example in this regard are nuclear arms produced and piled for security and defence purposes). (Max-Neef 1991) "Goods" are – as the word suggests – defined by their "value" within the framework of "bonum"; things become 'goods' when those things reflect a particular situation e.g., become a 'situation marker'. Meaning that those things – X – play an important role in defining the given situation, a situation which could otherwise not be defined. One example in this regard would be a heating system in a home or household, the heating defines this household situation; a Sat-Nav would be a similar marker in a car. The marker X represents the goods – bonum – for the person P; if P chooses a situation defined or marked by X rather than the situation without the marker X.

Entrepreneurial activities do not necessarily take place within a household but it cannot be carried out in an unbiased, non-judgemental environment, meaning it cannot be effected in a space where there is no value-system and which would have to be 'imported' from outside, i.e., one can only do business with 'goods' and their inherent values which makes them precious or prized or priceless. Trade and economies involve the production and transaction of goods whether they be services or commodities, but which will in some act on, react to and counteract the requirements, needs and demands made by human beings. This may remind us all of the bother and hassle of certain goods, such as the vacuum cleaner bag which was the bane of Manfred Sauer's life; he was confined to a wheel-chair having had tetraplegia since an accident in 1963. His predicament inspired James Dyson to develop and manufacture the first 'bagless' vacuum cleaner – the perfect product for wheel-chair users. Such a products -goods – are a direct response to a demand felt and expressed in managing life issues. The scale of values on which any management structure is based is the outward sign that management competence comprises capacities to co-ordinate and co-operate. (cf. Bröckling 2007, pp. 120–122)

Trade becomes inhuman when it ceases to make an active contribution to a 'decent life': a decent life being one in which worth and worthiness plays a key role, worth being synonymous with respect (of self and others). Similarly, the term 'decent economy' can only be engendered in a 'decent society' which in turn is practically aware of the concept of 'decent work'. In Avishai Margalit's view (1996), a 'decent' society is one with 'decent' institutions, decent in the sense of respectable and honest, where individuals are not humiliated or made to feel worthless. Humiliation is a rational and reasonable feeling brought on by one's sense of self having been injured or bruised in some way. Entrepreneurial activity in a decent economy requires awareness of an active steps to prevent humiliating patterns of malpractice entering the chain of management; humiliation often found in exploitation and immoral contracting. Any discussion about decent work within a decent economy has to comprise the notion of freedom as a key concept, the freedom to choose suitable work in reasonable conditions and commensurately rewarded and remunerated; working conditions that are safe and secure, and provide enough social protection for employees and their families. Factors such as equal opportunities and social dialogue are the prerequisites of such structures. (cf. Margalit 1996) A "decent economy" by sheer definition must be able to provide, or at least attempt to provide, "decent work" corresponding to basic human freedoms. Management firmly anchored within a framework of ethical value standards will ultimately create humane working conditions. Sustainable management satisfies real human needs with real, worthwhile goods. However, a decent economy is more than just a framework of management, it also has to do with the economy per se, existing to serve people and not exert pressure with 'non-products' as its prime objective. The term "non-products" stems from the term "non-disease" commonly used in medical sociology to describe classic conditions such as going bald or drooping eye lids, which become pathological due to social and cultural pressures (cf. Smith 2002, pp. 883–885) Non-products could also be described as 'would be' goods, which due to social and peer-group pressures are portrayed – advertised – as real, worthwhile goods, but which in actual fact do not improve or enhance quality of life any meaningful way, nor do they bring about a "decent life".

2   What Do We Mean by "Sustainable Entrepreneurship"?

A sustainable economy demands a form of management which knows that: (i) reserves must be put aside for 'lean years' – times of need, (ii) the construction and application of a tangible infra-structure has to be built with a sense of permanency in mind and transcends a 'here and now' blinkered perspective which managers might have. Thus, sustainable management is also a strategy in survival when the going gets really tough and a crisis management can be an opportunity for future-oriented management. Since the World Summit in Rio de Janeiro in 1992 (the largest summit conference in the twentieth century), "sustainable development" was at the top of the agenda in public debate. The idea put forward was to satisfy human needs, preserve the environment and at the same time maintain a just and sustainable economy. Sustainable development is feasible and viable long-term i.e., for future generations, without it having to overload ecological systems;

sustainable development embraces a structure of management which satisfies the needs of this and future generations, while preserving natural resources. Or to put it another way, sustainable economies strive to achieve a level of development which does not trespass upon the goods and options available to the here-and-now generation – "N" – or impinge upon so as to reduce them for successive generations – N + 1. The actual term sustainability may be hazy and blurred, making it difficult to argue and reason one the one hand, or be glorified to the extent of undermining it on the other, but the fact remains that the term and implications inherent in meaning *have* made it a relevant factor in public debate despite it seeming to bear no clout in the global political playing field.

Sustainable policy-making appears to lack clout or 'bite' because it is constantly falling into the trap of the "Tragedy of the Commons". This is a recurring tragedy now well known since Jared Diamond made an in-depth analysis of the collapse of societies and the key factors involved in their own ruin. (Diamond 2005: esp. Chaps. 1 and 14) Diamond defines the collapse of a society as one in which the total population has dropped drastically and/or the politico-economic-socio long-term impact persists over a wider geographical area. The five main contributory factors, Diamond believes are: damage to the environment, change in climate, hostile neighbours and friendly trading partners, and the societies' response to the same. Diamond then turns to the question of why societies watch the demise of their own society; to illustrate his point, he uses the example of development drive in the US state of Montana which has systematically levelled a hitherto pristine eco-system – but why? The players on this particular field have behaved in a manner which can only be driven by self-interest and the need to protect and maximize self-gain to the detriment of their 'commons' (common ground). Agricultural land was sold off to second-home property developers, the mining industry was fired by short term contracts with no regard as to long-term interests and/or impact, the federal state protected the interests of home-buyers and owners whose homes were at risk due to the heightened possibility of forest fires which in turn were more likely due to de-forestation and meant the priority to protect homes was given higher priority than protecting forests: clearly a vicious circle. These have become the familiar dynamics of world summit conferences: each participating nation wanting to further its own interests, rationally, reasonably and understandably and as a result was bound to undermine any 'common goods' they may share with other participating (and non-participating) nations. In other words a 'tragedy of commons': a tragedy brought about by rational thinking which can cause the complete breakdown and demise of common goods. Conflicting self interests can similarly be the result of past 'tragedies': short-term perspectives in any age must eventually lead to a long-term conflict of interests. The greatest obstacle and challenge in the way of sustainable management is solving the tragedy of 'our' commons.

Of course, one might be tempted to ask why we actually need to consider sustainable entrepreneurship at all and the answer would be to take a close look at our present situation. In the second half of the twentieth century two issues became key in this concern – the question: *how much will it cost*? And, *who is going to pay*? The whole issue of costs and payments became the set underlying frame-work for planning at any level. Niklas Luhmann did therefore see money as the universal symbol generating common communication between modern-day trading partners, enabling and driving the exchange of any 'goods' and any trading transactions. (Luhmann 2001, pp. 31–75) Money made everything – any transfer at any level – possible: it was its own universal language allowing free movement between states, beyond traditional (language) barriers. Money stood for control, complete control, because when the key issue of covering costs incurred can be settled, the spectre of catastrophe becomes a real and tangible item and dominant cost factor on the household planning agenda. If it can be settled – before it happens – who is to pick up the bill in the event of a catastrophe, the planning can go ahead to scheduled, since the cost factor has been ironed out and provided for. Simply by raising and answering these two questions, causal and eteological aspects have been resolved: there is no going back, there is no reverse gear. However legitimate these two questions about costs incurred may be, they do have their limits, limits which are very much part of the question of sustainability. If we think "Fukushima" or "New Moore Island" we might begin to realise such limits.

"Fukushima" and "New Moore Island", seen as abbreviations of "signs of the times", jolt our attention to the fact that there is a limit to what can be paid for: not everything. Fukushima, the site and centre of a nuclear disaster, and New Moore Island, the tiny rock island which disappeared in the Bay of Bengal because of rising sea levels, are two places which also represent the "topoi" of our times, and rather like 'non-products' can be regarded – literally – as 'non-places', places where it is not 'good' to be. Topoi as signature tunes of our times, battling against the increasing tide of risk, exclusion and segregation, and imponderability. Fukushima and New Moore Island share certain common features: irreversibility, urgency, inescapability. Irreversibility is the snowball effect which becomes an avalanche, the common chain of cause and effect in which there is no going back or reversal of outcomes; once the floodgates have been opened and allow the tide of events to flow through, it may be briefly stemmed, slowed or even stopped for a while, but there can be no going back to, re-establishing the original circumstances. Nuclear energy and climate change are a fact and, in their chain reaction, have caused facts which cannot be circumnavigated as if they did not exist: there is no going back to a non-nuclear or pre-global-warming state. They may not immedi-ately trigger a change in conditions but do create circumstances which can and will give rise to future change. One could say that irreversibility is one of many worthy human creations, and that irreversibility is an unavoidable risk-factor of life's forward journey; human beings have to constantly test the limits of what is theirs to control, often to their own pain. Reminiscent of the Sorcerer's Apprentice, the trickle becomes an uncontrollable torrent which will not be mopped up by mere buckets – a situation out of and beyond human control – only the intervention of higher powers can work the wonders wished. Urgency: Fukushima and New Moore Island's common message is: catastrophes and disasters demand immediate action and, in their own imminency, are a threatening proclamation of what must follow. Martha Nussbaum (2001) underlines the sense of 'urgency' in her Theory of

Emotions; emotions by nature express a sense of urgency and press for action to be taken. The two catastrophes being discussed here are laden with emotion because they changed and will continue to change the fate and fortune of many and much which cannot be argued away or overcome by rational reasoning – they demand decisions be taken and steps be taken to resolve the urgency. Such disasters do test our powers of reaction, a test which cannot be passed by sticking one's head in the sand and waiting for things to blow over. Precious time is ticking by, and if immediate action is not taken worse will ensue. Irreversibility: Urgency comes about exactly in those situations which cannot be easily and quickly solved, where the risks are great and cannot be solved locally by conventional means: issues pertaining to nuclear power and climate change are global issues. Similar to the omniscience of God described in Psalm 139 ("If I ascend to heaven, you are there; if I make my bed in Sheol, you are there". Ps 139,8), catastrophes are omnipresent – we cannot carve out a life-style niche which will not in some way be impinged by this topoi: they have become a signature tune of our time, are part of the fabric of fate which we have to learn to live and deal with since it cannot be reversed. Fukushima und New Moore Island show us the limits of cost-analysis, and in acknowledging these limits we can begin to 'negotiate' the limits of real and seeming problems. Or, bluntly speaking, when the world is no longer inhabitable due to radioactive contamination, (a consequence outlined by Michael Frayn in his dystopian novel *A Very Private Life*, 1968), a world in which the question: 'who's going to pay for it?' is superfluous to any needs: the situation is irreversible. I know a woman who lost her daughter in the Austrian Kaprun funicular railway inferno of 2000. She knows that no money will ever bring her daughter back, or ease the pain and the question such catastrophes raise is not: 'How can we foot the bill?'. In fact they show us close up the boundaries of what can be paid for and ticked off as settled. We all the know the now household saying: 'you can't eat money', and it is in situations like this that it comes home to roost, situations like New Moore Island and Fukushima. We are being forced to manage our 'households' sustainably because we are pushing our own limited resources to their tangible and intangible limits.

When there is a major shortage of resources (raw materials, fuel and energy. . .) we begin to realise the importance of other resources (knowledge, value systems etc.). Sustainable management also means conceding the primacy of intangible infrastructures over tangible ones. The intangible infrastructures of any society are the roots of its knowledge and wisdom, and the foundation stone of its value system, but it can also be sources and resources of identity. The Credit Suisse Research Institute thoroughly analysed and came up with the concept of an intangible infrastructure. (Natella et al. 2008) While material infrastructures pertain to roads, rails, water and electricity supplies and of course airports etc., intangible structures are concerned with connections and avenues between what is known and what is valued, things like education, technology, and health care. Developing knowledge based economies, increases the likelihood of intangible structures playing a lead role in future socio-economic prosperity. The above study identifies five interconnected corner stones of intangible infrastructures: education, health care, financial development, investment in technology and the spread and distribution of Business Services. It does have to be said that of the five, education is the key element. The study defines an infrastructure as: "the set of factors that develop human capability and permit the easy and efficient growth of business activity". (Natella et al. 2008, p. 7) According to this definition, human capabilities must make up the heart of such intangible infrastructures. Human capabilities depend on a benchmark value and education system for their nurture and growth. Human capabilities are dormant potential waiting to realise possibilities, capabilities are the means to transforming opportunity to reality turning an unwanted situation – A – into a desired situation – B. Intangible Infrastructures can be seen as a cluster of abilities implemented to transform reality using a system of knowledge and values. Such systems and such capabilities rely on cultural, social, legislative, political and economic factors; they depend on the social cohesion stemming from stable political policies, public safety, a strong tax system and the 'capacity' of institutional structures to support them. Katie Warfield, Erin Schultz und Kelsey Johnson came to similar conclusions in their conceptualisation of infrastructures in a framework comprising cultural aspects. (Warfield et al. 2007) The fundamental difference between them is that tangible infrastructures are manifest in physical constructs and intangible infrastructures characterised by soft skill factors which are not manifestly tangible. Intangible infrastructures are about human identity and our notion of 'self' and 'self respect' and makes an economy and the managing of that economy a plausible means to an end in striving for and fulfilling that identity. Intangible infrastructures are closely related to the family of 'non-economic' forms of capitalism e.g., social, cultural, symbolic. A sustainable market economy is an economy that is firmly rooted in dealing with other trade partners in other types of capital markets and is therefore a true reflection of the economy.

3   The Benedictine Monastic Order: A Model for Sustainable Management

In understanding how such a sustainable economy or a spirit of entrepreneurial activities can work and grow, we could take the Benedictine monastic order as an exemplary model. Benedictine monasteries have proved themselves to be economically resilient. Their economic structures have not collapsed in times of crisis and have in fact proved to be incredibly stable. Bruno Frey and Emil Inauen looked for and studied reasons which might explain why Benedictine monasteries in the German- speaking countries were economically resilient. Aspects they indentified included: *good governance*, particularly the democratic framework practised in choosing an Abbot and making major decisions, *monitoring* via regular external supervisors, *careful recruitment* and clearly defined *allocation* of work and tasks for the various offices and functions within the monastery e.g., Abbot, Prior, Cellerar, Master of Novices, etc.. Apart from these structural factors, a clear system of values was also seen to be key prerequisite for the unhindered management of such an institution tried and tested over centuries by generations of monks. A resilient enterprise needs 'bricks and mortar' but it also needs a benchmark system of values as mainstay of its intangible infrastructure which can be relied upon to withstand the onslaughts of tough times and crises. Benedictine monasteries have eminently proved their mastery of such benchmark systems.

Taking a look at the Benedictine tradition means taking a look at the sustainability of a unique way of life, taking a 'master class' in creating lasting living and livable structures. A number of studies e.g., Luy's study on health and monastic life style could be cited in supporting the claim that monks live longer (Inauen and Frey 2008) one reason perhaps being that they do indeed have a clearly defined rhythm of life embedded in a 'daily routine'; they are moderate in carrying out tasks and have a clear framework of values to work by: all within a community. Such factors can contribute significantly to the sustainable management of any institution: a strong sense of community and moderation in all things played a key role. We can learn a lot from the Benedictine tradition, such as the vital role of identity resources for a sustainable way of life. A monastic life focusing on a moderate approach in all things will prevent any erosion of self. The French sociologist, Alain Ehrenberg, describes the dynamics at work in the erosion of self in what he terms the "weariness of self" including mental health issues, burn-out and the other tell-tale signs of a loss of equilibrium; all linked in some way or form to the burden of self identity issues – situations in which individuals are forced to be something they are not, cast in a role of someone they are not and who they cannot inwardly identify with. The gap between external expectation of role-fulfilment and the internal dynamics at work against the same lead to cachexia anorexia. (Ehrenberg 2009) The clearly-defined foundation of identity in a Bene-dictine environment is the counter-weight to those qualities of ego which erode 'self'. For it is only be adhering to a strong sense of moderation that those other effects of self-erosion can be counteracted and prevented (even within a monas-tery), namely: avaritia and its brother acedia: destructive forces immobilising any feelings of contentment.

The key reference text in understanding the Benedictine tradition is of course that opus magnum "Regula Benedicti" generally acknowledged as having been written by Benedict of Nursia, who founded the monastic community of the Benedictine Order in the sixth century. (Doyle 2001) The text could be described as a manual with guidelines for different people from different backgrounds living together under one roof, long term. It is a collection of directives or instructions for a shared way of life, *not* a list of all those 'New Year's Resolutions' we crave to keep but fail to do so because we lack the necessary attitude but an approach based on a perception of life from the inside – the centre of being – out, in other words a life orientated towards moderation to achieve a set goal. Not only without harming inner self or outer environment (other selves), but actively striving to bring out the best in self and others. How can this possibly be a single goal in a shared community of many? The Rule of Benedict (RB) contain a wide spectrum of instructions which serve the everyday needs of any household management, regardless of age or era.

We have seen above that a decisive factor in sustaining a 'common culture' is dealing with the "Tragedy of the Commons" – the greatest challenge to sustainability. The issue of 'ownership' is referred to again and again in relation to collective decision-making (RB 3); Brothers should not stubbornly defend their own personal opinions (RB 5,12), concord is of uppermost significance in a culture of (re)conciliation. The *Regula* describes the strongest monks as being willingly cenobite, who live in monasteries under the 'rule' of an Abbot – living co-ordinating and co-ordinated lives. The resulting stability is both the stalwart and bulwark of the *Regula*, the very notion of sustainability is a corner stone of the Order, which is outlined in the Prologue through the Biblical metaphor of the house built upon a rock (RB Prologue, 33f, ref., Mt 7. 24f); sustainability is regarded as a key value in the recruitment process: only candidates can be considered for admission who are in no doubt as to the harshness and hardships involved in monastic life (RB 58,8). Above all, they must pass the test of steadfastness. These basic benchmark values and a real sense of living together provide the stability needed. But is not just a question of survival of the fittest, the 'survival' of the weakest is a prime factor not to be overlooked and demands both the attention and care of the others. The Abbot must make allowances for the 'weakest links' (so to speak) (RB 4,25), meaning that in what might be called a recruitment drive' a monastery should – first and foremost – consider the presence of aptitude and capabilities. Stability is, among other things, the reward of the independence sought in any monastery regardless of the upheavals and impacts from round about in the outside world. A Benedictine monastery is an autocracy: "where possible, a monastery should always be constructed so that it own daily needs can be met within the environs of the monastery itself, namely water supply, mill to grind its own flour and a herb garden, moreover, it should have enough trades and craftsmen to be able to manage its daily running" (RB 66,6); the reasoning behind this being: "Monks will have no need to go running round outside, something which does them no good" (RB 66,7). Through such measures and means, the monastery is its own closed micro-unit in which everything is manageable, focusing on common a landscape of culture and values which maintains its own dynamics of stability. Relationships with and to the world outside are also carefully 'regulated'; the gatekeeper on the boundary between the 'two worlds' – inside and out – has to be a man of maturity and wisdom (RB 66), similarly a brother assigned to receive and look after guests (RB 53). Similar regulations apply for journeys outside of the Order: brothers must be given permission (by the Abbot and having received strict instructions about time and practice of prayer according to the rules of the Order) before they can entertain the idea of travelling. Upon their return, they should be mindful not to take it upon themselves to relate happenings and/or events which they have seen or experienced on the travels, since this will be harmful effect on anyone listening (RB 67,5). The clash between life styles in inner and outer worlds will invariably jeopardise the stability attained by a closed community (worthy of discussion!)

The secret to Benedictine household management lies, however, in its bench-mark set of values. Interiority is of prime importance, the significance of this is even mentioned in the Prologue of the *Regula,* the importance prayer before setting out on any mission (RB, Prolog, 4). Spirituality as leading principle is reflected in a unwavering trust in God above all else, a direct reference to what is commended in Mt 6,33 and Lk 16,2. The Abbot should live up to his title – *superior* – reflected in his absolute trust in God and re-act accordingly: "he should not fret about a shortage or lack of monastic funds" (RB 2, 35). Nor is poverty something to be ashamed about (RB 48,7): "*And if the circumstances of the place or their poverty should require that they themselves do the work of gathering the harvest, let them not be discontented*"; which does not say that hard labour and poverty are not the rule but the exception, nevertheless, to be taken in the spirit of spirituality. Setting spiritu-ality above economy defines a culture of moderation: everything should be done, undertaken and accepted moderately: "*Let all things be done with moderation, however, for the sake of the faint-hearted*" (RB 48,9). In choosing an Abbot, the following should be guide choice: "*Merit of life and wisdom of doctrine should determine the choice of the one to be constituted*" (RB 64,17f). And the idea of a moderate life style is genuinely conceivable since any immoderate conduct is only haughty hubris in disguise, which knows no limits and thus at the complete mercy of avarice, envy and greed since it lacks a firm foundation of inwardness. "Avarice" is the haphazard pursuit of either, a dysfunctional sense of 'order' or the striving for (material) possessions; in all cases immoderate in nature and erroneous. One should not make the mistake of thinking that avarice is the natural conclusion of hunger or thirst satisfied because she does in fact know no boundaries, which therefore need to be set by defined values. Such boundaries are set with wisdom which will always focus on and perceive what is good. Moderation is reflected in prudence and discretion. The Thomistic philosopher Josef Pieper pinpoints the necessity of moderation for preserving self – for we destroy our very selves when we slither into immoderacy: when we can no longer see or say: 'enough is enough'. Enough applies mutatis mutandis in the Benedictine tradition both in work and prayer.

And, it is essential to remember that a monastic order is not a business under-taking which might have built-in millstone to contend with. A monastery is first and foremost a spiritual institution to serve and safeguard its members, only then is it an economic enterprise in the traditional sense. This is clearly a difference based on prioritizing; "On the artisans in the monastery" is section 57 of the *Regula.* Here we find out about those monks who are 'employed' as workers – artisans – who have entered the Order on having pursued a career or craft in the outside world. They are allowed to carry out their trade:, "*provided the Abbot has given his permission*"; but, "*if any one of them becomes conceited over his skill in his craft, because he seems to be conferring a benefit on the monastery, let him be taken from his craft and no longer exercise it*". (RB 57,2). Economic increase and financial gain are secondary to the spiritual well-being of the whole. Might the secret to sustainable success lie therein?

Moderation leaves no room for avarice; the cellarer must not in any way shape or form be driven by greed, (RB 31,12). Avarice by virtue (or lack of it!) invariably demands a 'more and more and more' policy of management, with growth and gain for its own sake. Such an attitude is not compatible with the Benedictine tradition: this holds true not only for maximisation of self-interest but also for maximising common or shared interests, which is verified in section 57 of the *Regula* on monks working as craftsmen and sell their handicrafts: "And in the prices let not the sin of avarice creep in, but let the goods always be sold a little cheaper than they can be sold by people in the world," (RB 57,7–8), and the reason? "that in all things God may be glorified (1 Peter 4:11)", (RB 57,9). If we stop to think about this one moment and remember that it is a monastery we are talking about, which can gain a 'good' income from the sale of its own products and produce; then we may be surprised: for although the idea of maximum 'gain' is manifest, the ever-present and overriding danger of "avaritia" is to be avoided at all costs, for it will stand in the way of spirituality and represents a far greater malum than the bonum gained by a 'good' price. Maximisation does have a brake which should be used when the need – temptation – arises. In the same vein, the admission of brothers should not be dependent on their having or giving the Order money, if they do have money at their disposal, then they are free to give it all away to the poor or bequeath it to the monastery: "*If she has any property, let her either give it beforehand to the poor or by solemn donation bestow it on the monastery*" (RB 58,24).

4   Good Management

(a) Effort

The significance of making an effort is pinpointed in various sections throughout the *Regula: "Run while you have the light of life, lest the darkness of death overtake you*" (RB Prologue 17 reference to, John 12:35); "Turn away from evil and do good; seek after peace and pursue it" (RB Prologue, 17, reference to Psalm 33); "*while we are still in the body and are able to fulfill all these things by the light of this life, we must hasten to do now what will profit us for eter*nity", and a second time in the Prologue: "*For if we wish to dwell in the tent of that kingdom, we must run to it by good deeds or we shall never reach it*" (RB Prologue, 22). We are obliged to make an effort but too much zealousness is not good either; the *Regula* mentions this: "*Just as there is an evil zeal of bitterness which separates from God and leads to hell, so there is a good zeal which separates from vices and leads to God*" (RB 72). An urgent sense of being prepared (even when asleep) is also conveyed: "*thus be always ready to rise without delay when the signal is given and hasten to be before one another at the Work of God*" (RB 22,6). The message of 'effort' is imbedded in memory in the supreme power of moderation which is omnipresent throughout the *Regula*, for effort in itself -or making too much effort- is not a desirable goal (cf. RB 39 on the measure of food, RB 40 and drink) and too much effort e.g., in evil zeal of bitterness should be abandoned (RB 72,1), for it cuts us off from God. A good culture of effort requires pains be taken and due care practised, as well; care taken to be mindful as we have already seen in Pachomius' instructions for cenobite monks living in one house. One can well imagine and believe that adopting such an attitude would indeed defend against the onslaught of a "Tragedy of the Commons" and contribute towards a healthy economy. Mindfulness is a basic virtue, particularly in Abbots, who should have a mental list at his finger-tips of all the property and possessions of the monastery; property of the monastery should always be easily traceable and clearly recorded (RB 32,3). Carelessness will not be tolerated: *"If anyone treats the monastery's property in a slovenly or careless way, let her be corrected"* (RB 32,4); the Abbot should entrust tools and utensils to those he can rely on to use and then return them; *"He shall return the utensils of his office to the cellarer clean and in good condition, and the cellarer in turn shall consign them to the incoming server, in order that he may know what he gives out and what he receives back,"* (RB 35,10). Tools and other implements are to be looked after and returned in good condition, or else: *"When anyone is engaged in any sort of work, (. . .)she commits some fault, or breaks something, or loses something,"* (RB 46,1–3) they will be severely punished. And: Good management of a monastery depends on

(b) Work

Physical hard work – as opposed to only mental labours – is a mainstay of the Benedictine ethos and gains in value and importance; traditionally, physical work had been down-graded and looked down upon in Greek and Roman philosophy but in the Benedictine tradition it is upgraded and attains higher, healthier spiritual value: " let no one be excused from the kitchen service", (RB 35,1). For: *"Idleness is the enemy of the soul"* (RB 48,1; RB 48,18), and care must be taken to avoid it since it can destroy a community. Even the sick or overly sensitive brothers must be provided with some mental occupation or exercise: *"Weak or sickly sisters should be assigned a task or craft of such a nature as to keep them from idleness and at the same time not to overburden them or drive them away with excessive toil. Their weakness must be taken into consideration by the Abbess"*. (RB 48,24). They should not have the time or opportunity to moan: *"that the brethren may do their work without just cause for murmuring"* (RB 41,5). If the task allocated to them should prove to be too much, it should be approached: modestly and not too fervently; it is up to the superior to decide if a task is reasonable or too strenuous, beyond the capacities of a fellow sister: *"If it happens that difficult or impossible tasks are laid on a sister, let her nevertheless receive the order of the one in authority with all meekness and obedience"* (RB 68,4f). Moderation in work is as important as moderation in other aspects of life – workaholics are not on the agenda – should not be encouraged in their work ethic (cf. RB 48).

It is interesting the way persons, responsible for some aspect of monastic life, are characterised: the Cellerar should be as already mentioned, (RB 31): *"mature character, sober, not a great eater, not haughty, not excitable, not offensive, not slow, not wasteful, but a God-fearing man who may be like a father to the whole community."* In other words a perfect fatherly role-model figure (RB 31,1). And: *"Let helpers be given him, that by their assistance he may fulfill with a quiet mind the office committed to him,"* ("aequo animo"; RB 31,17); administrative duties demand that "temporalia" be used wisely and sparingly with as little fuss and bother as possible. The cellerar is mentioned again at a later point in the text: *"Let him not vex the brethren"* ("Fratres non contristet"), but: *"should humbly give the reason for denying the improper request"*.("rationabiliter cum humilitate"; RB 31,7) and not give a: *"contemptuous refusal"*. For: *"he who has ministered well will acquire for himself a good standing"* (1 Tim. 3:13). He is responsible for care – cum omni sollicitudine- of the sick, guests and the poor (RB 31,9); which might otherwise seem to be contradictory to a growth-oriented management scheme aiming to reach the highest maximum returns. The Abbot: *"shall take the greatest care that they*(the sick*) suffer no neglect"* (RB 36). Clearly, maximization is not the ultimate objective in a Benedictine household economy, since interestingly: *"At the second hour let Terce be said, and then let all labor at the work assigned them until None"* (RB 48,12) work is interrupted at regular intervals for proper prayer; it is not a question of milking the cow for as much as possible, or pushing subordinates beyond their own physical or mental limits: prayer is more important than economic gain. Such a scale of values – prayer before economic gain should perhaps be considered, both at the individual level as well as the collective. Seen thus, the *Regula Benedicti* do present a working and workable sustainable management ethic which has now been practised – tried and tested – for almost 1,500 year.

5 Concluding Comments

What can we learn from this tradition – from the Benedictine code of sustainable management and a sustainable entrepreneurial spirit? I would say three things: a meaningful set of values applied and practised in moderation in everyday business; stability anchored temperance and orderliness; and last but not least, co-operation and co-ordination in overcoming the Tragedy of Commons as the greatest threat to sustainability.

Literature

Bröckling, U. (2007). Das unternehmerische Selbst. Frankfurt/Main: Suhrkamp.
Diamond, J. (2005). *Collapse. How societies choose to fail or succeed*. New York: Viking.
Doyle, L. J. (2001). Saint Benedict's rule for monasteries. translated from the Latin by Leonard J. Doyle OblSB.
Ehrenberg, A. (2009). The Weariness of the Self. Diagnosing the history of depression in the contemporary age, with a preface by Allan Young. Montreal: McGill-Queen's University Press 2009. (The French original was published 1998 and entitled "La Fatigue d'être soi".)
Elon, A. (1988). *Founder: A portrait of the first Rothschild and his time*. London: Penguin (1996).
Ferguson, N. (1999). *The house of Rothschild. Money's prophets 1798–1848*. London: Penguin.
Inauen, E., & Frey, B. S. (2008). Benediktinerabteien aus ökonomischer Sicht. Working Paper No. 388. Institute for Empirical Research in Economics, University of Zurich, Zürich.
Landes, D. (1999). *The wealth and poverty of nations*. London: Abacus.

Lewicki, R. J., & Wiethoff, C. (2000). Trust, trust development, and trust repair. In M. Deutsch & P. T. Coleman (Eds.), *The handbook of conflict resolution: Theory and practice* (pp. 86–107). San Francisco: Jossey-Bass.

Luhmann, N. (2001). Einführende Bemerkungen zu einer Theorie symbolisch generalisierter Kommunikationsmedien, in: Aufsätze und Reden. Stuttgart.

Margalit, A. (1996). *The decent society*. Cambridge: Harvard University Press.

Max-Neef, M. (1991). *Human scale development. Conception, application and further reflections*. New York: The Apex Press.

Natella, S., et al. (2008). *Intangible infrastructure: Building on the foundation*. Zurich: Credit Suisse Research Institute.

Nussbaum, M. (2001). *Upheaval of thought*. Cambridge UK: Cambridge University Press.

Smith, R. (2002). In search of "non-disease". *British Medical Journal, 324*, 883–885.

Warfield, K., et al. (2007). Framing infrastructure in a cultural context. Working Paper 3. Centre of Expertise on Culture and Communities. Burnaby, British Columbia.

Wilson, D. (1988). *Rothschild. The wealth and power of a dynasty* (pp. 38–44). New York: Scribner.

International Perspective on Sustainable Entrepreneurship

Liangrong Zu

1 Introduction

The increasing social and environmental issues and challenges have become an imperative for businesses, governments and international community to be addressed, and failure to address these issues and challenges jeopardizes their ability to create prosperity and to be sustainable in the long-term for businesses and society as a whole. Sustainability has long been on the agenda at many companies, but for decades their social, environmental, and governance activities have been disconnected from core strategy. Most of companies still take a fragmented, reactive approach – launching ad hoc initiatives to enhance their "green" credentials, to comply with regulations, or to deal with emergencies – rather than treating sustainability as an issue with a direct impact on business results. The result of McKinsey's survey on sustainability shows that only 36 % of executives say their companies have a strategic approach to it, with a defined set of initiatives (McKinsey 2011). The global stakeholders and constituents such as government, NGOs, and international community have become a major force in promoting social responsibility and environmental sustainability. They can provide integrated solutions across interconnected issues area such as economic, social, environmental, and security action. They can facilitate universal dialogue to arrive at joint solutions and mobilize new constituencies to join businesses, governments and international organizations to address global emerging issues and challenges, and they also can legitimize new norms, structures and processes for international cooperation on sustainable development. Now a global movement that aims to encourage businesses to pay closer attention to their social and environmental impact has gained momentum in recent years.

Significant achievement in sustainable development should be realized through the strategies and initiatives of corporations, governments, and international

L. Zu (✉)

International Training Centre of the ILO, Viale Maestri del Lavoro, 10, Turin 10127, Italy

e-mail: l.zu@itcilo.org

C. Weidinger et al. (eds.), *Sustainable Entrepreneurship*, CSR, Sustainability, Ethics & Governance, DOI 10.1007/978-3-642-38753-1_6, © Springer-Verlag Berlin Heidelberg 2014

67

community if there is a long-term commitment to change. The long road toward sustainability involves incorporating sustainable development principles into the design and development of new products, business processes and new technologies over time. Entrepreneurship is increasingly being recognized as a significant conduit for bringing about a transformation to sustainable products and processes. Entrepreneurship exists in large companies, where individual executives take the initiative to innovate and expand the business. Entrepreneurship also drives many civil society organizations, and it exists in government and public administrations. Individuals in these organizations have the drive to innovate and pursue opportunities with the passion and dedication of an entrepreneur. Entrepreneurship also flourishes most in small and medium firms with significant potential to grow and innovate. This dynamic segment is typically the hotbed of entrepreneurship and innovation. It can drive economic growth, create jobs and foster competition, innovation and productivity.

It has been only recently that entrepreneurship is emerging as a new forum within which sustainability issues are being addressed. Entrepreneurship is concerned with activities to promote social-economic progress, creating new values, and providing employment opportunities. Therefore, entrepreneurship has the potentials to create value within each of the three dimensions (social, economic and environmental) of sustainability while boosting innovation through new products, services, and business models. The global stakeholders and constituents have been advocating and promoting entrepreneurship as a panacea for many social and environmental concerns.

This section intends to explore how and what the sustainable entrepreneurship are addressed from an international perspective, especially to examine the initiatives and principles of social responsibility and environmental sustainability that the international community has developed for businesses to be considered and integrated in their sustainable business strategy.

2   From Sustainable Business to Sustainable Entrepreneurship

Before the discussion of sustainable entrepreneurship, we develop the management construct of sustainable entrepreneurship and its subset design for sustainability (see Fig. 1).

Figure 1 shows the logical progressions and an overview of this perspective. The changing business environment, driven by the forces of economic, social and environmental issues, stakeholders, technology, has changed the way that companies are doing businesses traditionally. Particularly the changes in the external business environment, for example, competition for resources, climate change, economic globalization, connectivity and communications, etc., have put pressure on companies to fine-tune their business strategies to tackle the new risks and

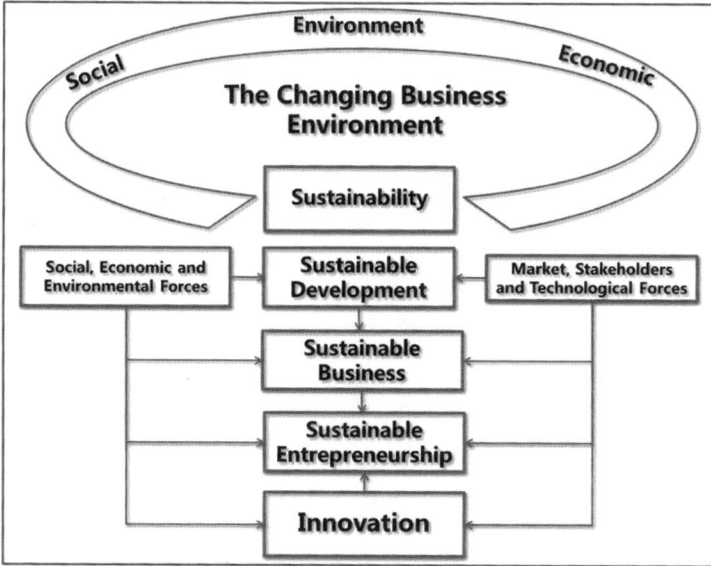

Fig. 1 The linkage between sustainability and Sustainable Entrepreneurship

challenges. More and more businesses will have to take a long-term strategic view of the issues by identifying and pursuing sustainability opportunities that hold the highest value potential. To respond to the forces and uncertainty, entrepreneurs must take proactive and innovative action to change repetitious and routine activities of an organisation and to explore new business opportunities through social responsibility and environmental sustainability program.

From the scholarly and academic perspective, sustainable entrepreneurship has been proposed as a "breakthrough discipline for innovation" (Fussler 1996), as a "source of creative destruction" (Hart and Milstein 1999), as well as the beginning of the "next industrial revolution" (McDonough and Braungart 1998; Levins and Hawken 1999; Senge and Carstedt 2001). In his influential book on Eco-Innovation, Fussler (1996) states that a majority of today's firms is not actively pursuing sustainability entrepreneurship as a strategy to create market share. However, he does not believe that this "innovation lethargy" (Fussler 1996) will persist in the years to come. Using a number of anecdotal case studies he shows that innovative firms can succeed in driving ecological innovation profitably, not by following current customer demand but by creating future market space. This notion that firms can actively transform market structures to make them more conducive to ecological innovation is also proposed by Dyllick (1999), Schaltegger and Wagner (2008) even propose that the ambition to transform an industry is a defining element of sustainable entrepreneurship, implying that sustainable entrepreneurial firms do not only see sustainability as central to core business activities, but at the same time aim for mass-market transformation beyond the eco-niche. On the social side of sustainability entrepreneurship the term "corporate social innovation" was first introduced by Kanter (1999) who argues that firms should use social issues as a learning laboratory for identifying unmet needs and for developing solutions that create new markets.

Thus, building on the above scholarly views, we argue that sustainable entrepreneurship is associated with the promise of more traditional concepts of entrepreneurship, but also has additional potential both for society and the environment. The topic of entrepreneurship for sustainable development lies at the nexus of innovation, sustainability concerns, and entrepreneurship as indicated in Fig. 1. Therefore, from the international perspective, in this section, sustainable entrepreneurship refers to the sustainable business initiatives, policies and strategies which companies are committed to the combination of economic, social and environmental value creation, and contribute to a sustainable future.

3 Social, Environmental and Sustainable Entrepreneurship

In recent years, the three different concepts of social, environmental and sustainable entrepreneurship have emerged in the management literature. Social, sustainable, and environmental entrepreneurship researchers attempt to understand the potentially positive outcomes for society that result from entrepreneurial actions. Academic work in each of these areas of research evolved from a variety of disciplines and methodologies, providing evidence that research in these areas of entrepreneurship is multidisciplinary. Whether these three concepts explore the same issues? If not, what are the differences among them? Thompson, Kiefer and York (2011) compare the differences and similarities for each of these areas by examining concepts, questions, methodologies, and disciplinary roots. They intend to clarify the terminology used in each of these literature streams to allow for distinction between each, arguing that social entrepreneurs focus mainly on problems that affect people today, environmental entrepreneurs are focused on creating simultaneous economic and ecological benefit, whereas sustainable entrepreneurs focus on a "triple bottom line" of people, planet, and profit. Social entrepreneurship is concerned with opportunities that relate to socially relevant issues and how altruistic motivations influence the exploitation of such opportunities. These opportunities may be exploited to for-profit or non-profit organizations, but regardless of their form, the organizations fostered by social entrepreneurs place primacy on their social mission, that is helping people. Environmental entrepreneurship is the simultaneous creation of economic and ecological profit by addressing environmentally relevant market failures. These organizations are for-profit firms, but they may be distinguished from traditional entrepreneurship by their focus on resolving environmental degradation through the creation of new products, services, and markets. Motivations are likely to be mixed, with a blend of environmental and economic ideology. These organizations focus on both creating an economic profit and provisioning environmental benefits. Sustainable Entrepreneurship examines opportunities to transition to a socially, economically, and environmentally sustainable society. Similarly to social entrepreneurship, these opportunities may be sought through organizations that create economic profit, or through non-profit means, but the organization must be economically self-sustaining. Again, this work focuses on noneconomic motivations of entrepreneurs to exploit these opportunities. These organizations balance the triple bottom line of people, planet, and profit, thus organizational design is a key consideration. This Section touches on all three concepts from the international perspective.

4 Sustainable Entrepreneurship: An International Perspective

The work and initiatives of the international community such as United Nations agencies, World Business Council on Sustainable Development (WBCSD) and other global stakeholders have become an important catalyst for the rapid emergence of sustainability and entrepreneurship. For example, the UN Global Compact, lunched in 2000, is a strategic policy initiative for businesses that are committed to aligning their operations and strategies with universally accepted principles in the areas of human rights, labour, environment and anti-corruption. Now over 10,000 corporate participants and other stakeholders from over 130 countries have joined Global Compact, it is the largest voluntary corporate responsibility initiative in the world. The Global Compact assists the private sector in the management of increasingly complex risks and opportunities in the environmental, social and governance realms, seeking to embed markets and societies with universal principles and values for the benefit of all. WBCSD, a Geneva-based consortium of more than 100 leading companies has spearheaded the business community's response to the 1992 Earth Summit conference in Rio de Janeiro, developing and initiating a series of programs and studies that demonstrate the business value of sustainability, and coordinating the definition of agendas for change in industries. In recent years, a number of WBCSD member corporations have established highly visible sustainability programs. For example, top management at BP, Dow, DuPont, Ford, General Motors, IBM and Royal Dutch Shell have gone public with ambitious commitments to generate shareholder returns while addressing the needs of humanity.

Table 1 lists some of the leading organizations that endorse and subscribe to sustainability, sustainable entrepreneurship, and identifies their working perspectives and definitions. It reveals that stakeholders in the global business environment typically view sustainable development and entrepreneurship in the context of the social, political, economic, technological and ecological implications of governmental laws, regulations, and action, and in terms of business operations, activities, and outcomes. The perspectives are often based on what the businesses

Table 1 Perspectives of selected leading international organizations on sustainable entrepreneurship

Organization	Perspectives	Background
International Labour Organization (ILO)	ILO is one of United Nations agencies, and its mission is to promote social justice and internationally recognized human and labour rights	The ILO was created in 1919. The ILO encourages the tripartism within its constituents and member states by promoting a social dialogue between trade unions and employers in formulating, and where appropriate, implementing national policy on social, economic, and many other issues
	ILO is the only UN body that brings together representatives of governments, employers and workers to jointly shape policies and programmes to promote sustainable entrepreneurship. ILO assists member countries to assess and adapt policies, laws and regulations with the goal of encouraging investment and entrepreneurship that balances the needs and interests of the enterprise – both workers and employers with the broader aspirations of society	Since 1919, a system of international labour standards has developed and maintained aimed at promoting opportunities for women and men to obtain decent and productive work, in conditions of freedom, equity, security and dignity. The international labour standards are an essential component in the international framework for ensuring that the growth of the global economy provides benefits to all
	ILO has developed sustainable entrepreneurship program to help entrepreneurs – including special target groups like youth and women to start and build successful enterprises, to support efforts to stimulate entrepreneurship through training, provision of business development services, access to information, technology and finance, and connecting enterprises to local value chains	The work of ILO focuses on promoting and realizing standards and fundamental principles and rights at work; creating greater opportunities for women and men to decent employment and income; enhancing the coverage and effectiveness of social protection for all, and strengthening tripartism and social dialogue
	One of the important ILO instruments for promoting sustainable and responsible businesses is the tripartite declaration of principles concerning multinational enterprises and social policy (MNE Declaration) declaration. It provides companies with the guidance on social policy and responsible labour practices	

(continued)

Table 1 (continued)

Organization	Perspectives	Background
United Nations Global Compact (UNGC)	UNGC is a United Nations initiative to encourage businesses worldwide to adopt sustainable and socially responsible policies, and to report on their implementation. UNGC is a principle-based framework for businesses, stating 10 principles in the areas of human rights, labour, the environment and anti-corruption The main work of UNGC is to assist the entrepreneurship and private sector in the management of increasingly complex risks and opportunities in the environmental, social and governance realms, seeking to embed markets and societies with universal principles and values for the benefit of all. Therefore, GC has developed and launched, in collaboration with other agencies, some of key principles and guidelines, including the principles for responsible investment; principles for responsible management education; supply chain sustainability: A practical guide for continuous improvement, and as well sustainability reporting system	UNGC was officially launched at UN headquarters in New York in 2000. It is supported by six UN agencies: the United Nations high commissioner for human rights; the United Nations environment programme; the international labour organization; the United Nations development programme; the United Nations industrial development organization; and the United Nations office on drugs and crime. Now over 10,000 corporate participants and other stakeholders from over 130 countries have joined UNGC. It is the world's largest corporate citizenship initiative with two objectives: "Mainstream the ten principles in business activities around the world" and "Catalyse actions in support of broader UN goals, such as the millennium development goals (MDGs)"
Office of the High Commissioner for Human Rights (OHCHR)	OHCHR is mandated to promote and protect the enjoyment and full realization, by all people, of all rights established in the charter of the United Nations and in international human rights laws and treaties One of the important policies for business and human rights is the framework for "Protect, Respect and Remedy", which was endorsed by the UN human rights council as the guiding principles in 2011. It provides a global standard for preventing and addressing the risk of adverse impacts on human	The office was established by the UN general assembly in 1993 in the wake of the 1993 world conference on human rights The office is headed by the high commissioner for human rights, who co-ordinates human rights activities throughout the UN system and supervises the human rights council in Geneva, Switzerland

(continued)

Table 1 (continued)

Organization	Perspectives	Background
	rights linked to business activity. For better managing business and human rights challenges, the policy framework helps businesses to identify and clarify standards of corporate responsibility and accountability for business enterprises with regard to human rights	
United Nations Environmental Program (UNEP)	UNEP's mission is to provide leadership and encourage partnership in caring for the environment by inspiring, informing, and enabling nations and peoples to improve their quality of life without compromising that of future generations	UNEP was founded as a result of the United Nations conference on the human environment in June 1972. It is an international institution that coordinates United Nations environmental activities, assisting developing countries in implementing environmentally sound policies and practices
	"Environment for development" underlines UNEP's vision of sustainability. It inspires, informs, and enables governments and people to improve their quality of life without compromising that of future generations	It is the voice for the environment, acting as a catalyst, advocate, educator and facilitator to promote the wise use and sustainable development of the global environment. Its offices develop policies and practices for mitigating the risk of degradation
	UNEP's work encompasses assessing global, regional and national environmental conditions and trends; developing international and national environmental instruments; strengthening institutions for the wise management of the environment; facilitating the transfer of knowledge and technology for sustainable development, and encouraging new partnerships and mind-sets within civil society and the private sector	It issues a biennial global environment outlook to provide an overview of the world-wide environment assessment process
	UNEP teams up with other organizations and industry partners to promote sustainable development thinking and practices and in production and in general business strategies.	UNEP has also aided in the formulation of guidelines and treaties on issues such as the international trade in potentially harmful chemicals, transboundary air pollution,

(continued)

Table 1 (continued)

Organization	Perspectives	Background
	One of the initiatives, launched jointly with society for environmental toxicology and chemistry (SETAC) is life cycle initiative (including life cycle management, and life cycle assessment)	and contamination of international waterways
United Nations Development Program (UNDP)	Since 1966, UNDP has been partnering with people at all levels of society to help build nations that can withstand crisis and drive and sustain the kind of growth that improves the quality of live for everyone	The UNDP was founded in 1965. It advocates for change and connects countries to knowledge, experience and resources to help people build a better life
	The UNDP report on "Unleashing entrepreneurship: Making business work for the poor" was issued in 2004. The report offers recommendations on how the major actors – governments, public development institutions, the private sector and civil society organizations – can modify their actions and approaches to significantly enhance the ability of the private sector to advance the development process. The objective of poverty alleviation leads UNDP to focus on developing businesses that create domestic employment and wealth by unleashing the capacity of local entrepreneurs	UNDP issues the annual human development report, focusing on the global debate on key development issues, provides new measurement tools, innovative analysis and often controversial policy proposals The UNDP works internationally to help countries achieve the millennium development goals (MDGs). Currently, the UNDP is one of the main UN agencies involved in the development of the post-2015 development agenda. In addition, UNDP also works in the areas of poverty reduction, democratic governance, crisis prevention and recovery, and environment and sustainable development
United Nations Industrial Development Organization (UNIDO)	The UNIDO's mandate is to promote and accelerate sustainable industrial development in developing countries and economies in transition. In recent years, UNIDO has assumed an enhanced role in the global development agenda by focusing its activities on poverty reduction, inclusive globalization and environmental sustainability	UNIDO was established in 1966 and became a specialized agency of the United Nations in 1985. It has 173 member states and is headquartered in Vienna, Austria. UNIDO works toward improving the quality of life of the world's poor by drawing on its combined global resources and expertise. It provides comprehensive and integrated packages of services which combine its operational activities with its analytical, normative and convening roles, both globally and locally

(continued)

Table 1 (continued)

Organization	Perspectives	Background
	UNIDO also promotes sustainable patterns of industrial consumption and production to de-link the processes of economic growth and environmental degradation. UNIDO is a leading provider of services for improved industrial energy efficiency and the promotion of renewable sources of energy. It also assists developing countries in implementing multilateral environmental agreements and in simultaneously reaching their economic and environmental goals	UNIDO's three priorities focus on poverty reduction through productive activities; trade capacity-building, and energy and environment
	UNIDO is also actively engaged in promoting entrepreneurship and private sector development in developing countries and transition economies through the provision of policy advice and institutional capacity-building services	
International Finance Corporation (IFC)	IFC, a member of the world bank group, is the largest global development institution focused exclusively on the private sector in developing countries	Established in 1956, IFC is owned by 184 member countries, a group that collectively determines the policies. Its work in more than a 100 developing countries allows companies and financial institutions in emerging markets to create jobs, generate tax revenues, improve corporate governance and environmental performance, and contribute to their local communities
	IFC has developed the sustainability framework to promote sustainable entrepreneurship. The IFC's framework articulates the corporation's strategic commitment to sustainable development. It comprises IFC's policy and performance standards on environmental and social	IFC's strategic priorities include strengthening the focus on frontier markets; addressing climate change and ensuring environmental and social sustainability; addressing constraints to private sector growth in infrastructure, health, education, and the food-supply chain; developing local

(continued)

Table 1 (continued)

Organization	Perspectives	Background
	sustainability, and IFC's access to information policy. The policy on environmental and social sustainability describes IFC's commitments, roles, and responsibilities related to environmental and social sustainability The sustainability framework along with other strategies, policies, and initiatives is used by IFC to direct the business activities of the corporation in order to achieve its overall development objectives	financial markets and building long-term client relationships in emerging markets
World Business Council for Sustainable Development (WBCSD)	The WBCSD is committed to sustainable development via the three pillars of economic growth, environmental protection and social equality. It galvanizes the global business community to create a sustainable future for business, society and the environment through the following ambitious activities and objectives: – be a leading business advocate on sustainable development; – participate in policy development to create the right framework conditions for business to make an effective contribution to sustainable human progress; – develop and promote the business case for sustainable development; – demonstrate the business contribution to sustainable development solutions and share leading edge practices among members, and – contribute to a sustainable future for developing nations and nations in transition	The WBCSD was created in 1995 in a merger of the business council for sustainable development and the world industry council for the environment and is based in Geneva, Switzerland The council provides a platform for companies to explore sustainable development, share knowledge, experiences and best practices, and to advocate business positions on these issues in a variety of forums, working with governments, non-governmental and inter-governmental organizations The council plays the leading advocacy role for business. Leveraging strong relationships with stakeholders, it helps drive debate and policy change in favour of sustainable development solutions WBCSD has been identified as one of the "most influential forums" for companies on corporate social responsibility and sustainability issues over the past years
Coalition of Environmentally Responsible Economics (CERES)	Ceres, a non-profit organization, an advocate for sustainability leadership. Ceres's mission is to mobilises a network of	Founded in 1989, Ceres has been working for more than 20 years to weave sustainable strategies and practices into the fabric and

(continued)

Table 1 (continued)

Organization	Perspectives	Background
	investors, companies and public interest groups to accelerate and expand the adoption of sustainable business practices and solutions to build a healthy global economy Ceres is committed to helping business leaders, investors and policymakers address sustainability the smart way – by developing practical strategies that are good for society, the earth, and the bottom line. Ceres bring investors, public interest groups and companies together to examine complex environmental and social issues. Among the many issues, Ceres works with companies, investors and policymakers on climate change, energy, water and supply chain Ceres launched "the Ceres roadmap for sustainability" in 2010, which serves as a vision and practical guide for integrating sustainability into the DNA of business. The Ceres roadmap is designed to provide a comprehensive platform for sustainable business strategy and for accelerating best practices and performance	decision-making of companies, investors and other key economic players. It leverages the power of their partners – leading investors, Fortune 500 companies, thought leaders and policymakers – to positively influence change. Ceres works with more than 130 member organizations that make up the Ceres coalition to engage with corporations and help advance the goal of building a sustainable global economy, and works with more than 80 companies across a broad range of sectors committed to engaging with diverse stakeholders, improving their performance on social and environmental issues and disclosing strategies and progress publicly. Ceres also works with investors worldwide to improve corporate strategies and public policies on climate change and other environmental and social challenges across the global economy
International Organization for Standards (ISO)	ISO is an international standard-setting body composed of representatives from various national standards organizations. international standards give state of the art specifications for products, services and good practice, helping to make industry more efficient and effective. Developed through global consensus, ISO helps to break down barriers to international trade ISO 26000, launched in 2010, is one of important standards on social responsibility and	Since it was founded in 1947. ISO works to facilitate the international coordination and unification of industrial standards, ISO has published more than 19500 international standards covering almost all aspects of technology and business. From food safety to computers, and agriculture to healthcare, ISO international standards impact all human lives Today ISO has members from 161 countries and 3,368

(continued)

Table 1 (continued)

Organization	Perspectives	Background
	sustainability. ISO 26000 provides guidance for businesses and organizations to translate principles into effective actions and shares best practices relating to social responsibility	technical bodies to take care of standard development
Social Accountability International (SAI)	SAI is a non-governmental, multi-stakeholder organization whose mission is to advance the human rights of workers around the world. It partners to advance the human rights of workers and to eliminate sweatshops by promoting ethical working conditions, labor rights, corporate social responsibility and social dialogue	SAI was established in 1997. SAI works to protect the integrity of workers around the world by building local capacity and developing systems of accountability through socially responsible standards
	SAI shared vision is of decent work everywhere – sustained by widespread understanding that decent work can secure basic human rights while benefiting business	The SA8000 standard for decent work, a tool for implementing international labor standards that is being used in over 3,000 factories, across 65 countries and 66 industrial sectors
	The SA8000 standard for decent work was developed in 1997. SA8000 is an auditable certification standard that encourages organizations to develop, maintain, and apply socially acceptable practices in the workplace	SAI is one of the world's leading social compliance training organizations, having provided training to over 30,000 people, including factory and farm managers, workers, brand compliance officers, auditors, labor inspectors, trade union representatives and other worker rights advocates
World Commission on Environment and Development (WCED)	The WCED views sustainability as the principle of ensuring that actions today do not limit the range of economic, social and environmental options open to future generations	The WCED is an international commission started in 1987 to study the connection between economics and the environment The commission wrote our common future (The Brundtland Report)
International Chamber of Commerce (ICC)	ICC is an international NGO established in 1919 to promote trade, investment, and the market economy. Its membership extends to more than 130 countries and includes thousands of business organizations and enterprises with international interests	The environment and energy commission (as of Oct 2007) comprises 227 members representing 75 multinational corporations from as well as representatives from 33 industry associations, and 52 ICC national committees that federate ICC members in their

(continued)

Table 1 (continued)

Organization	Perspectives	Background
	ICC is on the forefront in the development of ethics, anti-corruption and corporate responsibility advocacy codes and guidelines, providing a lead voice for the business community in this rapidly changing field The environment and energy commission makes recommendations for business on significant regulatory and market issues concerning energy and environment	countries. The commission examines major environmental and energy related policy issues of interest to world business via task forces and thematic groups. The commission usually meets twice a year though task forces and other thematic groups may meet more frequently. ICC formulated its Business Charter for Sustainable Development in 1990
World Economic Forum (WEF)	WEF includes sustainable development in its agenda	WEF is an annual meeting in Switzerland at which political leader and business executives discuss economic, social and environmental issues
	It discusses environmental issues as related to economic performance WEF encourages businesses, governments and civil society to commit together to improving the state of the world. Its strategic and industry partners are instrumental in helping stakeholders meet key challenges such as building sustained economic growth, mitigating global risks, promoting health for all, improving social welfare and fostering environmental sustainability	WEF is in close collaboration with the Schwab foundation for social entrepreneurship to provide unparalleled platforms at the regional and global level to highlight and advance leading models of sustainable social innovation. They identify a select community of social entrepreneurs and engage it in shaping global, regional and industry agendas that improve the state of the world
Watchworld Institute (WWI)	The WWI works to accelerate the transition to a sustainable world that meets human needs. The Institute's top mission objectives are universal access to renewable energy and nutritious food, expansion of environmentally sound jobs and development, transformation of cultures from consumerism to sustainability, and an early end to population growth through healthy and intentional childbearing WWI publishes the state of the world on an annual basis	WWI is a NGO, founded in 1974 to devote to global environmental concerns. WWI develops innovative solutions to intractable problems, emphasizing a blend of government leadership, private sector enterprise, and citizen action that can make a sustainable future a reality

(continued)

Table 1 (continued)

Organization	Perspectives	Background
World Resources Institute (WRI)	WRI's mission is to move human society to live in ways that protect Earth's environment and its capacity to provide for the needs and aspirations of current and future generations. WRI focuses on the intersection of the environment and socio-economic development, working globally with governments, business, and civil society to build transformative solutions that protect the earth and improve people's lives	WRI was launched in 1982 as a centre for policy research and analysis addressed to global resource and environmental issues
	The objective of WRI's sustainability initiative is to "learn by doing" and to apply their research to reduce the environmental footprint	WRI is a global environmental think tank that goes beyond research to put ideas into action. WRI has over 50 active projects working on aspects of global climate change, sustainable markets, ecosystem protection, and environmentally responsible governance
The International Institute for Sustainable Development (IISD)	IISD's mission is to champion innovation, enabling societies to live sustainably	IISD was founded in Canada in 1990. IISD is headquartered in Winnipeg and has offices in Ottawa, New York and Geneva. It has over 100 staff and associates working in over 30 countries. It is a non-partisan charitable organization specializing in policy research and analysis, and information exchange
	IISD promotes the transition toward a sustainable future, and seek to demonstrate how human ingenuity can be applied to improve the well-being of the environment, economy and society through the tools of policy research, information exchange, analysis and advocacy	IISD reports on international negotiations and disseminate knowledge gained through collaborative projects, resulting in more rigorous research, capacity building in developing countries, better networks spanning the North and the South, and better global connections among researchers, practitioners, citizens and policy-makers
	IISD contributes to sustainable development by advancing policy recommendations on international trade and	

(continued)

Table 1 (continued)

Organization	Perspectives	Background
	investment, economic policy, climate change and energy, and management of natural and social capital, as well as the enabling role of communication technologies in these areas	

and governments should be, rather than what they are doing. The views are also based on how they should interact with the business environment and the natural world.

The session will address the sustainable entrepreneurship from the international perspective by focusing on social and labour issues, business and human rights, sustainable and responsible supply chain, environmental sustainability, and as well as sustainability reporting and auditing.

4.1 Perspective on Social and Labour Issues

The social and labour issues and challenges, such as poverty, employment, gender equality, child mortality and child labour, safety and health, HIV/AIDS, malaria and other diseases, etc. have been put on the agenda of businesses and international community over the past decades. "Eradicating extreme poverty continues to be one of the main challenges of our time, and is a major concern of the international community. Ending this scourge will require the combined efforts of all, governments, civil society organizations and the private sector, in the context of a stronger and more effective global partnership for development "(BAN Ki-moon 2009).

The Millennium Development Goals (MDGs), adopted by world leaders in 2000 and set to be achieved by 2015, provide a framework for the entire international community to work together towards a common end. MDGs include goals and targets on income poverty, hunger, maternal and child mortality, disease, inadequate shelter, gender inequality, environmental degradation and the global partnership for development. The eight goals – along with a set of targets and indicators – serve as milestones against which to measure international and country progress towards the overall goal of reducing extreme poverty.

Business has the power and responsibility in the fight against poverty and other social issues, and thus plays an important role in achieving MDGs. The research in the area of the Growing Inclusive Markets Initiative, undertaken by UNDP over the past years, shows that the inclusive business models can contribute towards meeting the Millennium Development Goals, because they include poor people into value chains as producers, employees and consumers. Business has six essential roles to play: by generating growth, by including poor people into their value chains, by contributing knowledge and capabilities, by developing innovative approaches, by replicating those approaches across borders and by advocating for policies that will alleviate poverty. The private sector can alleviate poverty by contributing to economic growth, job creation and poor people's incomes. It can also empower poor people by providing a broad range of products and services at lower prices. Small and medium enterprises can be engines of job creation – seedbeds for innovation and entrepreneurship. But in many poor countries, small and medium enterprises are marginal in the domestic ecosystem. Many operate outside the formal legal system, contributing to widespread informality and low productivity. They lack access to financing and long-term capital, the base that companies are built on (UNDP 2008, 2010). In this context, an global business leadership platform, called Business Call to Action (BCtA), was launched in 2008, to accelerate progress towards the MDGs. BCtA challenges companies to develop inclusive business models that offer the potential for both commercial success and development impact. Now, worldwide, 63 companies have responded to the BCtA by making commitments to improve the lives and livelihoods of millions through commercially-viable business ventures that engage low-income people as consumers, producers, suppliers, and distributors of goods and services.

One of the ways to combat poverty is to create more jobs and decent work for all. However, the recent financial crisis has deteriorated the labour world and brought about mass unemployment, underemployment and cuts in wage earnings and social benefits in many countries. The ILO's statistics shows the dismal global picture (lLO 2013):

- Social and economic inequalities in their multiple forms are rising.
- Some 200 million women and men are unemployed.
- A further 870 million women and men – a quarter of the world's working people – are working but unable to lift themselves and their families above the $2 a day per person poverty line.
- Some 74 million young women and men have no jobs. Youth unemployment is at dramatic levels in a number of countries in Europe and North Africa. The length of time young people are remaining idle is increasing and the scars of youth unemployment can last a lifetime.
- Alongside jobless young women and men, child labour persists.
- So too does forced labour – in seeking to escape the traps of joblessness and poverty at home, many women and men are falling into the traps of human traffickers in modern forms of slavery.
- Eighty percent of the world's population lacks adequate social security coverage and more than half have no coverage at all.
- Discrimination in its many manifestations is holding back hundreds of millions, especially women, from realizing their potential and contributing on an equal footing to the development of our societies and economies.
- And in many countries working women and men seeking to exercise their right to organize freely to uphold justice and dignity at work are prevented from forming and joining trade unions.

Therefore, Mr. Guy Ryder, ILO Director-General sent the message to the business world at World Day for Social Justice in February 2013 to call for the international cooperation and policy coordination for recovery to transform into inclusive, equitable, sustainable global development, to focus on generating full and productive employment and decent work for all including through support for small and medium-sized enterprises. There must be the recognition that respect for fundamental rights at work unleashes human potential and supports economic development as do social protection floors. A commitment to building a culture of social dialogue also helps to generate just, balanced and inclusive policies. This is the underpinning of the legitimacy and sustainability of open societies and of the global economy.

ILO and other UN agencies have come up with different initiatives, principles, guidelines, and labour standards to advocate and promote social justice, sustainable entrepreneurship. One of the important initiatives is the promotion of sustainable enterprise development (SED). It was proposed at the June 2007 International Labour Conference. Promoting sustainable enterprises is about strengthening the rule of law, the institutions and governance systems which nurture enterprises, and encouraging them to operate in a sustainable manner. Central to this initiative is an enabling environment which encourages investment, entrepreneurship, workers' rights and the creation, growth and maintenance of sustainable enterprises. Policy frameworks must balance the need of enterprises to turn a profit with the aspiration of society for a path of development that respects the values and principles of decent work, human dignity and environmental sustainability (ILO 2007).

The social dimension of sustainable development from ILO's perspective includes "a commitment to promote social integration by fostering societies that are stable, safe and just and which are based on the promotion and protection of all human rights and on non-discrimination, tolerance, respect for diversity, equality of opportunity, security and participation of all people including the disadvantaged and vulnerable groups and persons". A central tenet of the social pillar of sustainable development is the generation of secure livelihoods through freely chosen productive employment.

The social responsibility for businesses is also embedded in the Principles of the UN Global Compact. UNGC is a principle-based framework for businesses, stating ten principles in the areas of human rights, labour, the environment and anti-corruption. UNGC requires companies to embrace, support and enact, within their sphere of influence, a set of core values in the areas of human rights, labour standards, the environment and anti-corruption. Of the ten principles, four principles are concerned with the labour issues:

- Principle 3: Businesses should uphold the freedom of association and the effective recognition of the right to collective bargaining;
- Principle 4: the elimination of all forms of forced and compulsory labour;
- Principle 5: the effective abolition of child labour; and
- Principle 6: the elimination of discrimination in respect of employment and occupation.

Multinational enterprises (MNEs) play even much more powerful and influential role in alleviating poverty and creating employment in the world. ILO laid down the guidelines to MNEs in the Tripartite Declaration of Principles concerning Multinational Enterprises and Social Policy (MNE Declaration) which was adopted by the ILO Governing Body in 1977. MNE Declaration requires businesses, governments and workers' organizations to apply, to the greatest extent possible, the guidelines in the areas of employment, training, conditions of work and life, and industrial relations.

Today, the prominent role of MNEs in the process of social and economic globalization renders the application of the principles of the MNE Declaration as timely and necessary as they were at the time of adoption. As efforts to attract and boost foreign direct investment gather momentum within and across many parts of the world, the parties concerned have a new opportunity to use the principles of the Declaration as guidelines for enhancing the positive social and labour effects of the operations of MNEs.

4.2 Perspective on Business and Human Rights

Human rights have been a concern for some companies since the anti-Apartheid divestment campaigns of the 1980s, but there has been no broad-based uptake of human rights as a business discipline. Relatively few companies have human rights in their corporate vocabulary. Therefore, during his serving as the UN Secretary General's Special Representative on Business and Human Rights, Professor John Ruggie has forged a working consensus among companies, governments and advocate that human rights are not just a business concern, but that both governments and companies have human rights responsibilities. This led to the initiative of the Ruggie, or UN, Framework of "*Protect, Respect and Remedy*". The Framework asserts that governments must protect against abuses by companies; companies must respect human rights; and victims must have access to remedies. It is a policy framework for better managing business and human rights challenges,

The Framework of Business and Human Rights was endorsed by the UN Human Rights Council as the Guiding Principles in its resolution for both governments and companies to meet their responsibilities under the Framework. It is for the first time that companies have a clear roadmap for making human rights part of their compliance and corporate responsibility efforts.

One of the main goals of the UN Framework 'Protect, Respect and Remedy' is to identify and clarify standards of corporate responsibility and accountability for business enterprises with regard to human rights. For example, the foundational principles for corporate responsibility to respect human rights include (HRC 2011):

- Business enterprises should respect human rights. This means that they should avoid infringing on the human rights of others and should address adverse human rights impacts with which they are involved

- The responsibility of business enterprises to respect human rights refers to internationally recognized human rights – understood, at a minimum, as those expressed in the International Bill of Human Rights and the principles concerning fundamental rights set out in the International labour Organization's Declaration on Fundamental Principles and Rights at Work.
- The responsibility to respect human rights requires that business enterprises:
  - Avoid causing or contributing to adverse human rights impacts through their own activities, and address such impacts when they occur;
  - Seek to prevent or mitigate adverse human rights impacts that are directly linked to their operations, products or services by their business relationships, even if they have not contributed to those impacts.
- The responsibility of business enterprises to respect human rights applies to all enterprises regardless of their size, sector, operational context, ownership and structure. Nevertheless, the scale and complexity of the means through which enterprises meet that responsibility may vary according to these factors and with the severity of the enterprise's adverse human rights impacts.
- In order to meet their responsibility to respect human rights, business enterprises should have in place policies and processes appropriate to their size and circumstances, including:
  - A policy commitment to meet their responsibility to respect human rights;
  - A human rights due diligence process to identify, prevent, mitigate and account for how they address their impacts on human rights;
  - Processes to enable the remediation of any adverse human rights impacts they cause or to which they contribute.

Therefore, for the executives and managers of companies, they should know and do ten things about human rights:

1. Corporate human rights responsibilities go beyond legal compliance.
2. Consider the full range of human rights and all of your company's activities and relationships.
3. Adopt a human rights policy.
4. Invest in human rights due diligence.
5. Act on the findings.
6. Track and communicate your human rights performance.
7. Ensure that corporate human rights initiatives contain effective grievance mechanisms.
8. The UN Framework is now the de facto human rights standard for companies and their stakeholders.
9. Stakeholders will use the UN Framework to hold your company accountable for respecting human rights.
10. Pay attention to how governments and companies connect the "Protect" and "Respect" pillars of the UN Framework.

Executives may be tempted to focus exclusively on the "Respect" pillar of the Framework, because it describes the responsibilities of companies. The connections among the pillars, however, are where the most challenging issues arise, and where there are likely to be significant developments going forward (HRC 2011).

The Guiding Principles also requires governments to foster corporate respect for human rights; and companies to encourage governments to protect human rights. States may establish legal liability for legal persons (corporations) and take steps to regulate firms operating abroad. Specific measures on the horizon are human rights reporting requirements, and mandatory human rights impact assessments. Financial reporting requirements may be the first place where required human rights reporting by redefining materiality to include human rights impacts. In the other direction, companies are increasingly raising human rights concerns with the governments of countries where they operate.

4.3 Perspective on Sustainable Supply Chains

The widespread concerns about poor social and environmental conditions in companies' supply chains have emerged over the past years. Weak implementation of local social and environmental regulation has forced companies to address issues that traditionally have been seen to lie outside of their core competencies and responsibilities. Moreover, public scrutiny of business behaviour has led to rising expectations that companies are responsible for the environmental, social and governance (ESG) practices of their suppliers. Failure to address suppliers' ESG performance can give rise to significant operational and reputational risks that can threaten to undermine any potential gains from moving into these markets. As a result, a company's overall commitment to corporate citizenship can be seriously discredited if low standards of business conduct are found to persist in their supply chain. Thus, supply chain sustainability is increasingly recognized as a key component of corporate responsibility. Managing the social, environmental and economic impacts of supply chains, and combating corruption, make good business sense as well as being the right thing to do.

The United Nations Global Compact (UNGC) and BSR (Business for Social Responsibility) launched a joint project in 2010 to develop strategic guidance materials for business on the implementation of the Ten Principles in supply chain programmes and operations, and to help companies overcome these challenges by offering practical guidance on how to develop a sustainable supply chain programme based on the values and principles of the Global Compact. The practical guide for Supply Chain Sustainability was developed in 2010, in the meantime, UNGC also provides companies with the Quick Self-Assessment & Learning Tool, and online the resources and practices of sustainable Supply Chains. This is the one-stop-shop for business seeking information about supply chain sustainability. It includes initiatives, programmes, codes, standards and networks, resources and tools, case examples of company practices. Companies can also access information UNGC designed to assist business practitioners in embedding sustainability in supply chains at UNGC homepage.

Supply chain sustainability is defined in UNGC guide as the management of environmental, social and economic impacts, and the encouragement of good governance practices throughout the lifecycles of goods and services. The objective of supply chain sustainability is to create, protect and grow long-term environmental, social and economic value for all stakeholders involved in bringing products and services to market (UNGC and BSR 2010).

The benefits of integrating the UN Global Compact principles into supply chain relationships, companies are able to advance corporate sustainability and to promote broader sustainable development objectives. There are numerous reasons why companies start a supply chain sustainability journey. Primary among them is to ensure compliance with laws and regulations and to adhere to and support international principles for sustainable business conduct. In addition, companies are increasingly taking actions that result in better social, economic and environmental impacts because society expects this and because there are business benefits to doing so. By managing and seeking to improve environmental, social and economic performance and good governance throughout supply chains, companies act in their own interests, the interests of their stakeholders and the interests of society at large.

Coalition of Environmentally Responsible Economics (CERES) is an advocate for sustainability leadership. Ceres mobilizes a powerful network of investors, companies and public interest groups to accelerate and expand the adoption of sustainable business practices and solutions to build a healthy global economy. Since it was founded in 1989, Ceres has been working for more than 20 years to weave sustainable strategies and practices into the fabric and decision-making of companies, investors and other key economic players. Sustainable supply chain is one of the key issues that Ceres deals with.

Ceres's research shows that for many companies, the largest opportunity for improving sustainability performances such as reducing carbon emissions, water use, toxic chemicals and addressing social and human rights concerns is in its global supply chain. For example, up to 60 % of a manufacturing company's carbon footprint is in its supply chain. By focusing not only on auditing and remediation but also on supplier education and engagement, companies can significantly reduce environmental and social impacts while also raising the standards of their suppliers and improving the bottom line (Ceres 2013b). Ceres requires that sustainable supply chain performance begins with companies establishing supplier policies and endorsing industry codes or practices that contain explicit references to social and environmental standards. Corporate policies and practices for their supply chain should not only cover where and how materials are sourced and the environmental impact of a product's life-cycle, but should also recognize the rights of supply chain workers as well as those directly employed by the company. By bringing sustainability improvements deep into supply chains, companies can better protect their reputations from human rights and environmental violations, increase productivity and save on costs related to energy, water use and reductions in waste and toxic chemical use. Ceres helps companies and investors understand the business case for improving supply chains and recognizing human rights, and provide guidance for partner companies.

In 2010, Ceres developed *"The Twenty-First Century Corporation: Ceres Roadmap for Sustainability outlines"*, which serves as a vision and practical guide for integrating sustainability into the DNA of business – from the boardroom to the copy room. It analyzes the drivers, risks and opportunities involved in making the shift to sustainability, and details strategies and results from companies who are taking on these challenges. The Ceres Roadmap is designed to provide a comprehensive platform for sustainable business strategy and for accelerating best practices and performance. The Ceres Roadmap also urges corporations to view human rights in terms of the overall supply chain and to use their influence to spread best practices – for example, by choosing suppliers and partners whose policies protect workers' rights (Ceres 2010).

Ceres also developed a self-assessment tool for companies to advance sustainable supply chains. *"The Supplier Self-Assessment Questionnaire (SAQ): Building the Foundation for Sustainable Supply Chains"* is provided for all companies to seek to strengthen their supply chain engagement, though it was designed with the industrial goods sector in mind, as well as those that are just beginning to address sustainability issues in their supply chains (Ceres 2012). Drawing on leading practices in the field, and addressing environmental, social, and governance issues, the SAQ is a "conversation starter" for companies to use with their suppliers as they begin to assess the sustainability risks in their supply chains. The goal is to help companies be more competitive and build resiliency in their supply chains by identifying, assessing, managing and disclosing supply chain sustainability risks. The SAQ is part of a broader strategy to raise the bar of supply chain sustainability performance across the global economy. Traditionally, supply chain management has focused on whether a particular supplier facility is complying with certain minimum standards or codes of conduct related to treatment of workers or environmental impacts. The SAQ is part of a broader strategy to raise the bar of supply chain sustainability performance across the global economy.

4.4 Perspective on Environmental Sustainability

Environmental sustainability is integral to and a key pillar of sustainable development. It is one of the Goals set in Millennium Development Goals (MDGs). Environmental sustainability is at the heart of the seventh goal (MDG 7). The global MDG framework contains targets and indicators that can be used to measure global progress towards achieving each of the goals (Table 2). In the case of MDG 7, the targets and indicators are illustrative of key global environmental issues and commitments. Because they are global in nature, they require responses from both developed and developing countries, with common but differentiated responsibilities. The framework assumes that improvements at the national level would impact regional and global trends through meeting the targets by 2015.

Table 2 MDG 7 global targets and indicators

MDG 7: Ensure environmental sustainability	
Targets	Indicators
Target 7.a Integrate the principles of sustainable development into country policies and programme and reverse the loss of environmental resources	7.1: Proportion of land area covered by forest 7.2: CO_2 emissions, total, per capita and per \$1 GDP (PPP) 7.3: Consumption of ozone-depleting substances
Target 7.b Reduce biodiversity loss, achieving by 2010, a significant reduction in the rate of loss	7.4: Proportion of fish stocks within safe biological limits 7.5: Proportion of total water resources used 7.6: Proportion of terrestrial and marine areas protected 7.7: Proportion of species threatened with extinction
Target 7.c Halve. By 2015, the proportion of people without sustainable access to safe drinking water and basic sanitation	7.8: Proportion of population using an improved drinking water source 7.9: Proportion of population using an improved sanitation facility
Target 7.d Achieve significant improvement in the lives of at least 100 million slum dwellers, by 2020	7.10: Proportion of urban population living in slums

Source: UNDP(2006), Making Progress on Environmental Sustainability: Lessons and Recommendations from Review of Over 150 MDG Countries Experiences. Environmental & Energy Group. Bureau for Development Policy, UNDP, October 2006. New York

UNDP and UNEP have been working to support countries in sound environmental management and, in particular, on achieving MDG 7 on environmental sustainability. Millennium Development Goal 7 contains three global targets – Target 9 to integrate the principles of sustainable development into country policies and programmes and reverse the loss of environmental resources; Target 10 to halve, by 2015, the proportion of people without sustainable access to safe drinking water and sanitation; and Target 11 to have achieved, by 2020, a significant improvement in the lives of at least 100 million slum dwellers – and eight global indicators that can be used to measure global progress. While these targets and indicators are a starting point for monitoring country-level progress towards ensuring environmental sustainability, they do not necessarily capture national and local priority issues.

In fighting against climate change, Ceres works at the intersection of business, investment and advocacy communities to address sustainability risks such as climate change in order to build a more sustainable global economy. Ceres works with more than 80 companies on a range of sustainability issues, including climate change. Working with their coalition of investors and advocacy groups to reduce the absolute greenhouse gas emissions (GHG) from business operations, supply chain, products and services and employees; disclose the financial and material implications of climate change as well as their plans and goals for mitigating that risk; put in place strong governance structures to manage risks at the board and CEO levels of the company, and develop products and practices that decrease GHGs and generate revenue for the company.

Ceres also leads the Investor Network on Climate Risk, a group of 100 investors representing more than $10 trillion in assets under management (Ceres 2013). Under Ceres' direction, this vast network is combating climate risk in their portfolios by engaging with corporations on their business strategies to reduce risk due to climate change through stakeholder engagements and shareholder resolutions; calling on the U.S. and other governments to pass strong climate and energy policies that will spur low-carbon investments, new jobs and transition us to a clean energy economy; integrating the financial and material risks of climate change into investment decisions. Ceres has organized businesses and investors nationwide and globally to call for strong climate and energy policies that will reduce carbon emissions, promote energy efficiency and renewable energy and increase investment in a clean energy economy.

Climate change presents business and investors with challenges and opportunity. Innovative products and new technologies are emerging to help mitigate pollution, reduce reliance on fossil fuels and limit the overall impact on the environment – all the while creating new markets, job opportunities and growth potential. Those companies and investors that seize these opportunities are best positioned to thrive in a resource-constrained economy.

Sustainable energy fuels sustainable development, and drives economic growth, expands social equity, and helps create a healthier environment for us all. Energy is central to nearly every major challenge and opportunity the world faces today. Energy enables social and economic development, from basic needs to advanced industrial activity. However, energy sourcing and usage also have a significant impact on the environment, and companies are under more scrutiny than ever about producing and consuming energy in a more sustainable manner.

To mobilize action and partnerships focused on sustainably meeting the increasing energy requirements of businesses and society, United Nations Secretary-General Ban Ki-moon has launched a global initiative. Called Sustainable Energy for All, the initiative has set three primary objectives, to be met by 2030: ensuring universal access to modern energy services; doubling the global rate of improvement in energy efficiency; and doubling the share of renewable energy in the global energy mix. Sustainable Energy for All strives to leverage the global convening power of the United Nations to mobilize people, organizations and countries on a broad scale and to facilitate a rapidly expanding, cross-sector knowledge and action network.

In 2011, United Nations launched the Sustainable Energy for All initiative with the goal of bringing together governments, businesses and civil society groups in an unprecedented effort to transform the world's energy systems by 2030. The three complimentary objectives for the initiative which have been set to achieve by 2030 are to ensure universal access to modern energy services; to double the global rate of improvement in energy efficiency, and to double the share of renewable energy in the global energy mix (UNGC 2012).

This initiative is one of the great opportunities for businesses to identify solutions for some of the toughest global challenges businesses and governments face such as poverty, inequality, energy security, climate change and environmental protection. These commitments demonstrate that businesses can make tremendous progress when all key stakeholders – developed and developing countries, private companies and civil society groups – work together in common cause for the common good. The initiative will require commitment and vigorous action from the private sector. It is obvious that many barriers – financial, political, and technical – will persist as the world works toward achieving these objectives, sustainable Energy for All is meant to remove some of these barriers – through its voice, convening power, and focus on mobilizing action and facilitating new public-private partnerships.

The Sustainable Energy for All initiative provides a clearly articulated, global, and shared vision for sustainable energy. It can leverage the unparalleled convening power and reach of the United Nations to build consensus and drive a common agenda. A significant challenge businesses face in advancing their strategic goals while pursuing a sustainable energy agenda is the complex ecosystem of players involved – internal company stakeholders, suppliers, customers, communities, governments, and more. Successfully driving value from action to advance access to energy, energy efficiency, and renewable energy, will require not only business innovation and investment, but also alignment and cooperation with relevant stakeholders. The UN Secretary General's vision for Sustainable Energy for All and the three objectives endorsed by governments and stakeholders from around the world, provide a common language and shared targets that companies can align with as they set strategic priorities, address the expectations of various stakeholders, and seek foundations to increase cross-sector collaboration.

Sustainable Energy for All also provides opportunities for peer companies within an industry to collaborate, when they are often justifiably reluctant to work together. The global objectives of the initiative can provide a shared target for companies who may typically compete with each other to focus the best of the their resources in partnership – in a fair, neutral, and respectful environmental where each stakeholder is viewed as an equal partner driving towards a common interest. By leveraging Sustainable Energy for All as a platform for new, "pre-competitive and safe" partnerships, businesses can work together to identify cost saving energy efficiency measures or develop innovative products and services.

4.5 Perspective on Sustainability Reporting and Social Auditing

The last decade has witnessed the rise of sustainability as a defining element of responsible business strategy and performance. As organizations work to address the sustainability challenges of the twenty-first century, we need to identify the better ways to assess sustainability performance.

Social and environmental performance has been viewed as one of the important indicators for companies to demonstrate the extent of their commitment to social responsibility and environmental sustainability, it also a key indicator for investors and other stakeholders to make investment decision. Therefore, hundreds of companies are trying to increase accountability to stakeholders by investing serious time and energy into reports disclosing their social and environmental impacts and performance. Today, we see many different reporting terms being used to describe sustainable performance reports issued on an annual or periodic basis by companies interested in getting their message out: CSR reports, social performance reports, corporate citizenship reports, sustainability reports, environmental, social and governance (ESG) reports. Most of these reports use methodologies that are less rigorous than the original idea of social audits. What these reporting processes have in common is that they make the public and stakeholders aware of their social and environmental programs, activities, and achievements. Some of the more advanced reports actually report company achievements relative to previous goals set by management. Others just report what the company has done during the previous reporting period.

The impetus for social and environmental performance reports in recent years has come from societal and public interest groups' expectations that firms report their achievements in the social responsibility and environmental sustainability arenas. Such reports typically require monitoring and measuring progress, and this is valuable to management groups wanting to track their own progress as well as be able to report it to other interested parties. Some companies create and issue such reports because it helps their competitive positions. Globalization is another driver for social and environmental performance reports. As more and more companies do business globally, they need to document their achievements when critics raise questions about their contributions, especially in developing countries.

One of the major impediments to the advancement of effective social performance reporting has been the absence of standardized measures for social reporting. Standardization of social and environmental performance reporting is a challenge for international community. Ceres gets a lot of credit for the interest in social and environmental performance reports during the past decades years. Ceres has been the pioneer in creating guidelines to help standardize social performance reporting globally. It is recognized by the business world as leadership in sustainability reporting. For example, Ceres initially developed globally applicable guidelines for reporting on the economic, environmental, and social performance of corporations, governments, and NGOs. It is called Global Reporting Initiative

(GRI), launched in 1997. GRI was spearheaded by Ceres in conjunction with the United Nations Environment Programme (UNEP). GRI includes the participation of corporations, NGOs, accountancy organizations, business associations, and other worldwide stakeholders. GRI is now considered the de facto international standard (used by more than 1,800 companies) for corporate reporting on environmental, social, and economic performance.

The GRI's Sustainability Reporting Guidelines were first released in draft form in 2000. They represented the first global framework for comprehensive sustainability reporting, encompassing the "triple bottom line" of economic, environmental, and social issues. The second generation of Guidelines, known as G2, was unveiled in 2002 at the World Summit on Sustainable Development in Johannesburg. UNEP embraced GRI and invited UN member states to host it. In the same year, GRI was formally inaugurated as a UNEP collaborating organization in the presence of then UN Secretary General Kofi Annan, and relocated to Amsterdam as an independent non-profit organization.

The uptake of GRI's guidance was boosted by the 2006 launch of the current generation of Guidelines, G3. Over 3,000 experts from across business, civil society and labour participated in G3's development. After G3 was launched, GRI expanded its strategy and Reporting Framework, and built powerful alliances. Formal partnerships were entered into with the UNGC, the Organization for Economic Co-operation and Development (OECD), and others. A regional GRI presence was established with Focal Points, initially in Brazil and Australia and later in China, India and the USA. Sector-specific guidance was produced for diverse industries in the form of Sector Supplements (now called Sector Guidelines). Educational and research and development publications were produced, often in collaboration with academic institutions, global centres of excellence and other standard-setting bodies. GRI's services for its users and network expanded to include coaching and training, software certification, guidance for small and medium sized enterprises in beginning reporting, and certifying completed reports. GRI's outreach was strengthened by its biannual Amsterdam Conference on Sustainability and Transparency, beginning in 2006; the third conference in May 2010 attracted over 1,200 delegates from 77 countries. In March 2011, GRI published the G3.1 Guidelines – an update and completion of G3, with expanded guidance on reporting gender, community and human rights-related performance (UNGC 2013).

Another specific initiative of Ceres has been its annual award for Sustainability Reporting. The initiative was launched in 2002 in conjunction with the Association of Chartered Certified Accountants (ACCA). The purpose of the awards program was to contribute to reporting on sustainability, environmental and social issues by United States and Canadian corporations and other organizations by rewarding best practice and providing guidance to other entities that are publishing or intend to publish sustainability reports. These awards have doubtless increased attention to the idea of social and environmental performance reporting. The initiative was suspended in 2012. However, Ceres and the Tellus Institute launched the new Global Initiative for Sustainability Ratings (GISR) in 2011. Its mission is to create a world-class corporate sustainability ratings standard as an instrument for transforming the definition of value and value creation by business in the twenty-first century in a way that aligns with the national and global sustainability agenda.

As companies develop enterprise-level strategies and corporate public policies, the potential for social responsibility and environmental sustainability reporting remains high. Social and environmental reporting is best appreciated not as an isolated, periodic attempt to assess social and environmental performance but rather as an integral part of the overall strategic management process as it has been described. Because the need to improve planning and control will remain as long as management desires to evaluate its corporate social and environmental performance, the need for approaches such as the social and environmental responsibility reporting will likely be with us for some time, too. The net result of continued use and refinement should be improved corporate social and environmental performance and enhanced credibility of business in the eyes of its stakeholders and the public. In terms of practice, social and environmental performance reporting has become more popular than the more complex task of social auditing. Both approaches serve much the same purpose and help to keep the organization on track with its social performance goals.

International Organization for Standardization (ISO) is the world's largest developer of voluntary International Standards. International Standards give state of the art specifications for products, services and good practice, helping to make industry more efficient and effective. Since it was founded in 1947 to facilitate the international coordination and unification of industrial standards, ISO has published more than 19,500 International Standards covering almost all aspects of technology and business.

In 2011, ISO published a new standard in the *ISO 14006 on environmental management systems – Guidelines for incorporating eco-design*, to help organizations reduce the adverse environmental impacts of their products and services. Every product or service has an impact on the environment during all stages of its lifecycle, from extraction of resources to end-of-life treatment. The goal of eco-design is to integrate environmental aspects into the design and development of products and services so as to reduce their environmental impacts and continually improve their environmental performance throughout their lifecycle. It gives "how to" guidance to product and service organizations on incorporating eco-design into any environmental, quality or similar management system. It will help organizations establish, document, implement, maintain and continually improve their management of eco-design as part of an environmental management system (EMS). It applies to those environmental aspects of an organization's products and/or services over which it has control or influence.

Integrating eco-design offers several advantages:

• Economic benefits, e.g. through increased competitiveness, cost reduction and attraction of financing and investments.
• Promotion of innovation and creativity, and identification of new business models

- Reduction in liability through reduced environmental impacts and improved product knowledge
- Improved public image
- Enhancement of employee motivation.

Another important standard is *ISO 26000: Social responsibility*. ISO 26000 was launched in 2010 to provide guidance on how businesses and organizations can operate in a socially responsible way. ISO 26000 provides guidance rather than requirements, it cannot be certified to unlike some other well-known ISO standards. Instead, it helps clarify what social responsibility is, helps businesses and organizations translate principles into effective actions and shares best practices relating to social responsibility globally. It is aimed at all types of organizations regardless of their activity, size or location. As part of the effort, ISO also established Memoranda of Understanding (MOU) with the ILO, UNGC and as well as OECD. The primary objectives of the MOU were to not amend their respective instruments in the social responsibility (SR) field, but to complement their work and provide authoritative, international voluntary guidance on the breadth of this subject for all organizations (e.g. not just corporate social responsibility - CSR). The standard specifically provides guidance on:

- Concepts, terms and definitions
- The background, trends and characteristics of SR
- Principles and practices relating to SR
- The core subjects and issues of SR
- Integrating, implementing and promoting socially responsible behaviour throughout the organization and, through its policies and practices, within its sphere of influence
- Identifying and engaging with stakeholders
- Communicating commitments, performance and other information related to SR.

ISO 26000 effectively provides a global context for social responsibility, a context in which various existing tools and initiatives already provide important solutions. In the area of "reporting", extensive input and expertise from key global players, such as GRI, has been provided throughout the development of ISO 26000. There are no fewer than 19 instances of the term "reporting" in the body of ISO 26000, intended to provide guidance on communicating results within the organization, with other stakeholders and with society as a whole. However, no requirements for reporting are indicated in ISO 26000, nor are there requirements on how this could be done in a manner that integrates financial and non- financial information. Thus the utility and complementarity of ISO 26000 and GRI's sustainability reporting principles and indicators is very important. It is expected that in the future, the International Integrated Reporting Committee will be established by the international community to raise awareness of the merits of measuring business success in a new, holistic and integrated manner.

Social Accountability International (SAI) has played a pivotal role in facilitating and promoting social auditing over the past years. SAI is a non-governmental, multi-stakeholder organization with its mission of advancing the human rights of workers around the world. It partners to advance the human rights of workers and to eliminate sweatshops by promoting ethical working conditions, labour rights, corporate social responsibility and social dialogue. SAI works to protect the integrity of workers around the world by building local capacity and developing systems of accountability through socially responsible standards.

SAI established one of the world's preeminent social standards – the SA8000 standard for decent work, a tool for implementing international labour standards that is being used in over 3,000 factories, across 65 countries and 66 industrial sectors. Many more workplaces are involved in programs using SA8000 and SAI programs as guides for improvement (SAI 2013).

The SA8000 standard is one of the world's first auditable social certification standards for decent workplaces, across all industrial sectors. The SA8000 standard is based on conventions of the ILO, UN and national laws. The SA8000 standard spans industry and corporate codes to create a common language for measuring social compliance. Those seeking to comply with SA8000 have adopted policies and procedures that protect the basic human rights of workers. The management system supports sustainable implementation of the principles of SA8000: child labour, forced and compulsory labour, health and safety, freedom of association and right to collective bargaining, discrimination, disciplinary practices, working hours, remuneration.

In addition to the standard setting, since 1997, SAI has also provided training of supply chain management and CSR to over 30,000 people worldwide, including factory and farm managers, workers, brand compliance officers, auditors, labour inspectors, trade union representatives and other worker rights advocates across a wide array of industrial sectors. SAI's training courses integrate compliance requirements with management systems to embed improvements in daily operations. SAI has provided training to executives from numerous industries over the past 5 years, including apparel, footwear, agriculture, electronic assembly and light manufacturing.

SAI consults with trade unions, companies and NGOs to provide interpretive guidance: to SA8000 auditors to verify compliance with the standard; and to managers and workers to implement SA8000 at their workplace. SAI has convened 20 national or regional multi-stakeholder consultations on the SA8000 system or on specific compliance issues that are particularly challenging in one region or industry.

The organization that keeps the most comprehensive data on social and environmental performance reports is CorporateRegister.com. CorporateRegister.com is a free directory of company-issued CSR, Sustainability, and Environment reports from around the world, and the site is continually updated with new reports and companies. The tremendous growth in CSR Reports can be seen by data collected by CorporateRegister.com. In the year 2012, more than 46,000 reports across about 10,000 companies were issued (CorporatRegister 2013).

5 Conclusion

It is recognized that significant achievement in sustainable development can only be realized through the concerted efforts and contributions that corporations, governments, NGOs and international community should make. Businesses and other stakeholders must work together to examine the risks and challenges in economic, social and environmental arena, and seek innovative solutions to a range of challenges in order to develop a truly sustainable society in the long-run.

The role of entrepreneurship in fighting social and environmental issues and achieving progress towards sustainable development is of critical importance. More and more international stakeholders and constituents, as I state in the Section, are actively engaged in promoting Sustainable Entrepreneurship and private sector development through provision of principles, policy advice and institutional capacity-building services. These include principles and measures to integrate sustainability into business strategy, global value/supply chains and enable them to achieve compliance with social and environmental standards prevailing in international markets. This section examines the creative initiatives and principles that some of the international organizations have developed over the past years to promote sustainability and entrepreneurship in the business world, particularly in the areas of social and labour responsibility, human rights and business, supply chain sustainability, environmental sustainability and sustainability reporting and social auditing. The section is unable to cover all areas of Sustainable Entrepreneurship; however, it is served as a window in which we can understand what efforts and commitments that international community has made to Sustainable Entrepreneurship.

Literature

Ban Ki-moon, United Nations Secretary-General. (2009). *The millennium development goals: Lifting the bottom billion out of poverty.* Retrieved February 20, 2013, from http://www.un.int/wcm/content/site/gmun/cache/offonce/pid/8510

Ceres (2010). The 21st century corporation: The Ceres roadmap for sustainability: coalition of environmentally responsible economics (Ceres).

Ceres (2012). Supplier self-assessment questionnaire (SAQ): Building the foundation for sustainable supply chains. http://www.ceres.org/resources/reports/supplier-self-assessment-questionnaire-saq-building-the-foundation-for-sustainable-supply-chains/view. Retrieved on 25 Feb 2013.

Ceres (2013a). Climate change. Ceres. http://www.ceres.org/issues/climate-change. Retrieved on 26 Feb 2013.

Ceres (2013b). Sustainable supply chains. Ceres. http://www.ceres.org/issues/supply-chain. Retrieved on 26 Feb 2013.

CorporateRegister.com. (2013). Corporate reporting. http://www.corporateregister.com/. Retrieved on 25 Feb 2013.

Davidsson, P. (2003). The domain of entrepreneurship research: Some suggestions. In J. K. a. A. C. Corbett (Ed.), *Advances in entrepreneurship, firm emergence and growth* (vol. 6, pp. 315–372). UK: Emerald Group Publishing Limited.

Dyllick, T. (1999). Environment and competitiveness of companies. In D. Hitchens, J. Clausen, & K. Fichter (Eds.), *International environmental management benchmarks*. Berlin: Springer.

Fussler, C. (1996). *Driving eco-innovation, a breakthrough discipline for innovation and sustainability*. Pitman: Financial Times/Prentice Hall.

Hart, S. L., & Milstein, M. B. (1999). Global sustainability and the creative destruction of industries. *Sloan Management Review, 41*(1), 23–33.

Hawken, P., Levins, A. B., & Lovins, L. H. (2000). *Natural capitalism: Creating the next industrial revolution*. Boston: Back Bay Books.

HRC. (2011). *Guiding principles on business and human rights: Implementing the United Nations "protect, respect and remedy" framework*. Geneva: Human Rights Council (HRC).

ILO. (2007). *Conclusion concerning the promotion of sustainable enterprises*. Geneva: International Labour Office.

ISO (2010). ISO 26000 – Social responsibility. International Organization for Standardization. http://www.iso.org/iso/home/standards/iso26000.htm. Retrieved on 26 Feb 2013.

ISO (2011). ISO 14006: Environmental management systems-guidelines for incorporating ecodesign. http://www.iso.org/iso/catalogue_detail?csnumber=43241. Retrieved on 26 Feb 2013.

Kanter, R. M. (1999). From spare change to real change: The social sector as a beta site for business innovation. *Harvard Business Review, 77*(3), 123–132.

Kelley, D. (2011). Sustainable corporate entrepreneurship: Evolving and connecting with the organization. *Business Horizons, 54*(1), 73–83.

Levins, A. B., & Hawken, L. H. (1999). *Naturalcapitalism: Creating the next industrial revolution*. Boston: Little Brown.

ILO (2013). Ryder, Guy, ILO Director-Genera, Message on World Day for Social Justice. http://www.ilo.org/global/about-the-ilo/who-we-are/ilo-director-general/statements-and-speeches/WCMS_205246/lang–en/index.htm. Retrieved on 25 Feb 2013.

McDonough, W., & Braungart, M. (1998). The next industrial revolution. *The Atlantic Monthly, 282*, 82–92.

McKinsey (2011) The business of sustainability: Mckinsey global survey results. McKinsey Quarterly 2011.

SAI (2008). SA8000 standard (labour standards). Social Accountability International (SAI). http://www.sa-intl.org/index.cfm?fuseaction=Page.ViewPage&pageId=1365. Retrieved on 25 Feb 2013.

Schaltegger, S., & Wagner, M. (2008). Types of sustainable entrepreneurship and conditions sustainability innovation: From administration of a technical challenge to the management entrepreneurial opportunity. In R. Wustenhagen, J. Hamschmidt, S. Sharma, & M. Starik (Eds.), *Sustainable innovation and entrepreneurship*. Cheltenham: Edward Elgar.

Senge, P., & Carstedt, G. (2001). Innovating our way to the next industrial revolution. *Sloan Management Review, 42*(2), 24–39.

Senge, P. M., Carstedt, G., & Porter, P. L. (2001). Innovating our way to the next industrial revolution. *MIT Sloan Management Review, 42*(2), 24–38.

Thompson, N., Kiefer, K., & York, J. (2011). Distinctions not dichotomies: Exploring social, sustainable and environmental entrepreneurship. In G. T. j. A. K. Lumpkin (Ed.), *Social and sustainable entrepreneurship*. UK: Emerald Group Publishing Limited.

UNDP. (2006). *Making progress on environmental sustainability: Lessons and recommendations from review of over 150 MDG countries experiences*. New York: Environmental and Energy Group. Bureau for Development Policy, UNDP.

UNDP. (2008). *Creating value for all: Strategies for doing business with the poor*. New York: United Nations Development Programme.

UNDP. (2010). *The MDGs: Everyone's business: How inclusive business models contribute to development and who supports them.* New York: United Nations Development Programme.

UNGC, & Accenture. (2012). *Sustainable energy for all: The business opportunity.* New York: UN Global Compact (UNGC).

UNGC, & BSR. (2010). *Supply chain sustainability: A practical guide for continuous improvement.* New York: United Nations Global Compact.

Yamada, J. (2004). A multi-dimensional view of entrepreneurship: Towards a research agenda on organisation emergence. *Journal of Management Development, 23*(4), 289–320.

Part II
Business Related Concepts

Sustainable Entrepreneurship: A Driver for Social Innovation

Thomas Osburg

1 Introduction

Sustainable Entrepreneurship and Social Innovation are not necessarily new concepts; they have existed for several years but seem to become increasingly relevant to companies, NGO's and governments throughout the world every year. Many companies take these developments very seriously, as they understand that they are part of a new way of doing business in the future and thus create their own capacity to endure. One of the key challenges is, however, a partially unclear understanding of the concepts, how they can be implemented and how an impact can be created. In addition, concepts like Social Entrepreneurship or Sustainable innovation exist and add to the complexity. As with most new concepts, this unclear understanding could ultimately hinder the development of a concept that, if applied seriously, might have a significant contribution on improving the way we collaborate, innovate and ultimately have a positive impact on the world.

2 Social Innovation as an Opportunity

2.1 Trust in Business as a Macro-Economic Driver

Since the beginning of the current economic crisis in 2008, we saw a slow decline in trust towards businesses and governments across the world, which since then has taken up speed. The year 2012 was a significant turning point, as trust in governments across the globe declined stronger than ever before. Only 43% of

T. Osburg (✉)
Intel GmbH, Dornacher Str. 1, Feldkirchen 85622, Germany
e-mail: Thomas.osburg@intel.com

C. Weidinger et al. (eds.), *Sustainable Entrepreneurship*, CSR, Sustainability, Ethics & Governance, DOI 10.1007/978-3-642-38753-1_7, © Springer-Verlag Berlin Heidelberg 2014

the world's population now trusts governments to do what is right, but 53% trust business to do what is right. Both numbers were at a comparable level the previous year (Edelman 2012). NGO's remain the most trusted institutions with 58% of people trusting them to do what is right. The relevance of this shift over the years has become significant and now offers companies a huge opportunity to assume leadership with the appropriate behavior and actions.

In order to gain leadership for trust, business mainly needs to shift to or continue exercising principle-based leadership instead of a rules-based strategy. Firms need to focus on what is really creating shared value, both to shareholders and society, and defocus on operating only on what they are legally permitted to do. While in the past operational factors were key to building trust (like products, brands, regulations and financial returns), this is changing now. Trust will be built around engagement-oriented behaviours of a more social beneficial nature (Edelman 2012). In other words: Social Innovation will increasingly become a key driver for companies to continue building leadership and gaining trust from societies.

2.2   The Concept of Social Innovation

We have seen an astonishing rise in the use of the concept of Social Innovation over the past years, both in theory and in practice (Osburg and Schmidpeter 2013). It's importance for companies, NGO's and the public sector can't be denied these days. One of the reasons clearly lies with the proactive approach and newness of solutions for firms, compared to responsible behavior and the partially reactive approach of similar concepts. The concept of Social Innovation usually implies a normative approach that something positive should be created for society. As Googins (2013) points out, innovation has always been in the DNA of firms, but it has not really been an integral part of CSR.

The EU Commission defines Social Innovation as "...Innovations that are both social in their ends and in their means. Social Innovations are new ideas (products, services and models) that simultaneously meet social needs (more effectively than alternatives) and create new social relationships or collaborations" (EU-Commission 2012a).

While the EU Commission has a strong focus on the results produced and the importance of collaboration, leading business schools also increasingly focus on the impact achieved with Social Innovations. For the European Business School (EBS), Social Innovations are "...new solutions that address societal challenges in a way that is contextual, targeted, and promotes common welfare" (European Business School (EBS) 2012). The INSEAD Social Innovation Centre defines Social Innovation as the "...introduction of new business models and market-based mechanisms that deliver sustainable economic, environmental and social prosperity" (INSEAD 2012).

2.3 Embedding Social Innovation in Company Functions

The above given definitions clearly show the broadness of the Social Innovation concept if understood professionally. No business unit or department alone can "own" Social Innovation; it is a true cross-business cooperation of several functions within a company. In addition, the current product- or solution centric view of a Social Innovation needs to be expanded to a more process-oriented strategy. Social Innovation should not be seen as an add-on, but rather an integral part of the firms offering. Some examples include:

- **Marketing** is a key function in the process of creating Social Innovation through identifying customer needs, designing solutions with stakeholders, communicating solutions and, ultimately, including them in the overall product or service portfolio of the firm.
- Social Innovation is also closely linked to the **Research & Development (R&D)** agenda that the company is following. The more Social Innovation is understood as a key element of the company's overall vision and Mission, the less a specific R&D budget for Social Innovation is needed.
- By shifting to Social Innovation, a new type of employee with different skill sets and beliefs is needed. **Human Resources (HR)** has already experienced for several years now a rising interest from potential candidates on the firm's ethical behaviour, what products are produced and how, as well as the impact on society generated through Corporate Social Responsibility or Social Innovation.
- The company's **Innovation Process** in general is usually the key underlying conceptual framework for Social Innovation (Osburg 2013). Innovation always was and will always be the key driver for companies to thrive. However, innovation as such is neither positive nor negative, it is just new. Framing and guiding this innovation towards a direction that benefits society as much as the company is the key to arriving at Social Innovation. Section 2.4 will specifically look at this critical integration.
- And ultimately, Social Innovation needs to be financially measured like any other investment or innovation. This is certainly more challenging to undertake than for traditional innovations, as the bottom-line impact is tougher to detect and the external impact can often only be seen after several years.

Social Innovation, if done in a meaningful way, thus affects the whole organization and this is also where it's full potential lies. Social Innovation has the power and capacity to transform the largest organizations in order to increase their bottom-line result and create societal value at the same time.

2.4 Social Innovation as Part of the Innovation Process

Mainly and foremost, as Social Innovation needs to be embedded in the innovation processes of a firm, it is helpful to take a look at innovation as it is the key underlying conceptual framework for Social Innovation. Sometimes, little attention is paid to known and proven concepts of innovation when Social Innovation is discussed. Ultimately, Social Innovation can be considered to be traditional innovation, when a normative social component is added. Wanting to do good is not enough. Social Innovation needs to be a process driven by innovation and adding a new goal and value system to it leads to true sustainability.

Based on the works of the Austrian economist Joseph A. Schumpeter, innovation in general can be understood as a new combination of production factors (Schumpeter 1982). Innovation is the creation and adoption of something new that creates value for the organization that adopts it (Baldwin and Curley 2007). It can be a specific instrument of **entrepreneurship**, the act that endows resources with a new capacity to generate wealth (Drucker 1985). Contrary to the mere Invention, the concepts of innovation include the process of transforming an idea or an invention into a solution that creates value for stakeholders such as customers, shareholders or societies. Thus, innovation should not be confused with Invention.

Innovation as a term is rather ambivalent and this, as we will see later, is one of the root causes of different understandings of Social Innovation. Schumpeter holds that innovation focuses on the types (product, process, market), the dimensions (objective or subjective), the scope of change (radical, incremental, reapplied) or how it was created (closed or open innovation) (Stummer 2010). All of these differentiations are highly relevant to concepts of Social Innovation as well.

- **Types of Innovation** – Product and Service innovations are certainly a major area to focus on for companies, as these innovations are typically very visible and shape the reputation of a firm. However, process innovation (i.e. a new form of production that saves emissions and resources) or market innovation (i.e. creating new markets for social solutions) is often as important as product innovations.
- **Scope of Change** – It is commonly understood that innovation needs to be something big and ground-breaking. However, most innovations are not this way. The radical or disruptive innovation fundamentally changes markets and daily lives of people. Often, they are closely related to the inventor and bear high opportunities but also high risks. Incremental innovations rather build on the constant improvement of disruptive innovations; they are more related to the organisation and less to the inventor. In general, they offer a high potential for economic success. A third area to review is reapplied innovation. These often include existing concepts that are successfully implemented in a new area (Baldwin and Curley 2007).

- **Sources of Innovation** – Closed innovation processes strongly focus on the intellectual capacity and property of the organization; inventions and innovations are developed in-house and then results are shared with external stakeholders. Open innovation, on the contrary concerns "...the use of purposive inflows and outflows of knowledge to accelerate innovation. With knowledge now widely distributed, companies cannot rely entirely on their own research, but should acquire inventions or intellectual property from other companies when it advances the business model..." (Chesbrough 2003).

2.4.1 Open Innovation

Open innovation as source for creating new solutions is a key concept to consider, as it calls for significant stakeholder interaction to achieve results. While the relevance of Open innovation for business is steadily increasing, Open innovation is a *must* for Social Innovation. Even more than in business, solving problems today in society requires constant collaboration between all sectors to determine the most burning problems and approaches to resolve them. There are no serious issues today that can be solved by any single sector alone.

The concept of Open innovation has two different focuses using knowledge sharing that are significant for cross-sectorial collaboration in Social Innovation:

- **Outside-In Processes** integrate external knowledge into the innovation process and thus is enhances a company's internal knowledge base through the integration of external stakeholder knowledge. This can be through a loose collaboration or formal agreements. Through the Outside-In Process, external Social Knowledge is brought into the company.
- **Inside-Out Processes** are focused on the externalization of knowledge, which is far less common than Outside-In. Here, companies can license or provide technology or knowledge to capitalize on potential economic benefits outside the firm. It can also be used to run processes of joint development.

Both directions of Open innovation require significant collaboration between the stakeholders and, in Social Innovation, also among the different sectors. Open innovation is a critical concept for Social Innovation as it requires companies to internalize external knowledge, but also to externalize internal knowledge, leading to new cross-sectorial partnerships that go far beyond traditional approaches of Public-Private Partnerships.

2.4.2 Crossing the Chasm

The theory of crossing the chasm relates to the difficulties that exist when trying to develop a great idea or invention into a scalable and long-lasting success (Moore 1999). Originally developed for the HighTech Industry, the concept can be applied

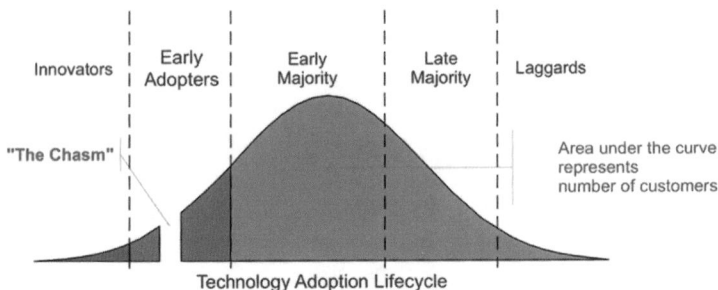

Fig. 1 The chasm in a typical lifecycle curve (Moore 1999)

to all innovation processes, as the underlying findings can be generalized. It is not sufficient to invent and have a great idea; the challenge is how to implement it so that it really has an impact. This is where a lot of great ideas ultimately fail.

Mostly, different personalities are required to work on an innovation during the early phases and then later on by scaling and mainstreaming the solution. The Chasm is simply the point in a typical lifecycle curve where a lot of great ideas fail for various reasons: either the customer doesn't see the value, the resources are not sufficient to scale, competitive solutions were not considered, etc. Particularly evident in the field of Social Innovation, which is often dominated by passionate and enthusiastic Social Entrepreneurs who *burn* for their solutions, little attention is given to next steps – how to implement, how to find needed resources and how to build a network of collaborating organizations to help scale (Fig. 1).

2.5 The State of Social Innovation

The picture of how widely Social Innovation concepts are disseminated today is rather unclear, despite a lot of public focus and communication. Social Innovation is not yet mainstream, partially because of a lack of clarity for a majority of firms. As a result, only few leading companies like Intel Corp. or HP have engaged in serious efforts to uplevel Social Innovation by linking it to the business side. For example, Intel has a clear innovation approach with a significant focus on the social impact of their business solutions (i.e. the World Ahead Program) to make it a more sustainable company. Similarly, HP has a Global Social Innovation group where the focus is to use innovation to make a positive difference in the world.

In academia, especially in the leading (business) schools, Centres for Social Innovation have been created over the last years. Examples are the European Business School (EBS) in Germany, INSEAD in France or Stanford in the US. Leading research on this topic is nowadays done in most business schools and a strong collaboration between universities and leading companies in this area, like Intel or IBM, is underway. However, at the vast majority of Higher Education Institutions in Europe, Social Innovation is not yet a significant research focus.

At the political level, some scattered activities in about half of the European countries can be observed. In Germany, the Federal Ministry of Education and Research (BMBF) announced in 2011 its funding support for a 2-year basic research project that looks into "Social Innovation in Germany". The project is carried out jointly by the World Vision Institute and its university partner, the European Business School, who for this purpose have collaborated to create the Centre for Social Innovation and Social Entrepreneurship.

At EU levels, the Directorate General (DG) Enterprise and Industry is leading the Social Innovation efforts, clearly linking it to Enterprise innovation and the Europe 2020 Agenda. Among others, a "Social Innovation Europe" (SIE) initiative was created in 2011 with three aims: research and publication of reports and recommendations, hosting an online exchange platform and one to two events in Europe per year. At the end of 2012, a European Social Innovation Competition was launched by Commissioner Barroso to showcase current success and encourage more Social Innovation in Europe.

The current challenges for Social Innovation can be seen in these two areas: "Social Innovation is *little known as a concept:* many social Innovations take place without them being known under that term, causing problems when asking for evidence in surveys and interviews. The second issue is the *cross-cutting nature* of social Innovation. Social Innovation is not a specific sector; it is not an easily defined activity. Statisticians have yet to develop an agreed approach, and so we lack reliable measures of spending on social Innovation and indicators of its scale of activity." (EU-Commission 2012b).

3 Sustainable Entrepreneurship

"Currently there is business hype in sustainable entrepreneurship. Every self-respecting company tries to brand itself as a sustainable entrepreneur. Business schools and employers' organizations devote whole conferences to the topic" (Crals and Vereeck 2007). This quote from a leading Business School illustrates the importance that the concept of *Sustainable Entrepreneurship* has gained over the previous years in the business world.

Sustainable Entrepreneurship stands for a unique concept of sustainable and entrepreneurial business strategies that focus on increasing social as well as business value at the same time. This means that it is, in essence, the realisation of sustainability innovations aimed at the mass market and providing benefit to the larger part of society. Sustainable Entrepreneurship can be understood as an entrepreneurial approach to developing business solutions to address the most urgent social and ecological challenges. It is an ongoing commitment by businesses to contribute to economic development while at the same time improving the quality of life of societies and the environment (World Business Council for Sustainable Development 2012).

Sustainable Entrepreneurship today is an approach that is applied mostly by large, often industrial companies, while small and medium enterprises (SMEs) have not yet embraced the concept. This is mainly an issue of available time, and less of lacking financial resources (Crals and Vereeck 2007).

The term Entrepreneurship in this sense needs to be understood in a much broader context than just in the sense of a start-up company. Entrepreneurship does not only need to be the creation of a new venture, business or company. Hence, according to this understanding, Entrepreneurship can also be embedded into larger organizations that actively promote and foster an entrepreneurial culture. Such an understanding of Entrepreneurship mainly focuses on the capability to create a competitive advantage by linking inventions with market success and is highly influenced by personal characteristics of the leader (Schaltegger and Wagner 2011).

3.1 Sustainable Entrepreneurship or Social Entrepreneurship?

The concept of Sustainable Entrepreneurship needs to be distinguished from Social Entrepreneurship, as there is one significant difference. While the key focus of Social Entrepreneurs is on the importance of using Inventions to achieve social change, they don't take "no" for an answer, but see it as a challenge to try harder. They manage to be extremely resourceful, making something out of seemingly nothing. In addition, almost all social entrepreneurs have been called crazy by their immediate friends and family (Schoening 2013). A main characteristic of Social Entrepreneurs is their explicit focus on social value creation (Weber 2007).

Building on this concept, Sustainable Entrepreneurship concentrates on creating both business and societal value, or shared value as Porter and Kramer describe it (Porter and Kramer 2011). This approach argues that by including a societal perspective into the strategic direction of the company, these solutions can serve as a source for attaining a competitive business advantage, while at the same time contributing to societal progress. Or in other words, while Social Entrepreneurship contributes to solving societal problems and creating value for society, Sustainable Entrepreneurship contributes to solving societal and environmental problems through the realization of a successful business venture (Schaltegger and Wagner 2011).

A key prerequisite for this is a focus on the core competencies of the firm. Whilst Social Entrepreneurs concentrate on finding a solution to solve a problem that they feel requires a solution, sustainable entrepreneurship is linked to creating shared value and thus has to be close to the core business of the company by definition.

Fig. 2 The impact-driven
approach of entrepreneurship

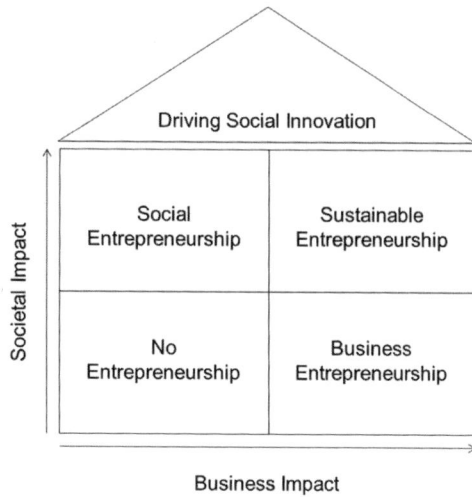

Ultimately, this helps companies focus their efforts in areas where they have a core competency and thus can be sustainable and innovative.

Both, Social Entrepreneurs and Sustainable Entrepreneurs can create Social Innovations, the difference lies in the impact for the company (Fig. 2).

3.2 Success Factors and Benefits

In order to embed Sustainable Entrepreneurship in a company to successfully drive Social Innovation that lasts and has a societal impact, some key success factors have emerged over the last years. All of them need to be part of a wider strategic direction for the company and can be influenced by company policy and leadership. Despite the many challenges outlined earlier, there is solid evidence in public discussion today that implementing Social Innovation will be key for companies in the coming decade and Sustainable Entrepreneurship will be the main driver. However, we have a long way to go.

First and possibly above all, **true Leadership** is needed to move the company towards Social Innovation. The concepts need to be part of the company's DNA, so to speak, and should not be questioned all the time. If this thinking is embraced and lived by the firm's leadership team, all business units and functions will become involved and work in this direction. It has to be part of the overall vision of the company to offer products and services that have both a business and societal value. This will ultimately ensure the needed management buy-in across all geos and functions and is also important to make sure that the ownership does not remain within any CSR or Sustainability Team. A good example is the current vision of Intel Corp., connecting the core business to societal value. *In this decade, we will create and extend computing technology to connect and enrich the lives of every person on earth.*

Social Innovation needs to be considered like any other **Business Innovation**, but enriched with a societal goal. It needs to link strongly to Corporate innovation Initiatives and ultimately become the driver of it. If there is no Business innovation behind new ideas and approaches, the long term viability of the innovation will not be present. As of today, most Social Innovation discussion focuses on doing good for society. While this is a noble cause, it fails to tackle the core essence of Social Innovation, which is still innovation. The *Social* component is adding a *triple-bottom-line* thinking to already established innovation Strategies within firms. Social Innovation have to be planned, evaluated, tested, introduced and measured like all other innovation. In order to do this, all company functions (Marketing, Sales, Research, HR, Finance,...) need to be aligned.

Building coalitions and cross-sectorial partnerships is more and more coming into the focus of businesses, for all the right reasons. Successful Social Innovation depends on all sectors collaborating by increasing the focus on **Open Innovation**, a close collaboration of knowledge sharing inside and outside of the firm is crucial for success.

The whole company culture needs to breathe an **Entrepreneurial Spirit** if Sustainable Entrepreneurship is to thrive. Trying to introduce Sustainable Entrepreneurship into a non-entrepreneurial company is a road to failure. In the end – it's all about Entrepreneurship driving innovation, but now with a sustainable and socially beneficial direction. Similar to all entrepreneurial endeavours – a proactive search for future trends and business opportunities is at the core of Sustainable Entrepreneurship.

CSR Managers are possibly in the best position to become real **Change Agents** in leading their company towards Sustainable Entrepreneurship and Social Innovation. This is a massive change from today, where most CSR departments are add-ons or only slowly moving to align with the business. Speaking in innovation terms, this is a radical innovation on its own. CSR Managers today need to be the drivers helping companies realize the potential for a sustainable company future.

Crossing the Chasm will be a key challenge in achieving Social Innovations through Sustainable Entrepreneurship. A lot of initiatives and strategies driven by companies have a noble and honest background, but they will not lead to the needed results (both for Business and Society) if the scaling and diffusion is missing. The concept from innovation theory of *how to cross the chasm* to reach a significant target audience is completely missing from today's discussion on Social Innovation. However, it is critically important for firms to become sustainable companies. As cynical as it sounds – doing good is not enough. It has to be scaled and thrust to a lasting level to become truly sustainable and impactful.

There needs to be a **strong fit** between the societal strategy and the company's overall objectives. Sustainable Entrepreneurship and Social Innovation can only be successful if there is a strong link to the core business. For example, supporting sustainable entrepreneurs in an area completely detached from the firm's core business does not create shared value and is not very credible. It might have a philanthropic angle but it is not a sustainable Social Innovation. Only by being close to the core business of the firm does the creation of company-specific benefits become possible.

Successful Social Innovations that were driven by Sustainable Entrepreneurship need to be properly **communicated** by showing the societal impact of the product or service. Being able to articulate and talk about the achieved impact will significantly drive credibility and reputation.

An excellent showcase on how to bring it all together can be observed in Intel's **World Ahead Program**. It starts with the global vision of Intel Corp., where Entrepreneurship is part of the DNA. "This decade we will create and extend computing technology to connect and enrich the lives of every person on earth". It then continues with a strategy on one hand while sustainability goals are described as one of four overall company goals. Based on this mindset, the World Ahead initiative was created. To align with stakeholders from the telecommunications industry, from the energy sector, local NGO's and governments to create real innovations by bringing technology to underserved parts of the world. While this is closely linked to Intel's core business and creates business value, the major focus here is clearly in creating societal value by bringing knowledge and technology to underserved areas of the world. Training modules for teachers are a key component to the program as well as local manufacturing of PCs which create jobs and wealth. For a higher impact and credibility, the Intel World Ahead program is running in nearly 100 countries worldwide and the Intel Teach program has reached more than 12 million teachers in over 70 countries so far.

If Sustainable Entrepreneurship is implemented in a proper way to drive Social Innovation, the benefits for the company are abundant:

- By communicating social impact, companies can build significant trust and credibility among key stakeholders.
- This external acknowledgement of the company, plus the entrepreneurial leadership spirit that is needed to achieve this praise, will result in higher motivation of employees and a higher skilled staff.
- Through the cross-sectorial collaboration with other businesses and external Institutions like NGO's or governments, a new level of knowledge sharing is achieved for all to benefit.
- There are significant positive impacts for the societal license to operate, which means agreement from society to perform the business operations as planned.
- Through constant innovation of societally accepted products or services, it might potentially open up new markets and thus contribute to the economic bottom-line. It will also help the business to stay ahead of its competition.

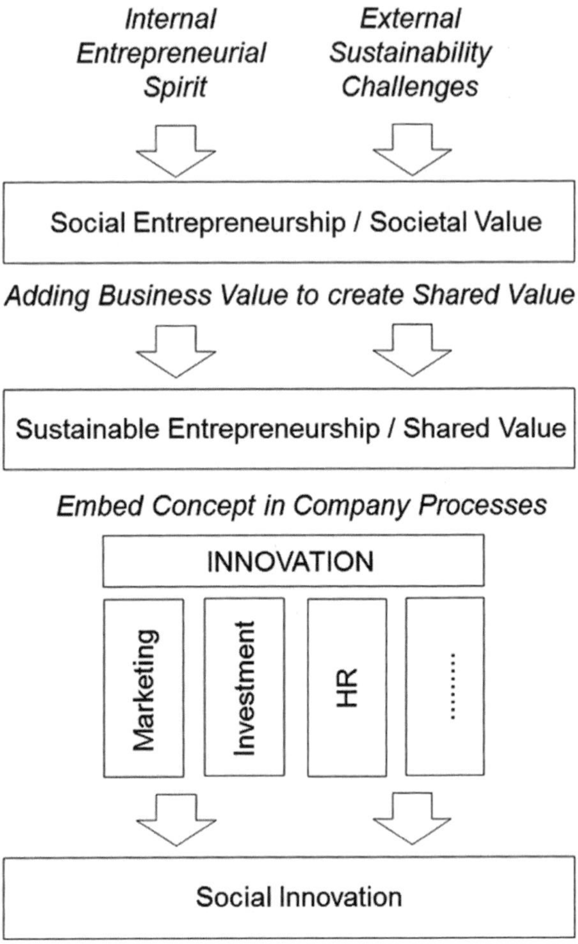

Fig. 3 Moving to social
innovation through
Sustainable Entrepreneurship

Internal   External
Entrepreneurial   Sustainability
Spirit   Challenges

Social Entrepreneurship / Societal Value

Adding Business Value to create Shared Value

Sustainable Entrepreneurship / Shared Value

Embed Concept in Company Processes

INNOVATION

Marketing   Investment   HR   ....

Social Innovation

- Driving Social Innovations through Sustainable Entrepreneurship might also lower the risk and burden from changes in legislation, because leading companies in this area usually meet all the requirements already.

Figure 3 summarizes the construct of how Sustainable Entrepreneurship can drive Social Innovation.

To conclude, Social Innovation is closer to the core business of what is generally thought of and what is the key for companies to achieve Corporate Sustainability and thus meet the needs of triple bottom-line reporting. Sustainable Entrepreneurship is the key driver to achieving this Social Innovation. But it will need work to transform current organizations; the business value has to be clear to reach this goal. The companies who will fully embrace Sustainable Entrepreneurship to ultimately drive Social Innovations will be the ones leading the decades to come. We are only at the very beginning now.

Literature

Baldwin, E., & Curley, M. (2007). *Managing IT innovation for business value* (IT Best Practice Series). Santa Clara: Intel Press.

Chesbrough, H. (2003). *Open innovation: The new imperative for creating and profiting from technology.* Boston: Harvard Business School.

Crals, E., & Vereeck, L. (2007). Sustainable entrepreneurship in SME's. Theory and Practice. http://www.inter-disciplinary.net/ptb/ejgc/ejgc3/cralsvereeck%20paper.pdf. Accessed 09 Dec 2012.

Drucker, P. F. (1985). *Innovation & entrepreneurship.* New York: Harper Trade.

Edelman (Eds.). (2012). Edelman Trust Barometer 2012. http://trust.edelman.com/. Accessed 04 Jan 2013.

EU-Commission (Eds.). (2012a). Social innovation http://ec.europa.eu/enterprise/policies/Innovation/policy/social-Innovation/index_en.htm. Accessed 09 Dec 2012.

EU-Commission (Eds.). (2012b). Strengthening social innovation in Europe – a journey to affective assessment and metrics. http://ec.europa.eu/enterprise/policies/Innovation/files/social-Innovation/strengthening-social-Innovation_en.pdf. Accessed 09 Dec 2012.

European Business School (EBS) (2012). http://www.ebs.edu/socialinnovation.html?&L=1. Accessed 09 Dec 2012.

Googins, B. (2013). Transforming corporate social responsibility: Leading with innovation. In T. Osburg & R. Schmidpeter (Eds.), *Social innovation – solutions for a sustainable future.* Heidelberg: Springer.

INSEAD (2012). What is "social innovation"? http://www.insead.edu/facultyresearch/centres/isic/home/about_us.cfm. Accessed 09 Dec 2012.

Moore, G. (1999). *Crossing the chasm.* New York: HarperCollins.

Osburg, T. (2013). Social innovation to drive corporate sustainability. In T. Osburg & R. Schmidpeter (Eds.), *Social innovation – solutions for a sustainable future.* Heidelberg: Springer.

Osburg, T., & Schmidpeter, R. (2013). Social innovation – Quo vadis. In T. Osburg & R. Schmidpeter (Eds.), *Social innovation – solutions for a sustainable future.* Heidelberg: Springer.

Porter, M. E., & Kramer, M. R. (2011). Creating shared value. How to reinvent capitalism – and unleash a wave of innovation and growth. In Harvard Business Review January-February 2011.

Schaltegger, S., & Wagner, M. (2011). Sustainable entrepreneurship and sustainability innovation: Categories and interactions, Wuerzburg. http://www.bwl.uni-wuerzburg.de/fileadmin/12020800/_temp_/Wagner_-_Sustainable_Entrepreneurship_and_Sustainability_Innovation.pdf. Accessed 04 Jan 2013.

Schoening, M. (2013). Social entrepreneurs as main drivers of social innovation. In T. Osburg & R. Schmidpeter (Eds.), *Social innovation – solutions for a sustainable future.* Heidelberg: Springer.

Schumpeter, J. A. (1982). *The theory of economic development: An inquiry into profits, capital, credit, interest, and the business cycle.* New Brunswick: Transaction Publishers.

Stummer, C., Guenther, M., & Köck, A. M. (2010). *Grundzuege des Innovations-und Technologiemanagements.* Vienna: Facultas.

Weber, M. (2007). *Towards sustainable entrepreneurship. A value creating perspective on corporate societal strategies.* Lueneburg: University of Lueneburg, Centre for Sustainability Management.

World Business Council for Sustainable Development (2012). www.wbcsd.org. Accessed 10 Dec 2012.

Entrepreneurship: Introducing Shared Innovation into the Business Model

Bradley Googins and Manuel Escudero

1 Introduction

All of the great American Entrepreneurs from Henry Ford, Thomas Edison, David Packard and Steve Jobs, had a driving theme to their entrepreneurship: a burning desire to contribute something great to society. It defined their entrepreneurism and fueled their passion and relentless pursuit of their dreams (Isaacson 2011). Today's business model focuses primarily on shareholder returns, which characterizes much of the prevailing model of capitalism around the globe. This would be the least interesting driver for these visionaries, and in most cases they had to fight hard to keep their vision from the clutches and narrow interests of their investors. For the true entrepreneur there is little distance between the business model and the role of business in society. In fact the ultimate measure of a business is its contribute to society. This vision seems a far cry from the trajectory of the reigning business model. Even more telling is the relentless pursuit of a bottom line measured primarily by profit and loss statements.

However noticeable strains are confronting the existing order, and the shifting roles of government, business and civil society have created a much different environment within which business operates. Creating sustainable business today is increasingly being challenged by a new set of realities of the ground that are exposing serious fissures in global capitalism and its shortcomings in addressing persistently troubling social and environmental issues (Accenture 2010). Over the past few decades business has operated largely within its own domains, relatively

---

B. Googins (✉)
Boston College Carroll School of Management (Ret), 140 Commonwealth Avenue, Chestnut Hill, MA 02467-3809, USA
e-mail: googinsb@bc.edu

M. Escudero
Director General Deusto Business School, Bilbao Spain

C. Weidinger et al. (eds.), *Sustainable Entrepreneurship*, CSR, Sustainability, Ethics & Governance, DOI 10.1007/978-3-642-38753-1_8, © Springer-Verlag Berlin Heidelberg 2014

free from the concerns and dynamics of the societies within which they operated. Issues pertaining to societal welfare, conditions in the natural environment, the health and work-life concerns of employees, human rights, child labor and global supply chains, have been for the most part relegated to other sectors such as the government. Similarly, business was largely unaffected by activists and shareholder resolutions, the threat of protests and boycotts, not to mention calls for greater transparency and the dramatic increase in exposure provided by the Internet. Those days are long gone. Creating sustainable business today is directly tied to a model where the issues of environment and society have been brought into the business sphere, not by any choice or deliberate calculation by business, but as the result of shifting expectations, the growth of business influence and power in global capitalism, and the resultant diminution of the public sector. Consequently it is important for global business to better understand this new operating environment, adjust their business model to provide a sharper risk/opportunity analysis, and incorporate these new realities to insure their sustainability.

To address these issues some of the leading companies are rediscovering innovation as a key to creating sustainable impacts on social issues. This chapter will focus on the concept of shared innovation as a means by which business can unlock its core asset of innovation and create a new strategy along with key stakeholders to build a more sustainable model of business in today's society.

2 Sustainable Business in a New Era

As the twenty-first century unfolds, political, social and economic upheavals are now rolling out on a continuous basis, constituting a new normal, and a not so comfortable operating environment for business. This is an environment characterized by a great deal of turmoil, uncertainty and insecurity, exactly the opposite climate needed for sustainable business and society. But the upheaval is indisputable and inevitable. Several new significant global trends are driving these changes (Escudero et al. 2012a).

- *The "soft power" of granting legitimacy is changing hands:*
   If the social contract relates to legitimate governance, legitimacy, or the moral license to operate, comes fundamentally from public opinion (Habermas 1984). It is public opinion that sanctions whether an action with public consequences, a policy, a public authority or a company has legitimacy or not. However, in our days, public opinion is not only generated by the media but also, increasingly, by social media: an increasing number of "de-mediated", rank-and-file citizens are a new powerful source of reflexivity (Giddens 1984), rapid creation of global awareness, and legitimacy.

– *Social media are becoming powerful facilitators of civic participation in the public agenda:*

There is a related, complementary new trend at play in all societies: social media are revealing themselves as a multiplier for massive participation of normal citizens in the public agenda. The last year has shown us plenty of examples: the Arab spring, the 15M movement in Spain, gun violence in America, the massive demonstrations in Tel Aviv, the recent riots in Britain, and the civic response to them. Thus, the intervention in the public agenda (that goes beyond the political agenda) has been opened up to the citizens as never before.

– *Public powers are showing their limits:*

Today, public powers are shrinking. A big part of our daily problems (global terrorism, climate change, humanitarian crises, and the sovereign debt problem of Europe or the USA) cannot be dealt with single-handedly by any Nation-state on its own. This begs the problem of the limits of democracy in a globalized world: it is frustrating for citizens to vote for any political option, knowing that the capacity of national politicians to solve problems is severely curtailed. Also, during the second half of the twentieth century we have learnt that the unlimited growth of the public sector is no "silver bullet" to solve all social problems. That, of course, does not mean that public intervention is unnecessary. But it is equally necessary to prevent public abuses (imperium), through a much more deliberative and disputative democracy.

– *Corporate power is on the rise:*

While public powers have shown their limitations, corporate power has increased over the last decade, owing to globalization, deregulation and privatizations. This brings about a totally new dilemma: to whom are global companies and their new global power accountable?

– *There is a new social regulation of corporate power in the making:*

As a response to the increase of power of global corporations, a new trend of social regulation of companies by their stakeholders has emerged internationally. Various movements have arisen in this regard from Sustainability, Corporate Social Responsibility, or Corporate Citizenship. Many global companies have started to respond to demands for accountability and transparency on social, environmental and governance issues, thus paying attention to some international standards of responsible behavior, such as the ten principles of the UN Global Compact, GRI or ISO 26000. This trend heralds the emergence of a new paradigm of the firm in the twenty-first century, as an economic institution with positive impact in society and in the global agenda.

– *Institutional investors have acquired an inordinate dominance in the global economy:*

The trends mentioned until now, have combined with the economic-financial crisis of 2008, adding more arguments to rethink the social contract. In the light of the crisis we now realize that the conditions of a globalized financial market, an increased volume of international flows of money looking for short-term profits, and the exponential growth of financial products like derivatives and futures, have given institutional financial investors an excessive dominance in the global economy, never seen before. Sooner or later, as a core part of a new social contract, these private financial investment institutions will have to adjust to new international parameters, standards and codes of conduct, if they want to regain their lost legitimacy and trust from public opinion.

– *Income inequalities are at their highest:*

Budgetary cuts and unemployment have combined over the last 4 years in many developed countries, hitting millions of people and causing the greatest income inequality in our lifetime. At the same time, global companies remind us daily that they are recording large profits in spite of the crisis. New practices, legal or voluntary, to introduce a fairer redistribution of income and strategic "social-giving-back" will have to be incorporated sooner or later to the responsible code of conduct of the firm, as part of a new social contract.

– *We live now in a multipolar world:*

The crisis has revealed another new and crucial trait: we live in a multipolar world in which emerging countries will increasingly consolidate their presence and decision-making power. This will lead, sooner or later, to new multilateral global governance arrangements, both at the economic and political level. But this new multilateralism will not be necessarily a comfortable situation for those who live at the epicenter of globalization: India, China, Brazil or Russia could bring to the table ideas and suggestions not very familiar to us.

– *We live in a crowded planet:*

The last and unavoidable new trait of current reality is that there are other crises, less mentioned but, probably more relevant in the longer run. Over the last 10 years we have also learnt that we live in a crowded planet (Jeffrey Sachs) and that the four pillars of human growth – food, water, energy and climate – are under constant and increasing stress, starting to suffer periodic bottlenecks of supply. The need for a new sustainability paradigm concerning infrastructures, communications and life styles will be at the basis of the bulk of public and private investment over the next decades.

– *We have entered the knowledge economy:*

Today the rate of technological innovations is exponential, from software to nanotechnology. In this new era, value creation does not depend only on the transformation of raw materials, the transformation of inputs into manufactured goods, or the plethora of services added to industrial production. Beyond all that, in our times the most powerful driver of value and growth is human capital and the ever increasing knowledge it encompasses. The result is the constant breakdown of established value-chains, the successful competition of small knowledge-based companies against giant corporations, and, most fundamentally, the rise of innovation and entrepreneurship as fundamental assets of the growing company.

3 Implications for Business

These trends are transforming the he role of business in society and more directly the core business models:

- From a business perspective the line between social issues and business issues has blurred considerably. Trying to address environmental and social issues under the previous social contract or from the paradigm of an industrial society is no longer viable. If critical breakthroughs to the complex social and environmental issues are going to occur, social innovations on the magnitude of those found in technical innovations will have to be developed.
- Corporate legitimacy in the twenty-first century is achieved by embedding social, environmental and governance concerns into the strategy and operations of the company. Companies will maximize their rights by adding new opportunities for value creation, and they will be better equipped to fulfill their obligations and responsibilities though a smarter approach to risk management (Escudero et al. 2012a).
- An even broader set of issues are emerging across the globe, as strains in existing global capitalism, as manifest by the increased frequency of global warming, Arab Spring, and the failure of fiscal institutions . Consequently business will now have to broaden their understanding of sustainable business in the twenty-first century to encompass pressing problems in the societies in which they operate, such as the perspective of income redistribution and the notion of strategic social investment in the communities, as important concerns of the legitimate company.
- Rethinking the social contract highlights the call for companies to act as global citizens and global problem solvers, in terms of the challenges of development, the Millennium Development Goals, and the need to collaborate in smart global governance systems. Business currently is excessively embedded in a short term and narrow conception of its relationship to society so that companies still have not generated relevant positive impacts concerning some of the crucial challenges of the global agenda (Escudero et al. 2012a).

For business, this constitutes a new perspective on value creation, one that circles back to the driving forces of the early entrepreneurs, creating a closer link between the purpose of business and its ties and contributions to society. The very power of commerce and the business sector that has unleashed such prosperity and established higher standards of living over the past 50 years, are now being reexamined in a broader context. Issues of the environment and a growing list of social ills are now in the province of the corporation, creating a symbiosis (however uncomfortable this is at present) between business and the larger society. Creating a sustainable business and society will require a business sector more grounded in its values and purposes, more aligned with the needs and demands of society, more focused on its key assets of innovation and entrepreneurism and more willing to use its influence for achieving widespread prosperity and growth. For the business sector to bring their unique assets to this new context, a major reset and adjustment of the current model is necessary.

4 Shared Innovation: Unlocking Sustainable Entrepreneurship for Business

Innovation continues to serve as the key ingredient for business success-the driver of growth and sustainable value creation. Businesses looks to disruptive innovation as essential for continuous improvement, healthy competitiveness, and to insure through new products, or processes, a constant renewal and relevancy. It is now time for companies to use these same innovation frameworks, processes and principles to address the social and environmental issues that confront it across the globe.

Mirvis et al. (2013) coined a term Corporate Social Innovation as "a strategy that combines the unique set of corporate assets (entrepreneurial skills, innovation capacities, managerial acumen, ability to scale, etc.) in collaboration with the assets of other sectors to co-create breakthrough solutions to complex social, economic, and environmental issues that impact the sustainability of both business and society." This begins to address the potency and opportunity for business to reinvent its role in society by leading with innovation, something it is both comfortable and well equipped to do.

Given the centrality of innovation to business, it is curious that traditional approaches and strategies that companies employ in addressing societal issues have largely ignored innovation. Existing business strategies, focused on the role of business and society, have outlived their usefulness and now have minimal value for the business, and are only superficially helpful to the issues that business needs to address in the twenty-first century. Corporate philanthropy, community relations and most public- private partnerships reflect this misfit and consequently do not have the potency of innovation so central to the rest of the business enterprise.

New approaches to integrating innovation into societal and environmental roles within the company are opening up new concepts such as "Shared Innovation" (Escudero et al. 2012b). The concept of shared innovation is an attempt to redefine the interplay between business and societal needs as a major engine for corporate growth and social prosperity. Innovation here is shared in that it transcends the current notion of stakeholders, taking it from a reactive to a proactive position: stakeholders are there not only to prevent corporate misbehavior, but to enlighten companies about the needs and aspirations of citizens concerning new products and services. It is also shared innovation because the dominant mechanism currently in play si largely dysfunctional. Partnerships are neither shared, except on a conceptual level, nor innovative, in that they are not grounded in the one asset that is most likely to produce the breakthrough necessary to result in sustainable impacts and solutions – innovation.

Sustainability within a business context has to be understood as the co-creation of value both for the company and society, as Porter suggests (2011). But shared value goes beyond this view, in that it connects value creation with one of the core competitive advantages of successful companies: innovation. Transforming a company's approach to sustainability through innovation presents a unique opportunity to tie more closely business values and strategy with the company's approach and response to the social and environmental issues that are increasingly blending into the business itself. Developing catalytic innovation creates both new and more effective pathways linking innovation around social, environmental issues to sustainable business strategy. Disruptive change and catalytic innovation could open up new and refreshing ways of finding simpler and better solutions.

> "If shared innovation applies fundamentally to the innovation in existing companies, it also applies to the creation of new business, to social entrepreneurship. This is why social entrepreneurship, - a hybrid of entrepreneurism in the knowledge economy and the need to attend to societal needs-, should gain momentum as part of the new answer to fill the cracks of social integration in societies, in the form of a variety of new societal/business ventures, ranging from bottom-of-the-pyramid approaches or micro-finances to innovative business models where companies, public authorities and communities work together in partnerships" (Escudero et al. 2012a).

By resetting and transforming traditional approaches to CSR and sustainability, shared innovation becomes a critical building block of the new sustainable business model.

5   Embedding Shared Innovation in Companies

Embedding shared innovation into companies requires new frameworks and new approaches for transforming their role in society. A number of leading companies have already begun this process through some fundamental first steps (Googins and Mirvis 2012b):

– *Enact a social vision for your company.*
    ShoJi Shiba, the MIT professor who won Japan's fabled Deming Prize for individuals in 2002, says that transformative change can occur only when "noble purpose exists. A person wants to know: what is the contribution to society or the planet?" Research finds that social entrepreneurs are fired by noble purpose and their social innovations are intended to make a meaningful contribution to society. Paul Light amplifies, "The underlying objective of virtually everyone in the fields of social entrepreneurship and social enterprise is to create social value. ..... People have embraced these fields because they are new ways of achieving these larger ends" (Light 2011).
– *Bring employees to the center of the effort.*
    Intel focuses on rewarding and awarding ideas. The Intel Environmental Excellence Awards recognize employees or employee groups that have created

an environmental innovation. In 2010, there were 11 winners of Excellence Awards that in total had created $136 million in estimated cost savings in addition to their environmental benefits. The company also offers Sustainability-in-Action Grants to allow employees to get funding for an innovative sustainability idea or project. The grants offer a few thousand dollars, and perhaps more importantly frees up employee's time to develop their sustainability-in- action project.

Signing on to this program in Intel India, Sonia Shrivastava designed a low-cost hardware utility that helps visually challenged people communicate and access daily information. With Intel's financial and technical support, Sonia managed a team of internal and external experts who customized a set of freeware applications and utilities on a low cost Intel® Atom™ based netbook computer and created a solution that was 85 % less expensive than any other solution in the marketplace.

– *Nurture social intrapreneurship.* Founder Jo da Silva has created an International Development consultancy within Arup – the professional service firm that designed the Sydney Opera House and Pompidou Center in Paris. Her group provides technical advice and practical solutions to reduce poverty and address social and environmental health in developing countries. Hundreds of the company's consultants have been engaged as "social intrapreneurs" to develop solutions for clients that can be spread across continents.

There is a growing body of literature on how companies can engage their employees as social intrapreneurs. Accenture, for example, created a unique business model that was based around a three-way contribution: Accenture provides access to its high performers free of profit and corporate overhead; these employees voluntarily give up a substantial percentage of their salary; and non-profits would cut a check to Accenture for consulting and technology services at significantly reduced rates, with no reduction in quality for those services.

The Accenture Development Partnership went through a pilot phase, a development phase with a few social sector partners, to its institutionalization in the business. In 2011, Accenture Development Partnerships led 126 different projects for clients around the world, bringing the total since it began to 640. Of these 126 projects, 48 involved cross-sector collaboration and came under our Partnership Services offering. Total number of hours spent on projects in the last 12 months exceeds 157,000. In its last financial year, employees devoted an average of 1,344 man hours per project. (Googins and Mirvis 2012a)

– *Reset CSR to innovation.* The Shell Foundation used to be the philanthropic arm of the parent corporation. Now it is funding and developing commercially viable business models that can achieve sustainable social impact. Says, Jurie Willemse, one of Shell's NGO partners, "For us it was always about developing a business model that you can scale-up and replicate in numerous countries and regions and which sustainably addresses the needs of start-up and growing businesses – a solution of global value for emerging economies rather than just a few countries in Africa" (Shell 2010).

– *Engage a broad spectrum of interests using connective technology and social media for innovation.* Today Nokia runs a social innovation lab for scaling the good works of innovative NGOs. Also reaching out Dell sponsors a social innovation challenge for college students and Studio Moderna leads a Challenge: Future competition that spans over 200 countries, 15,000 schools, and over 23,000 innovators. This is all about using social media and networks to drive social innovation. Meanwhile, companies like Best Buy use social media to spark and shape programs such as the company's innovative reuse and recycle program for electronic equipment.

New social media and communication tools seem to be at the center of corporate social innovation.

6 Summary

Achieving real change in solving the social issues of our time has proven to be as elusive as challenging. As is often the case with "wicked" problems, solutions become institutionalized and approaches remain mired within narrow professional alleys and fragmented disconnected activities. Shared Innovation both in the public and the private sphere opens up new approaches, and a search for new solutions with a potential combination of private and public drivers – competitive advantage and disruption. While innovation in public and private spheres have a different set of drivers and motivators, both essentially disrupt and unfreeze current approaches that are often stuck and overly rigidified.

The world that captivated the early entrepreneurs reflects much of the vision and the purpose found in the context of business in the twenty-first century. While technology has brought incredible prosperity across the globe, it has outstripped much of human and community development. The very innovation that led to technological breakthroughs must now be applied to the more difficult social and environmental issues that are impeding our promise of achieving a sustainable society. For business, a shared innovation approach capitalizes on the true assets of a company and unlocks the potential for creating both a new business model and a more sustainable role for business.

Literature

Accenture, (2010). A new era of sustainability – UN Global Compact- Accenture CEO Study.
Escudero, M., & Googins, B. (2012a). Towards a new era for sustainable business: Shared innovation Deusto University.
Escudero, M. et al. (2012b). Rebuilding the social contract: Frontier issues for sustainable business in a new era Deusto University.
Giddens, A. (1984). *The constitution of society*. Cambridge: Polity Press.

Googins, B., & Mirvis, P. (2012a). Where is your Moonshot US Chamber of Commerce Blog, Aug 2012.
Googins, B., & Mirvis, P. (2012b). Corporate social innovation, US Chamber of Commerce Blog, April 2012.
Habermas, J. (1984). *The theory of communicative action, reason and the rationalization of society* (1). Boston: Beacon.
Isaacson, W. (2011). *Steve jobs*. New York: Simon & Schuster.
Light, P. (2011). *Driving social change*. New York: Wiley.
Mirvis, P., Googins, B., & Kiser, C. (2013). Corporate social innovation, Babson College.
Porter, M., & Kramer, M. (2011). Creating shared value. *Harvard Business Review, 61*, January–February, 1–17.
Sachs, J. (2008). *Economics for a crowded planet*. London: The Penguin Press.
Shell Foundation (2010). Enterprise solutions to scale: Lessons learned in catalyzing sustainable solutions to global development challenges. www.shellfoundation.org

The Evolution of CSR from Compliance to Sustainable Entrepreneurship

René Schmidpeter

1 Economic, Social and Ecological Challenges in Economy

Nobody would deny any more that we find ourselves in times of global change and high uncertainty. Financial crisis, scarcity of resources, climate change, demographic development, political upheaval and technological progress are becoming the main forces of our social development. This affects the thinking and behaviour of people as well as the prevalent political systems. The framework of national and international economic systems and enterprises' ability to compete are changing dramatically as well. More and more managers are realising that operating sustainably is becoming the central challenge for their enterprises as well as for the economy as a whole. The financial, the energy and the automotive sectors have already been challenged hard by the current developments – and some enterprises were able to survive thanks to massive governmental support (assumption of liabilities, scrapping premium) only. In other sectors of the economy as well it can already be observed that only the enterprises that cope best with the current challenges, i.e. actively take the opportunities inherent to the crisis, will be tomorrow's winners. The current ecological, social and economic challenges and the connected market changes will therefore create winners and losers. This means for all success-oriented enterprises that they have to increase innovation (product, process, management and social innovation) and react to current challenges with proactive management approaches.

This chapter is based on the German article "Unternehmerische Verantwortung - Corporate Social Responsibility" In: Sedmak, C./Kapferer, E./Oberholzer, K. (Eds.)(2013) Marktwirtschaft für die Zukunft. Edition Verantwortung. Wirtschaftskammer Salzburg.

R. Schmidpeter (✉)
Centre for Humane Market Economy, Salzburg, Austria
e-mail: rene.schmidpeter@gmx.de

C. Weidinger et al. (eds.), *Sustainable Entrepreneurship*, CSR, Sustainability, Ethics & Governance, DOI 10.1007/978-3-642-38753-1_9,
© Springer-Verlag Berlin Heidelberg 2014

Repositioning of Business in Society as Social Innovation

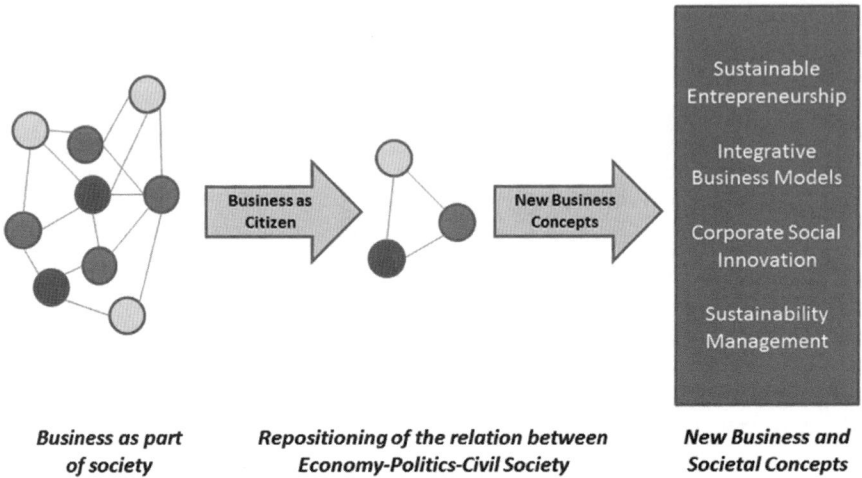

| Business as part of society | Repositioning of the relation between Economy-Politics-Civil Society | New Business and Societal Concepts |

2 Management by Sustainability: Innovation and Holism

This realignment of business models and processes can, according to recent management pioneers, only succeed if they are understood as parts of a greater whole and the current dichotomy between economy and society is overcome productively.[1] Viewing enterprises as part of the solution and not as part of the problem is a big opportunity within the crisis – and presumably our only chance.

---

[1] See Porter and Kramer (2011), Laszlo and Zhexembayeva (2011), Senge (2008) and the contributions in Schneider and Schmidpeter (2012).

Business as Driver for Social Innovation

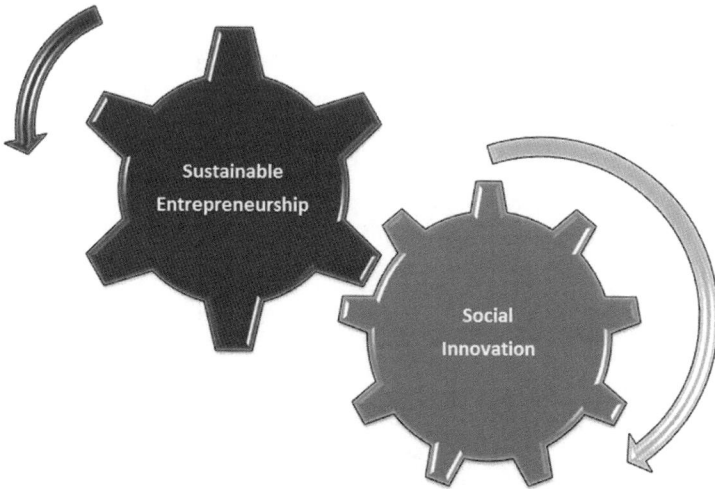

In order to view enterprises as drivers of social innovation, it is necessary not to reduce economic considerations to questions of business administration, but to always underlay them with socio-political reflection. The question of CSR as Sustainable Entrepreneurship thus becomes the central strategic question for every enterprise.[2] In the past there have been many misunderstandings and incorrect interpretations of CSR. Therefore, the goal of the following paragraphs is to highlight the key points of modern CSR in its new interpretation of Sustainable Entrepreneurship.

3 Integrative Management: Connecting Social Case and Business Case

For a long time fundamental critics of our economic system have argued with defenders of narrow profit-oriented management systems about whether economic reasoning has priority over ethics or vice versa. The participants of the discussion usually tend to think in stark contrasts, a very one-dimensional view that cannot be solved constructively and leads to never-ending conflicts. The answer to this problem caused solely by intellectual deliberations, however is relatively simple if the prevalent either-or approach is turned into pragmatic both-and thinking.

---

[2] For an overview to the different corporate divisions and functions see Schmidpeter (2012).

This means an integrated view on economic, social and ecological questions, instead of playing them off against each other as isolated issues. This is where the strength of the current discussion of Sustainable Entrepreneurship lies, wherein a new, productive view on the contribution of corporate responsibility to both one's own business and social progress takes form.[3] It is based on the assumption that entrepreneurship can only be reconstructed adequately if both the individual component of 'profit' (business case) and the social role of 'create added value for society' (social case) of the enterprise are considered equally. Sustainable entrepreneurship then has the goal to create added value for both society and the enterprise itself.

Sustainable Entrepreneurship: Added Value for Business and Society

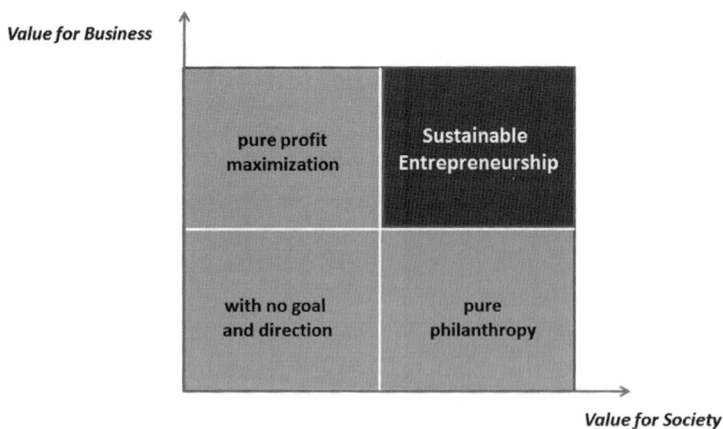

The actions of successful businessmen in the times of industrialisation may serve as an example of this view. Responsible businessmen have reacted to the huge social and ecological challenges by developing, for example, corporate pension funds, employee participation, education concepts and health or social projects. These success models practised in the enterprises were taken up by the state and pension funds, and dual education and public healthcare arose.

Like in the times of industrialisation, today we are once again dependent on enterprises that actively shape our country's framework and support the government in further developing social market economy by integrating the concept of sustainability. Enterprises are like laboratories in which new ideas are developed, tested and optimised. Often only through these experiences does it become possible to spread innovations to the whole sector or even society as a whole. In order to

---

[3] See contributions in Schneider and Schmidpeter (2012).

achieve that, social, ecological and economic questions and concerns have to be integrated systematically into management systems, thereby becoming part of the central DNA of the enterprise.

4 Strategic Management: Innovation Instead of PR and Compliance Alone

It is clear that Sustainable Entrepreneurship cannot and must not be mere PR or greenwashing (as the concept of CSR has often been accused of by its critics). A modern understanding of sustainable management implies defining enterprises as part of society and systematically identifying actual and potential areas of conflict between the enterprise and its environment. These conflicts are then reduced or solved by intelligent management approaches or by product and process innovation. This aims at adapting the business model to the ecological, social and economic framework in a way that generates added value from the given resources both for the enterprise and for society.

The necessary new problem-solving approaches require entrepreneurial innovation.[4] Thus, it is evident that Sustainable Entrepreneurship is much more than mere compliance. One could call compliance the compulsory and CSR in the meaning of Sustainable Entrepreneurship the voluntary exercise. Realising that, many enterprises now face the challenge of turning a defensive, compliance-oriented form of responsibility into a proactive, opportunity-oriented view of Sustainable Entrepreneurship. To accomplish this, enterprises as citizens have to redefine both their business goals and their relations to politics and society by systematically examining the current ecological and social challenges as well as their stakeholders' interests and integrating them into their business model. Integrated management that implements responsibility into all management processes is therefore increasingly becoming a prerequisite of economic activity.

5 Proactive Management: Turning Implicit Action into Explicit Strategy

Some enterprises still fail to see the necessity of working systematically on the topic of responsibility. Especially small and medium-sized businesses maintain that they, owing to their strong connection to their employees and their environment, work

---

[4] See Grieshuber (2012) on innovation.

responsibly by default. This may be true in many individual points, but working this way is no substitute for an explicit management approach, which takes up and permanently develops the opportunities provided by responsible economic activity. The following paragraph will therefore show the advantages of explicit CSR management compared to implicit responsibility:

First, explicit CSR approaches allow for a stronger involvement of employees and a better incorporation of scientific findings in taking responsibility. In this way, not only are existing innovation potentials better utilised, but the identification of employees and managers with taking responsibility is enhanced. Second, an explicit CSR approach facilitates the continuity of an existing culture of responsibility when transferring a business, as a strategy of responsibility that has been discussed and developed explicitly with the successors can be passed on to the next generation with less friction. Third, an explicit strategy of responsibility can be used to convey one's position to external partners (international customers, suppliers etc). Fourth, especially with a dynamic development of the enterprise, it is often indispensable to develop the enterprise's role in society as well. For example, if a business is growing successfully, in most cases it is necessary for the responsibility that was envisioned by the company's founder to 'grow' too. Additionally, increasing professionalism is demanded from large-scale enterprises in dealing with responsibility. This has consequences for small and medium businesses as suppliers, as they are confronted with new requirements/criteria on the part of the large-scale enterprises.

6 Conclusion: The Development of CSR into a Management Concept of Sustainable Entrepreneurship

Especially as general confidence in the economy is declining, it is becoming increasingly important for enterprises to report their position on responsibility openly and to communicate with relevant target audiences. To accomplish this, more transparency is needed but is not sufficient on its own. In the sense of the idea of social market economy, entrepreneurial activity always has to be 'approvable'. Entrepreneurial activity will be judged increasingly by the extent to which it considers the interests of society. Therefore, without an explicit CSR strategy, enterprises lag behind the potential benefits of their responsible economic activity, this means the resulting opportunities are not taken to their full extent.

Stages of Socio-Economic Management Thinking

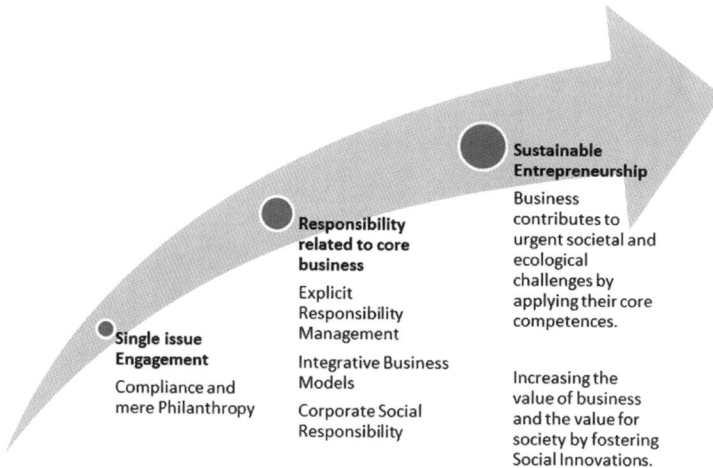

Single issue Engagement
Compliance and mere Philanthropy

Responsibility related to core business
Explicit Responsibility Management
Integrative Business Models
Corporate Social Responsibility

Sustainable Entrepreneurship
Business contributes to urgent societal and ecological challenges by applying their core competences.

Increasing the value of business and the value for society by fostering Social Innovations.

This is why CSR has developed in recent years from isolated engagement (sponsoring and donations) and legal compliance to explicit responsibility management in core business, along the triple bottom line of economic, social and ecological factors (CSR 2.0).[5] Also, the question of how profit is used is no longer the central point, but rather how profit is made. With the next step the general question of the contribution of enterprises to social innovation comes into focus. This conscious strategic positioning in society (business in society) aims at increasing social and entrepreneurial added value at the same time (shared value). Businesses as responsible corporate citizens are no longer viewed as a problem, but provide solutions for the pressing challenges of our time. This realignment of enterprises (in the sense of Sustainable Entrepreneurship) is the actual and fundamental contribution of economy to a sustainable development of our society. It is an investment in the competitiveness of enterprises and the wellbeing of future generations at the same time.

Literature

Grieshuber, E. (2012). CSR als Hebel für ganzheitliche innovation. In A. Schneider & R. Schmidpeter (Eds.), *Corporate social responsibility*. Berlin/Heidelberg: Springer Gabler Verlag.
Laszlo, C., & Zhexembayeva, N. (2011). *Embedded sustainability*. Stanford: Stanford University Press.

[5] On generations model and understanding of CSR see Peters (2009), Visser (2011) and Schneider (2012).

Lorentschitsch, B., & Walker, T. (2012). Vom integrierten zum integrative CSR-Managementansatz. In A. Schneider & R. Schmidpeter (Eds.), *Corporate social responsibility*. Berlin/Heidelberg: Springer Gabler Verlag.

Peters, A. (2009). *Wege aus der Krise. CSR als strategisches Rückzeug*. Gütersloh: Bertelsmann Stiftung.

Porter, M., & Kramer, M. (2011). Creating shared value. Harvard Business Review, 61, January–February, 3–17.

Schmidpeter, R. (2012). Unternehmerische Verantwortung – Hinführung und Überblick. In A. Schneider & R. Schmidpeter (Eds.), *Corporate social responsibility*. Berlin/Heidelberg: Springer Gabler Verlag.

Schneider, A. (2012). Reifegradmodell CSR – Eine Begriffsklärung und -abgrenzung. In A. Schneider & R. Schmidpeter (Eds.), *Corporate social responsibility*. Berlin/Heidelberg: Springer Gabler Verlag.

Schneider, A., & Schmidpeter, R. (Eds.). (2012). *Corporate social responsibility*. Berlin/Heidelberg: Springer Gabler Verlag.

Senge, P. (2008). *The necessary revolution: How individuals and organizations work together to create a sustainable world*. New York: Doubleday.

Visser, W. (2011). *The age of responsibility: CSR 2.0 and the new DNA of business*. Hoboken: Wiley.

Sustainable Entrepreneurship: Next Stage of Responsible Business

Mara Del Baldo

1 Introduction

The theme of CSR (Corporate Social Responsibility) is widely diffused in the literature and the debate among scholars of different disciplines has produced a rich set of contributions. In the last decade, studies of CSR have intersected and merged with those on sustainability, broadening the concept, as recently reviewed by the European Commission itself.

CSR is broadly defined as the extent to which firms integrate on a voluntary basis social and environmental concerns into their ongoing operations and interactions with stakeholders (Godoz-Diez et al. 2011). Nevertheless, there is no single, commonly accepted definition of the concept of CSR (Carrol 1999, 2008) and many different ideas, concepts, and practical techniques have been developed under the umbrella of CSR research including corporate social performance, corporate social responsiveness, corporate citizenship, corporate governance, corporate accountability, sustainability, triple bottom line and corporate social entrepreneurship.

Following the European Commission, CSR can be defined as a concept whereby "companies integrate social and environmental concerns in their daily business operations and in their interaction with their stakeholders on a voluntary basis" (EC 2001, Green Paper, p. 6) and, more recently, as "companies' responsibility for their impact on society" (EC 2011). CSR is placed at the base of the renewed European strategy for 2020, which is oriented toward a "smart growth" (based on innovation and knowledge), a "sustainable growth" (based on a more efficient use of resources and so-called green energy), and an "inclusive growth" (based on the employment development and on the social and territorial cohesion). EC also clearly states that certain types of enterprise, such as cooperatives, mutuals, and family-owned

M. Del Baldo (✉)
University of Urbino Carlo Bo, Department of Economics, Society and Politics, Via Saffi, 42, Urbino (PU) 61029, Italy
e-mail: mara.delbaldo@uniurb.it

C. Weidinger et al. (eds.), *Sustainable Entrepreneurship*, CSR, Sustainability, Ethics & Governance, DOI 10.1007/978-3-642-38753-1_10,
© Springer-Verlag Berlin Heidelberg 2014

businesses, have ownership and governance structures that can be especially conducive to responsible business conduct.

Since the 1990s, environmental and social factors have become increasingly important strategic considerations for enterprises of any size. Sustainable development or sustainability is defined as 'meeting the needs of the present generation without compromising the ability of future generations to meet their own needs' (WCED 1987). According to Elkington (1999), there are more than 100 definitions of sustainability and sustainable development. A number of scientists have elaborated on this theme (Ranganathan 1998).

The strategy of a sustainable enterprise has been defined as 'the process of aligning an enterprise with the business environment to maintain a dynamic balance' (WCED 1987). The phenomenon of sustainable entrepreneurship is associated with a number of other phenomena, e.g. social responsibility of entrepreneurs, corporate morality, and business ethics (Bucar et al. 2003). Sustainable entrepreneurship is a combination of creating sustainable development – on the one hand – and entrepreneurship, on the other. Particularly, sustainability strategies create many synergistic effects for SMEs (small and medium-sized enterprises) working collaboratively, as well as systemic benefits for the commons. By adding a sustainability lens within the framework of SME's strategies, SME's development seeks to balance resilience and growth. Integrating sustainability into their competitive strategy, and thereby obtaining greater profitability for SMEs through adoption of intentional sustainable strategies, can help them to optimize their rate of sustainable change (Giddings and O'Brien 2002).

Departing from these assumptions, the present work offers the hypothesis that evolutionary harmony between businesses and their surroundings is correlated with the firms' orientation toward "holistic development" (Sorci 2007) which represents a deeper and further development of a CSR-oriented strategy. This perspective is based on the literature from the disciplines of business administration and entrepreneurial management, which specifically refer to "territorial firms" (Del Baldo 2010, 2012) that are deeply rooted in their socio-economic, cultural, and relational surroundings and can be conceived both as community-based businesses (Peredo and Chrisman 2006; Jenkins 2006), spirited businesses (Lamont 2002) and collaborative enterprises (Tencati and Zsolnai 2008).

This analysis intends to answer the following questions: What are the entrepreneurial behavioral models and the inspired values of holistic development, and in which forms of sustainable businesses are they most fully expressed? What are the characteristic traits of those companies that are distinguished by a 'special' relationship with their surrounding territories – relationships that are never obscured or forgotten? What are the mutually beneficial pathways that characterize a firm's rootedness in its surroundings, and what are the roles played by their agents?

The responses to these questions are encapsulated within the following analysis that, embracing the systematic perspective, helps sketch out a model of development in which the firms selected as case studies are protagonists.

The work is developed both deductively and inductively (the so-called abductive approach) (Denzin 1978). The first part (Sect. 2) presents the theoretical framework, while the second (Sect. 3) is centered on a qualitative analysis of two "territorial" companies (multiple case studies; Yin 1994; Eisenhardt 1989; Eisenhardt and Graebner 2007), the Distilleria Varnelli spa and the Loccioni Group, both located in the Marche a central Italian region characterized by a diffuse entrepreneurial fabric and an elevated level of social cohesion (Brusco 1982; Becattini 1990; Bagnasco and Trigilia 1990; Leborgne and Lipietz 1991).

The reflections that emerge (Sect. 4) are intended to suggest that holistic development can be a route upon which both corporate strategies and industrial politics should converge. These reflections follow a bottom-up logic that considers the specificity of each unique local context and of the individual actors that animate them, and are aimed at the recuperation of competitiveness and of sustainability of the country. Linking the perspectives of corporate responsibility (Freeman et al. 2010; Garriga and Melé 2004; Zadek 2004) and business ethics with that of local development, and evaluating the anthropological dimension of business culture in its relationship with its surrounding territorial context, the work utilizes a new analytical framework to enrich the research on the relationship between sustainable entrepreneurship and local development (Eisenschitz and Gough 1993; Lechner and Dowling 2003). It does so by emphasizing the role of companies (and particularly small and medium-sized companies) deeply rooted in their regions (Enderle 2004). These are often little noted in the literature, but they are protagonists of a route toward sustainable development that runs counter to the model of "turbo-capitalism" (Matacena 2010), centered as they are on the soft elements of an intangible nature (social, intellectual, and relational capital), which are sedimented in the surrounding territory's *genius loci* (Granovetter 1985; Coleman 1988; Fukuyama 1999; Lin et al. 2001; Nahapiet and Ghoshal 1998; Adler and Kwon 2002).

2 Holistic Development of Businesses and Their Territories

"At the heart of studies concerning entrepreneurship and business management, scenarios emerge that correctly accept development in all of its dimensions, which can be called holistic development" (Sorci 2007, p. 12). Typically (but not exclusively) rooted in the Italian entrepreneurial tradition, in recent decades this paradigm has been closely analyzed and applied to international business models (Alford and Naugthon 2002). It has also been linked to the concepts of sustainable entrepreneurship and of sustainable development – a form of development that involves economic, ethical, environmental, and social dimensions (Elkington 1994).

A business's holistic development is founded on a system of universal ethical values that are embedded in the local environment, which are actively practiced by entrepreneurs, managers, and corporate stakeholders. Without these, the vision of development would often become crippled or "nomadic". Holistic development is the fruit of a managerial/entrepreneurial strategic orientation that is transferred throughout the organization, nurturing a successful entrepreneurial formula and a strong business culture.

Holistic development is by nature multi-dimensional; that is, it is able to be translated into diverse socio-competitive contexts and can create shared value (Porter and Kramer 2011), contributing to the common good and to collective progress, starting with the local community in which these businesses are an integral part (Spence 2000; Spence et al. 2003, 2004; Argiolas 2006; Zamagni 2007). The firms that follow such objectives are protagonists of a particular economic model, which is beginning to attract the attention of researchers, institutions, and practitioners. They see it as possibly capable of overcoming the present economic crisis and confronting the challenge of globalization, since the model is sustained primarily (but not exclusively) by small and medium-sized firms that live in a profound relationship with their territories (Harvey et al. 1991; Cornwall 1998; EC 2002; EU 2004; Jenkins 2004; Del Baldo 2009; Moore and Manring 2009; Del Baldo and Demartini 2012a).

In the perspective of holistic development, every decision and action is the result of an orientation aimed at producing economic results, but, at the same time, it is sensitive to the impact that decisions have on the company's stakeholders, and the system of values, on the development of understanding, on the professionalism of individuals and groups, on social cohesion, and on the socio-economic and environmental system surrounding the company (Spence and Schmidpeter 2003).

This perspective can be extended to every type of business (public or private, small or large, domestic or transnational), and brings to light two questions.

First, what are the values and behavioral models that it both expresses and synthesizes? Second, what kind of "atmosphere" – that is, what substrate of endogenous factors (referring to the firms) and exogenous factors (the particular and general contexts of which the firm is a part) – nurtures a form of management and entrepreneurship oriented toward holistic development objectives?

The response to the first question lies in positing that people – inspirers of holistic development – stand at the center of the company's work and of its being. Indeed, it must begin with understanding the unique system of values of the corporate system's key actors and collaborators (Melé 2002; Argandoña 2003). This, therefore, involves considering values as the principle factors of production and of recognizing the value of relationships, through which individual objectives can be harmonized with those of the firm, of the community and of the surrounding environment. This also emphasizes the link between the socio-economic context of the firm as well as its business culture, and understands that the behavioral codes of the economic subject is the expression of shared values among the community of individuals inside and outside of the business. In light of the position that interprets businesses as living organisms that have "souls" – that is, they are a community of people capable of perceiving and structuring a set of values that affect their activities (Lamont 2002) – it is possible to distinguish between growth and development. Growth is understandable only in quantitative terms; it regards the specific size of its productions of worth, and refers to single carriers of interests. Development, however, regards the company as well as all its stakeholders, and refers to the capability of an organization to develop over time by taking into consideration the economic, social, and environmental dimensions of its processes and performance. The conditions that favor the diffusion of holistic objectives into the economic fabric are cemented by social cohesion inside and outside of the firm.

The second question finds its response in the perspective that sees corporate organisms as creators of value for themselves and for their surrounding contexts, which open the moment they obtain full understanding of the need for holistic development. Alongside companies that are motivated by economic objectives – in which their understanding of their complex social role matures only through time – this perspective posits that there exist certain forms of "extraordinary" businesses. Some are already born with this understanding, and with a clear social vocation (for example, the Grameen Bank); others are the manifestation of a particular charisma, such as the firms belonging to the project of Economy of Communion (Gold 2010); and still others from particular conceptions of the dignity of work or of the founder's clients, which then develop all of the other dimensions. In this latter category are firms characterized by a solid working philosophy (i.e. Olivetti, Siemens), a strong corporate culture, and a deep-rootedness in the socio-economic context from which the firm comes. The last category denotes "territorial" businesses, whose fundamental features emerge in following through the analysis of the two cases in the second part of this work.

In conceptualizing the presuppositions of holistic development, it is necessary to look at the entrepreneurial and managerial profile (in terms of values, modes of understanding reality, their own objectives, their own roles and that of the firm), and in the way in which these are translated into a system of concrete decisions. The creation of lasting values for the corporate and environmental systems (i.e., "holistic success") is the fruit of a virtuous cycle that produces phenomena of accumulation of resources, mostly of an immaterial and intangible nature, such as categories of understanding, deduction, cohesion, and credibility. Along with the willingness to respect and valorize people, productive correctness, and informational transparency, one of the values inherent in a coherent model of holistic development is a profound sense of responsibility toward the socio-economic context. The development of a territory cannot emerge without the holistic development of the businesses that operate in it. The growth of a territory's wellbeing is, in fact, in large measure the result of the "values" that the businesses, of every type, are able to create. These values are intended not only in the economic sense (capital, profits, salaries, stipends, taxes, etc.), but also in other different ways: competitive products, effective services, material wellbeing, moral satisfaction, security for the future, cultural growth, social development, etc., created not only to benefit themselves, but also to benefit all stakeholders (shareholders, citizens, public institutions, etc.).

2.1 The Relational Dimension of the Territory and the Role of "Territorial Businesses"

If one considers a company to be a socio-economic system open to the "territorial" dimension of holistic development, one finds elements in common with the systematic perspective, which argues that the co-evolution of both the firm and its surrounding environment is an essential component for lasting development (Minguzzi and Passaro 2000). This interpretative-key considers firms and territories as systems that are partially or relatively open, dynamic, contextualized in time and space.

From the many national and international contributions that analyze the reciprocal dependence between businesses and territories (theories of international commerce, of districts and of local systems; international marketing studies), it is possible to distinguish a line of thinking which analyzes "territorialized innovation", and "external relations" that are based on "local and global space" (knowledge networks) (Moulaert and Sekia 2003; Cedrola et al. 2011). Firm growth depends not only on its network (egocentric perspective) but also on the relational environment in which it is inserted (socio-centric perspective) (Lechner and Dowling 2003). In this interpretative framework, the themes of innovation, local and international development, and competition are accompanied by concepts of rootedness and embeddedness, of networks, social capital, knowledge and reputational networks, which are the presuppositions of the concept of holistic development of the firm and of the territory.

Referring specifically to local development in Italy, which is centered on the model of the industrial districts and the prevalence of small and medium-sized firms, a number of contributions have emphasized the role of the territory as a site of the production of specific understandings and social interaction (networks of interpersonal relations) and illustrate how firms' behaviors contribute to the activation of virtuous circuits of effective collective dynamics which, though social, cultural, historical and productive components, determine the firm's behavior (Bagnasco 2004; Fuà 1988; Sabatini 2006). Here, the territory is not an external accompaniment, but an active force, the totalization of intangible resources such as knowledge, art, design, and creativity, which influence the characteristics of the business and foster a mode of operating through the reciprocal exchange of material and immaterial elements (Realacci 2012). It is the relational space, complex and difficult to replicate elsewhere, that attracts resources and skills that are then metabolized to the advantage of the actors who take part in it. Their actions, furthermore, nurture a collective process of social evolution.

The last step in the perspective in which the firm lives in a relationship of evolution with its territory – wherein the cognitive, spatial and anthropological dimensions are aligned – is to identify the "territorial" firm, which, from its rootedness in the local environment and from the quality of its relationships with it, creates diverse conditions of excellence. In territorial contexts in which a reciprocal relationship thrives between firm and territory entrepreneurs are "takers" of resources (human, physical, and immaterial, such as the traditions and the culture of the place), but, at the same time, they are particularly sensitive to giving back – and they do so through reciprocal exchanges, reputation, trust, and identity. This link emerges in particular in the theoretical framework that is focused on the approach to social responsibility and sustainability of small firms (MORI 2000; Lepoutre and Heen 2006; Vyakarnam et al. 1997; Perrini et al. 2006; Del Baldo and Demartini 2012b).

In this perspective, the orientation toward sustainability passes from the level of the individual to the plural, from the territorial firm that designs and enhances the network of its relationships with diverse carriers of interests to the development of networks of relations that take part in the local community. These networks also contribute to social capital – the sum of intangibles, trust, and shared norms and values that regulate their coexistence and promote cooperation, incorporated in authentic social and personal relations (Granovetter 1985; Coleman 1988; Putnam 1993; Fukuyama 1999; Lin et al. 2001). Territorial businesses oriented toward holistic development enhance forms of "territorial governance", which are the result of deeply rooted socio-economic and cultural contexts. It is opportune to deepen the knowledge of such a model of healthy "convergence" (from the Latin *cum vergere*, in the purest sense as the coming together) through an analysis of best practices.

3   The Case Studies: The Varnelli Distillery and the Loccioni Group

3.1   Methodology

The empirical analysis is based on a qualitative study aimed at identifying the values and the entrepreneurial behaviors (policies, strategies and actions) at the base of the coevolution of the firm and the context in which it is found.

The case study method allows for a deep analysis and is based on a multidimensional approach to the phenomenon investigated, in which the firms' networks and relationships are analyzed (Eisenhardt 1989; Yin 1994; Naumes and Naumes 2006). This method is often employed in studies on social responsibility and of sustainability both for theoretical objectives and for suggesting concrete routes of action, since it gives voice to the experiences of successful entrepreneurs. Qualitative research utilizes the sampling model technique (in particular of theoretical sampling) that does not readily allow for the generalization of its results, because the nature of its analysis favors the particularities of the cases rather than their representativeness (Glaser and Strauss 1967; Flick 2009). Nevertheless, it is important for generating theoretical propositions that can be tested through more ample quantitative research designs.

Varnelli and Loccioni are both located in small centers in the provinces of the Marche, an important region in the so-called "Third Italy" (Fuà 1988), which boasts leaders in the production of "traditional" goods under the *made in Italy* brand. The choice of the territorial context is linked to the distinctive nature of the Marchegian situation: it has a diffuse entrepreneurial fabric that is the result of a centuries-old artisanal tradition and a rural culture; the presence of numerous cases of best practices that have been recognized at both the national and international levels; and it is marked by a high level of social cohesion and a wealth of social capital, to which the local governments also contribute (Bonomi and Savignon 2011). Local governments are involved in the promotion of the territory through the lens of sustainability through projects aimed at the creation of a system of responsible Marchegian businesses which were begun from with the involvement of a number of territorial actors (Del Baldo and Demartini 2012a, b). The choice of these two firms, selected among those in the association Confindustria Marche, is already part of a field that has been utilized in an earlier exploratory study aimed at analyzing "spirited businesses" (Unioncamere 2003; Del Baldo 2012) and has been drawn up based on two characteristics: (a) a strategic approach to holistic development that is integral to its mission statement (value asset), as well as its governance (instruments and modalities of governance marked by transparency and to the sharing of decisions with its stakeholders) and in accountability; and (b) the systemization of activities such as stakeholder dialogue, engagement, and commitment with carriers of interests, initiated by those who come from the local context.

The deeper analysis, undertaken in 2011 and concluded in May 2012, is based on the triangulation of sources (semi-structured questionnaires, direct interviews, documentary analysis, focus groups, and participant observation) and was undertaken in two phases. In the first phase, a questionnaire was emailed to the owners and managers of the businesses (CEOs, human resources managers, etc.), followed by interviews (which lasted on average an hour; these were audiotaped, transcribed, and marked) during visits to the firm (three for each company) that involved six people (three for each business). The primary source data were added to secondary data that came from the analysis of the sites, internal documents, external publications, newspaper articles, and direct interaction with the entrepreneurs at conferences, seminars, and workshops.

3.2 Data Analysis

The following is a brief profile of the two companies (Table 1).

3.2.1 Loccioni Group

Founded by current president Enrico Loccioni, the group is composed of five companies that propose solutions for "tailoring technologies". Flexibility, relational

Table 1 Profiles of companies

Name, headquarters, year created, sector(s), primary activities, number of employees, sales figures (for the year 2011), proprietary assets	Instruments of implementation and communication of social responsibility and of sustainability (year of introduction)
Gruppo Loccioni, Angeli di Rosora (AN); 1968 Integrated solutions (technological systems) of automation, assembly, testing and quality control, environmental monitoring, taking place in the following sectors: Industry; Community; Home; Automotive; Environmental; Health; 350 employees; 60,000,000 euro (of which about 50 % are exports) Family business (first and second generation); not listed on the stock exchange; open (a minority of shareholders are non-family members)	List of values, 1969; Ethical code 1996; Social balance, 1997; Intangibles impact, 1997; Cause Related Marketing, 1999
Distilleria Varnelli Spa; Pievebovigliana (MC); 1868 Food and beverage sector: production of anise, infused liquors and herbal drinks; 11 employees; class of sales figures: 10,000,000 euro; Family business (third and fourth generation), not listed on the stock exchange	List of values, 2007; ISO 14001, 2007; SA8000, 2008; OHSAS, 2010

capacity, united to a distinctive core competencies, nurture for over 40 years a continual national and international development (its clients are spread across 40 countries). This development is sustained by a cohesive culture and a marked sensibility towards enhancing the common good, which finds expression in a model (Table 2) whose distinctive values (Table 3) and foundational principles (Table 4) are summarized below.

Enrico Loccioni was born in the Marchegian countryside into a family of farmers at the crossroads of three important Benedictine abbeys. The cultural and religious heritage, and that of his family, together with the link to the land, forged a model of the values of this primogenital firm, which he created when he was only 19 years old.

The group's strategic formula notes three aspects of its orientation toward holistic development and its rootedness: the centrality of its values and of the human element; an attitude toward creating and consolidating networks of relationships; and a privileged relationship with its surrounding territory.

From the rural culture we have learned the importance of traditional values; the transmission of trust through a handshake; the attitude of working in the uncertainty of the seasons; diversification to reduce risk; the strength to always start over. Actions, even those everyday actions, need a profound moral commitment. The firm is a container of values and not just of capital. Values are the identity of the Group: they provide a common language, they give strength to our businesses and guide them as they adapt to the marked towards success that benefits all (interview with Enrico Loccioni, 3 July 2012).

Table 2 Distinctive aspects of the model Loccioni Group's development

From rural farmer	To networked firm
From the dominion of nature	To the diffusion of knowledge
From the sweat from its brow	To products of the mind and to the market of ideas
From values as a foundation	To the ethics of development
From a microcosmic producer	To networked firm

Table 3 Loccioni Group's list of values

Imagination	To be able to create
Energy	To dream and to realize our dreams
Responsibility	For the air that we breathe, the land that we walk on, the resources that we utilize, the trust that we receive
Tradinnovation (Tradition + innovation)	To give form to the future learning from the past

The term "tradinnovation" synthesizes the capacity to "metabolize" the roots from which the firm and its founder draw their core values: a strong work ethic, dedication, a strong will, sobriety and parsimony, simplicity, solidarity, integrity, the sense of family and of community. The Group's orientation toward knowledge, innovation, and openness to international markets are based upon these values, which have produced a successful entrepreneurial and managerial synthesis.

The principles of the Marche's rural traditions (heritage value), whose stability and universality are linked to its ethics, constitutes a sort of corporate meta-culture. Promoted by the founder and interiorized in its organizational plan, they are projected outside of the company through network relations. Inspired by the entrepreneurial model of Adriano Olivetti and H. von Siemens, the Group is an evolved firm, a lab for continual betterment, and a vector of intangible factors, which follows Loccioni's dream "that doing business one can do something else" (Bartocci 2011).

Creativity and reciprocal opportunities for growth run the gamut from pre- to-post entry: *first*, through the articulated project Bluzone, an educational laboratory for young people that mediates between school and work (800 students doing coursework, over 7,000 hours of coursework and classes, partnered with 28 schools, 20 universities, five Master's programs), *during*, with the project Redzone, aimed at developing talent and generating entrepreneurial activities (business incubators and spin-offs); and *after*, with the project Silverzone, aimed at accompanying young people (from the Group or from other businesses) with the "wisdom" of those who had already gained important professional experiences. Loccioni Group has been formally regognized by the Regional government as Training Agency.

Relationality makes the Group a diffuse entrepreneurial network that is culture-driven and is expressed in the particular organizational network model: U-net (multi-disciplinary model of universities and research centers); Crossworlds (network of large international groups aimed at stimulating the transfer of automotive technologies to other industries); Nexus (a multi-sector network of local

Table 4 Loccioni Group's identity card

Attention to human resources	55 % of its collaborators have high school diplomas; 45 % have college degrees; average age 33 years old
	7 % of resources are dedicated to education
	Best Work Place Italia Award, from 2002 al 2007, for excellence in the work environment and the satisfaction of its collaborators
	Ernst &Young "Entrepreneur of the year" award, 2007, in the "Quality of Life" category
	"Olivettiano Entrepreneur" "award for 2008"
	Picus del Ver Sacru award, for the style of participatory command, 2012
Attention to research	4 % of resources invested in R&D
	12 patents and 7 research projects in application
	European recognition for the research project/DG XII, EU "MEDEA" (quality control in the appliance sector)
	Best Application, Automotive Forum 2008 Award (Progetto "MEXUS")
	"Marchigian of the year" award (2008) for technological innovation
	International Award Leonardo da Vinci (Associazione Italiana Progettisti Industriali), 2012
	Home Lab, first Italian consortium/pole of "domotica" founded by the Loccioni Group, among eight leaders chosen by Indesit, 2012
	Innovation Award ICT Lazio attributed to Loccioni Human Care, 2012
Attention to social responsibility and to sustainability	Sodalitas Social Award 2005 Finalist in the category "Internal Processes of CSR" and for "Metalmezzadro" project in the knowledge-based business
	Sodalitas Social Award 2008 Finalist in the category "Sustainability Projects"
	"Impresa e Cultura" ("Business and Culture") award for the project "Bluzone"
	Sodalitas Social Award 2009, Finalist in the category "sustainable initiatives" for the LOV project, The Land Of Values
	Leaf Community Project (first sustainable and ecologically integrated Italian community) – Leaf Energy and Future (partner with the European Commission's "Sustainable Energy Europe Campaign"), awarded by Legambiente
	Good Energy Award 2012 (producer category) for the firm Green Oriented

Source: Social balance sheet 2011 and the company's website, 2012

entrepreneurs – 550 collaborators and 28 businesses – who interact, developing the territory and creating synergies); and Land of Values – LOV (a network of businesses in the hospitality and food service industries representing local wine and gastronomy).

Finally, Loccioni's desire to leave a mark and valorize the territory "in which everyone was born" stands out. He thinks of his firm as a vehicle of communication and of development, not only under the industrial profile, but also cultural, social and touristic, and effects this through programs such as the project Land of Values, centered on the concept of welcoming various stakeholders and on local identity, realized in collaboration with small tourism firms from the most important Marchegian towns.

> LOV is a project aimed at giving the experience of the Group a unique and unforgettable moment: it permits everyone who visits to share in the same encounter with professionalism and conviviality, to small the perfume and the atmosphere of our culture. Our added value is innovation and hospitality (interview with J. Tempesta, MKTG & Communication, 5 June, 2011).

3.2.2   Varnelli Distillery

Varnelli is a company with very old traditions, as it has lived for over 100 years (1896) cultivating anise and producing bitters and liquors with traditional artisanal methods. Residing in a small center in the heart of the Sibillini Mountains, the founder, Girolamo Varnelli, knew how to take advantage of the observations and responsible use of nature, as well as having an understanding of popular customs to push his products. Guided by a set of shared values (Table 5), the Varnelli family has continuously operated the Distillery; today, the company is in its fourth generation of Varnelli, and is run entirely by the women in the family (the mother and her three daughters are the primary shareholders).

The company's mission statement links tradition with innovation, competition and social cohesion, and a strong local identity with a dynamic approach to global markets. This cultivates the company's belief in growth that can take place in a context of both development and the valorization of interested parties, in harmony with the quality of the territory, the culture, and the human relationships that characterize them. To achieve these objectives, which have received important recognition (such as Labor Value Award, given to high quality entrepreneurs in the Marche region), the company invests in intangible capital: human resources, technical skill, research, links with the community, youth education, and care for the environment, under the equation "brand = family = territory".

> We have never thought of selling the company - and there have been numerous requests to do so - because this business can have a future only if it maintains its historic link with the territory in which it was born. Girolamo Varnelli was convinced of the business's identification with the territory. In every occasion, he manifested the desire and the capacity to contribute to the emancipation of the entire district (Interview with O.M. Varnelli, Managing Director, 3 May, 2011).

The firm invests more than 3 % of its income in the territory, which is always referenced in the company's communications, synthesized with the motto "together for valorizing marvelous places". The map of local and extra-local relationships is exceedingly rich, and it provides the opportunity for reciprocal growth, starting

Table 5 The cardinal principles of "Varnelli's house"

Respect for the rules and continuous improvement	Social responsibility that goes beyond respecting the legislative prescriptions and standards in environmental material, security, and workplace health
Sustainable development	Programs of improvement aimed at the conservation of natural resources, reducing/eliminating negative environmental effects and risks associated with their products and services, to guarantee socially responsible products and services to its clients
Satisfaction of all interested parties	Transparency, communication (website, newsletter, direct contacts); dialogue, constructive relations
Respect for the individual	Through all of the activities and relationships with their stakeholders
Education, information, training and understanding	Shared objectives and outcomes
Collaboration with interested parties	Collaborations with providers, contractors, suppliers, public institutions, organizations, and local research centers to identify opportunities for improving social and environmental performance
Goal-orientation	Monitoring and analysis of data regarding the satisfaction of all parties interested, as well as the company's system, processes and services
Efficiency and dedication to the reduction of inequalities	Promotion of a sustainable orientation toward the company's decisions toward its stakeholders

Source: company's website

with neighboring communities, other local businesses (including the Loccioni Group), the "visitors to casa Varnelli" (such as students, associations and groups of Varnelli liquor enthusiasts, more than 1,000 a year), the Girolamo Varnelli Foundation, and research and educational institutions (such as the Adriano Olivetti Institute and the Symbola Foundation).

3.3 Discussion

Both of these companies share in common elements that illustrate their orientation toward holistic development, their rootedness and reciprocal relationship with the territory. These are synthesized in the following aspects (Table 6):

- Traditional values that mark the rural tradition of the Marche, which reinforce and nurture cohesion and a sense of belonging to the territory;
- Relationality, which is translated into the capacity to activate networks of multiple interlocutors and of actualizing concrete projects of firm-local context development;
- The will to "infect others" (in the positive sense), with such a sustainable orientation.

Table 6 Distinctive attributes common to Varnelli Distilley and Loccioni Group

A strong value system shared inside the family, the company, and the community from which the businesses come: the Marchegian culture as a virus and an emblem

Orientation toward holistic development at the top of the company, authentic and charismatic; best practices

Stakeholder commitment: vision, objectives, and clear goals that are constantly reinforced through communication, network relations, organizational practices, processes of governance (transparency, sharing, democracy), both internal and external to the company, through a number of forms of stakeholders' dialogue and engagement

Decision-making process based on trust, collaboration, participation, and sharing

A cohesive organizational climate. Flexible and integrated organizational structure

Presence of instruments of accountability

Development of intangible capital

Desire to demonstrate, communicate, and share in best practices

The strong moral and ethical bases that shape their existence and their business operations never lose site of people and has their roots in their territory. They nurture a model of family-based capitalism that expresses sustainable entrepreneurship. This is a genuine commitment inspired by these value-driven businesses; they illustrate their socially responsible orientation and relationships (Tencati and Zsolnai 2008) through the formation of networks with which the businesses interact with local and extra-local contexts; these networks are rich in intangible resources embedded in relational fabrics, and through them, they exchange "social goods" such as prestige, reputation, and friendship. "Weaving and pulling at the thread" of development and innovation, even in social terms, Loccioni and Varnelli play the role of stimulators and catalyzers. Projecting their values outside of their firms' walls, their relation-based logic flows together with the construction of sustainable value networks. These networks, in turn, nurture forms of collaborative local governance (Zadek 2006) centered on multi-stakeholder partnerships and projects, which can sustain coevolution through dynamic and reciprocal relationships (Fig. 1).

4 Conclusions

Both the theoretical and empirical planes reveal the fact that holistic development is strongly linked to a shared orientation constructed in conjunction with diverse actors from the socio-economic and institutional contexts. Effective instruments and processes of sustainability, as well as multi-dimensional development, can occur on solid bases if they derive from their own roots, and from internal cohesion around common values and objectives.

In terms of policy, these reflections are meant to point to a possible route for concretizing structural micro-economic policies aimed at sustaining the qualitative growth of the companies (innovation, education, and sustainability) and the local

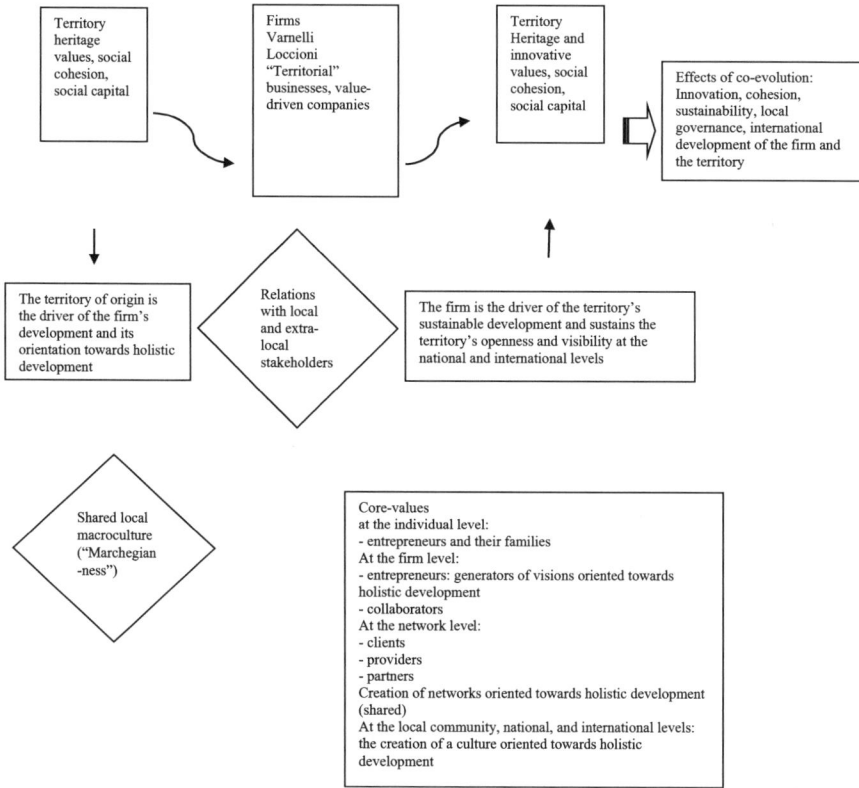

Fig. 1 The co-evolutionary model of holistic development (Source: author's adaption of Cedrola et al. 2011)

economy by tapping integrated systems of interlocutors, and directed at reinforcing the identity of the territories. These policies must be aimed toward the creation of shared values and that re-searches for relationships with the local community (Porter and Kramer 2011).

Such a model of development – that is based on tradition, territory, technological innovation and research, linking competition, respect for the environment and humanity, and social cohesion – requires a culture change. Culture change is necessary for re-personalizing modernity, re-imbuing the production of value with peoples' intelligence and of the sustainability and uniqueness of the territory. The challenge of sustainable modernity can be overcome by giving value to common goods (commons and connective links) and intervening on four levels: people (organizational behaviors characterized by the culture of doing business and of the value of work conceived of as an instrument of promoting the quality of life of oneself and others), ideas (innovations sustained by public and civic institutions, as well as businesses), relations, and values (common inspiration).

An example of this change can be found in companies based on holistic models of development (such as those presented in this study) capable of constructing the future by starting with their differences, defending entrepreneurial vocations, and generating social capabilities and innovation. The orientation toward holistic development therefore manifests itself when the dialogue between economics and ethics is restored, as well as the positive identity of these businesses and of many precious territories.

Literature

Adler, S., & Kwon, S. (2002). Social capital: Prospects for a new concept. *The Academy of Management Review, 27*(1), 17–40.

Alford, H., & Naugthon, M. (2002). Beyond the shareholder model of the firm: Working toward the common good of business. In S. A. Cortright & M. Naugthon (Eds.), *Rethinking the purpose of business. Interdisciplinary essays form the Catholic social tradition* (pp. 27–47). Notre Dame: Notre Dame University Press.

Argandoña, A. (2003). Fostering values in organizations. *Journal of Business Ethics, 45*, 15–29.

Argiolas, G. (2006, October 5–7). *The good management. Drivers of Corporate Social Orientation towards a multidimensional success.* Paper presented at the International Conference The Good Company, Rome.

Bagnasco, A. (2004). Trust and social capital. In K. Nash & A. Scott (Eds.), *The Blackwell companion to political sociology* (pp. 230–239). Oxford: Blackwell.

Bagnasco, A., & Trigilia, C. (1990). Entrepreneurship and diffuse industrialization. *International Studies of Management and Organization, 20*(4), 22–48.

Bartocci, M. (2011). *Animal spirits in Vallesina: Enrico Loccioni and the enterprise as a game.* Soveria Mannelli: Rubettino.

Becattini, G. (1990). The Marshallian industrial districts as a socio-economic notion. In F. Pyke, G. Becattini, & W. Sengenberger (Eds.), *Industrial districts and inter-firm cooperation in Italy* (pp. 37–51). Geneve: International Institute for Labour Statistic.

Bonomi Savignon, A. (2011, June 1–4). *Social capital and the government performance: Evidence from Italian regions.* Paper presented at the Euram Conference 2011, Tallin.

Brusco, S. (1982). The Emilian model: Productive decentralization and social integration. *Cambridge Journal of Economics, 6*(2), 167–184.

Bucar, B., Glas, M., & Hisrich, R. D. (2003). Ethics and entrepreneurs. *Journal of Business Venturing, 18*(2), 261–281.

Carrol, A. B. (1999). Corporate social responsibility. *Business and Society, 38*(3), 268–295.

Carroll, A. B. (2008). A history of corporate social responsibility: Concepts and practices. In A. Crane, A. McWilliams, D. Matten, J. Moon, & D. S. Siegel (Eds.), *The Oxford handbook of corporate social responsibility* (pp. 19–46). New York: Oxford University Press.

Cedrola, E., Cantù, C., & Gavinelli, L. (2011, June 1–4). *Networks, territory and firm development. The perspective of Italian family firms.* Paper presented at the Euram Conference 2011, Tallin.

Coleman, J. S. (1988). Social capital in the creation of human capital. *The American Journal of Sociology, 94*, 95–120. Chicago: The University of Chicago Press.

Cornwall, J. R. (1998). The entrepreneur as a building block for community. *Journal of Developmental Entrepreneurship, 3*, 141–148.

Del Baldo, M. (2009). Corporate social responsibility and corporate governance in Italian SMEs: An analysis of excellent stakeholders relationship and social engagement profiles. In C. Jayacharndran, U. Subramanian, & U. Rudy (Eds.), *Striving for competitive advantage and sustainability: New challenges of globalization* (Vol. 3, pp. 1515–1524). Montclair: Montclair State University, Comenius University in Bratislava.

Del Baldo, M. (2010). Corporate social responsibility and corporate governance in Italian SMEs: Toward a "territorial" model based on small "champions" of CSR. *International Journal of Sustainable Society, 2*(3), 215–247.

Del Baldo, M. (2012). Corporate social responsibility and corporate governance in Italian SMEs: The experience of some spirited businesses. *Journal of Management and Governance, 16*(1), 1–36.

Del Baldo, M., & Demartini, P. (2012a). Small business social responsibility and the missing link: The local context. In W. D. Nelson (Ed.), *Advances in business management* (Vol. 4, pp. 69–94). New York: Nova Science Publishers. Chapter 3.

Del Baldo, M., & Demartini, P. (2012b). Bottom-up or top-down: Which is the best approach to improve CSR and sustainability in local contexts? Reflections from Italian experiences. *Journal of Modern Accounting and Auditing, 8*(3), 381–400.

Denzin, N. K. (Ed.). (1978). *Sociological methods – A source book.* New York: McGraw-Hill.

Eisenhardt, K. M. (1989). Building theories from case study research. *Academy of Management Review, 14*(4), 532–550.

Eisenhardt, K. M., & Graebner, M. E. (2007). Theory building from cases: Opportunities and challenges. *Academy of Management Journal, 50*(1), 25–32.

Eisenschitz, A., & Gough, J. (1993). *The politics of local economic development.* New York: Macmillan.

Elkington, J. (1994). Towards the sustainable corporation: Win-win-win business strategies for sustainable development. *California Management Review, 36*(2), 90–100.

Elkington, J. (1999). *Cannibals with forks.* Oxford: Capstone.

Enderle, G. (2004). Global competition and corporate responsibility of small and medium sized enterprises. *Business Ethics A European Review, 13*(1), 51–63.

European Commission (EC). (2001). Commission of the European Communities, Green Paper for Promoting a European Framework for Corporate Social Responsibility, COM (2001)366 final. Brussels.

European Commission (EC). (2002). *European SMEs and social and environmental responsibility* (7th observatory of European SMEs, Vol. 4). Luxemburg: Enterprise Publications.

European Commission (EC). (2011, October 25). Communication from the Commission to the European Parliament, the Council, the European Economic and Social Committee and the Committee of the Regions. A renewed EU strategy 2011–14 for Corporate Social Responsibility. COM(2011) 681 final. Brussels.

European Union (EU). (2004, May 1–26). European Multistakeholder Forum on CSR: report of the round table on fostering CSR among SMEs', Final Version, 3.

Flick, U. (2009). *An introduction to qualitative research* (4th ed.). London: Sage.

Freeman, R. E., Harrison, J. S., Wicks, A. C., Parmar, B. L., & De Colle, S. (2010). *Stakeholder theory: The state of the art.* Cambridge: Cambridge University Press.

Fuà, G. (1988). Small-scale industry in rural areas: The Italian experience. In K. J. Arrow (Ed.), *The balance between industry and agriculture in economic development* (pp. 259–279). London: Macmillan.

Fukuyama, F. (1999, October 1). *Social capital and civil society.* Paper presented at IMP Conference on Second Generation Reforms, Lasar, Universitè de Caen, Miméo.

Garriga, E., & Melé, D. (2004). Corporate social responsibility theories: Mapping the territory. *Journal of Business Ethics, 53*(1), 51–71.

Gidding, B. H., & O'Brien, G. (2002). Environment, economy and society: Fitting them together into sustainable development. *Sustainable Development, 10*(4), 187–196.

Glaser, B. G., & Strauss, A. L. (1967). *The discovery of grounded theory: Strategies for qualitative research.* New York: Aldine.

Godoz-Diez, J. L., Fernandez-Gago, R., & Martinez-Campillo, A. (2011). How important are CEOs to CSR practices? Analysis of the mediating effect of the perceived role of ethics and social responibility. *Journal of Business Ethics, 98*, 531–548.

Gold, L. (2010). *New financial horizons. The emergence of an economy of communion.* New York: New City Press.

Granovetter, M. (1985). Economic action and social structure. *The American Journal of Sociology, 91*(3), 481–510.

Harvey, B., Van Luijk, H., & Corbetta, G. (1991). *Market morality and company size.* London: Kluwer.

Jenkins, H. (2004). A critique of conventional CSR theory: An SME perspective. *Journal of General Management, 29*(4), 37–57.

Jenkins, H. (2006, October 26). *A 'business opportunity' model of corporate social responsibility for small and medium sized enterprises.* Paper presented at the EABIS/CBS International Conference, Integration of CSR into SMEs business practice, Copenhagen Business School, Denmark.

Lamont, G. (2002). *The spirited business: Success stories of soul friendly companies.* London: Hoddes and Stoughton.

Leborgne, D., & Lipietz, A. (1991). Two social strategies in the production of new industrial spaces. In G. Benko & M. Dunford (Eds.), *Industrial change and regional development* (pp. 27–49). London: Pinter Publisher/Belhaven Press.

Lechner, C., & Dowling, M. (2003). Firm networks: External relationships as sources for the growth and competitiveness of entrepreneurial firms. *Enterpreneurship & Regional Development, 15*(1), 1–26.

Lepoutre, J., & Heene, A. (2006). Investigating the impact of firm size on small business social responsibility: A critical review. *Journal of Business Ethics, 67*(3), 257–273.

Lin, N., Cook, K., & Burt, R. S. (Eds.). (2001). *Social capital. Theory and research.* New York: Aldine de Gryter.

Matacena, A. (2010). Corporate social responsibility and accountability: Some glosses. In M. G. Baldarelli (Ed.), *Civil economy, democracy, transparency and social and environmental accounting research role* (pp. 7–59). Milano: McGraw-Hill.

Melé, D. (2002). *Not only stakeholder interests. The firm oriented toward the common good.* Notre Dame: Notre Dame University Press.

Minguzzi, A., & Passaro, R. (2000). The network of relationship between the economic environment and the entrepreneurial culture in small firms. *Journal of Business Venturing, 16*(2), 181–207.

Moore, A. B., & Manring, S. L. (2009). Strategy development in small and medium sized enterprises for sustainability. *Journal of Cleaner Production, 17*, 276–282.

MORI. (2000). SME's attitudes to social responsibility. Research study conducted for business in the community's impact on society taskforce, UK, January–February.

Moulaert, F., & Sekia, F. (2003). Territorial innovation models: A critical survey. *Regional Studies, 37*(3), 289–302.

Nahapiet, J., & Ghoshal, S. (1998). Social capital, intellectual capital and the organizational advantage. *Academy of Management Review, 23*, 242–266.

Naumes, W., & Naumes, M. J. (2006). *The art and craft of case writing* (2nd ed.). London: ME Sharpe.

Peredo, A. M., & Chrisman, J. (2006). Towards a theory of community-based enterprise. *Academy of Management Review, 31*(2), 309–328.

Perrini, F., Pogutz, S., & Tencati, A. (2006). Corporate social responsibility in Italy: State of the art. *Journal of Business Strategies, 23*(1), 65–91.

Porter, M. E., & Kramer, M. R., (2011). Creating shared value. *Harvard Business Review, 89*(1/2), 62–77.

Putnam, R. D. (1993). *Making democracy work. Civic tradition in modern Italy.* Princeton: Princeton University Press.

Ranganathan, J. (1998). *Sustainability rulers: Measuring corporate environmental and social performance*. Washington: World Resources Institute.

Realacci, E. (2012). *Green Italy*. Milano: Chiarelettere.

Sabatini, F. (2006). *The empirics of social capital and economic development: A critical perspective*. Roma: Fondazione Eni Enrico Mattei.

Sorci, C. (Ed.). (2007). *Lo sviluppo integrale delle aziende*. Milano: Giuffrè.

Spence, L. J. (2000, January 7). *Towards a human centred organisation: The case of the small firm*. Paper presented at the 3rd Conference on Ethics in Contemporary Human Resource Management, Imperial College, London.

Spence, L. J., & Schmidpeter, R. (2003). SMEs, social capital and the common good. *Journal of Business Ethics, 45*(1/2), 93–108.

Spence, L. J., Schmidpeter, R., & Habisch, A. (2003). Assessing social capital: Small and medium sized enterprises in Germany and the U.K. *Journal of Business Ethics, 47*(1), 17–29.

Spence, L. J., Habisch, A., & Schmidpeter, R. (Eds.). (2004). *Responsibility and social capital: The world of small and medium sized enterprises*. Palgrave: MacMillan.

Tencati, A., & Zsolnai, L. (2008). The collaborative enterprise. *Journal of Business Ethics, 17*(3), 311–325.

Unioncamere. (2003). *Models of corporate social responsibility in Italy. Executive summary*. Rome: Italian Union of Chambers of Commerce, Industry, Craft and Agriculture.

Vyakarnam, S., Bailey, A., Myers, A., & Burnett, D. (1997). Towards an understanding of ethical behaviour in small firms. *Journal of Business Ethics, 16*(16), 1625–1636.

World Commission on Environment and Development. (1987). *Our common future*. Oxford: Oxford University Press.

Yin, R. K. (1994). *Case study research: Design and methods* (2nd ed.). Thousand Oaks: Sage.

Zadek, S. (2004). The path to corporate responsibility. *Harvard Business Review, 82*(12), 1–9.

Zadek, S. (2006). Responsible competitiveness: Reshaping global markets through responsible business practices. *Corporate Governance, 6*(4), 334–348.

Zamagni, S. (2007). *L'economia del bene comune*. Roma: Città Nuova.

Opening the Door to Opportunities: How to Design CR Strategies that Optimize Impact for Business and Society

Michael Fürst

1 Introduction

Let's reflect for a moment about the staggering fact that more than 3.3 billion people are living in countries with a pro capita income of less than 3.900 USD a year: Are they just poor and therefore not attractive for companies from a commercial point of view or do parts of this income segment represent a distinct class of consumers and the responsibility of a modern, socially responsible corporation is to provide opportunities to them in form of most needed goods and services through inclusive business models or social innovations that can finally help them to uplift within their income segment? And how should companies contribute to solutions that are addressing the needs of the approximately 1.4 billion people living in huge poverty with an income of less than 1.25 USD? Is philanthropy or caritas the right intervention type? Or take a moment and think about issues such child labor, bounded labor or other human rights violations, corruption, or environmental pollution caused by industrial production or by excessive consumption of these products. How realistic is it that these issues affect a company of whatever size in today's world that is highly interconnected and virtually mobilized and expects a high degree of transparency? In a world that seems to become a fishbowl, should we still assume that what happens in Vegas stays in Vegas? Are these issues more relevant for academic consideration that professors and writers are busy with as they need to entertain their students and readers with "spooky" ideas? Is this just some random noise that is barely perceived in the global headquarters and in the

When not otherwise indicated, statements made in this article represent the personal opinion and perspective of the author and not of the affiliated organisation.

M. Fürst (✉)
Novartis International AG, CH-4002, Basel, Switzerland
e-mail: michael.fuerst1@gmail.com

C. Weidinger et al. (eds.), *Sustainable Entrepreneurship*, CSR, Sustainability, Ethics & Governance, DOI 10.1007/978-3-642-38753-1_11, © Springer-Verlag Berlin Heidelberg 2014

C-suites? Or do news about the involvement of a company in child labor, bonded labor, extortion or the corporate ignorance of the legitimate needs of people in the base of the pyramid travel faster around the globe than this sentence will be finished and negatively impact the reputation and the performance of a company?

Given the questions raised it may prove worthwhile to analyze with a theoretical view whether the success of a company or business in general can be secured and driven by mainly respecting and enforcing narrow classical economic parameters as still sometimes assumed in some parts of economic theory and of business or whether success does not require the consideration of two essential parameters: Firstly, it needs a values-based governance through which economic transactions can be ethically managed and successfully completed. This argument primarily refers to the policies and procedures by which a company meets its societal role – namely the provision of goods and services, and notably not the maximization of shareholder value (e.g. Drucker 1973; Heracleous and Lan 2010)[1] – by adhering to ethical standards that companies are measured against. Secondly, society expects and demands an increasingly visible and greater involvement of business in solving societal challenges that are outside of – though connected with – the scope of the traditional business models which often emphasizes short term success and which typically carries a narrow definition of a company's sphere of influence and responsibility.

The former speaks for the integration of ethical values such as integrity, fairness, trustworthiness, respect and loyalty into the corporate governance of a company. The latter for the fact that the society demands – as a prerequisite for the allocation of legitimacy – that companies take responsibility in areas where e.g. market failure, lack of political governance or weak infrastructure exists and where huge societal needs occur but are not equipped with sufficient purchasing power that is required to make the traditional business models work. That is, in other words, precisely in those areas where companies have not seen a strong sense of obligation or area of involvement as they didn't and couldn't see relevant commercial opportunities in these socio-economic segments of the global income pyramid. C. K. Prahalad has pointed out very lucidly that people in low income segments of the global income pyramid have a certain level of purchasing power which is indicating opportunities for companies if they understand to develop catalytic business models that meet the needs of the poor and provide them with social opportunities (see Prahalad 2004; Prahalad and Hammond 2002; Hammond et al. 2007). To sum it up: Ethical and legitimate behavior and social investments of companies are not longer to be seen as an option – it is a sheer necessity to satisfy the expectations of relevant stakeholders, to be seen as a legitimate organization that operates in line ethical standards and to successfully operate in a globalized economy.

Seen from a theoretical point of view, this is based on the simple-sounding but theoretically elaborate assumption that ethics has economic consequences and

---

[1] The work of Peter Drucker may serve here as an important reference. Of a more current date see the short but instructive remarks of Heracleous et al. in Harvard Business Review that is titled "The myth of shareholder capitalism".

economics has ethical consequences (Wieland 2004: 80ff). On the one hand this means that the successful realization and stabilization of economic transactions and cooperative relationships constitutively requires ethical behavior. On the other hand, the theoretically-led attempt to decouple economic action from its ethical consequences is doomed to failure as companies represent a specific form of societal cooperation that has a normative nexus to the legitimacy of its goals and behavior (Wieland 2009).

In practical terms this means that companies are held responsible for the avoidance of negative effects on society that are caused by corporate activities even if these are done within the existing legal framework. Furthermore this relates to the question whether and in which way companies are working together with other stakeholders within innovative and cooperative frameworks in order to solve societal problems such as lack of healthcare provision for the people living in the middle and base of the income pyramid. In principle philanthropic interventions are a conceptual approach to deal with this problem. However, philanthropy has its natural limitation in terms of scale and sustainability in an economic context as it is not generating any, or at least not sufficient income that can be reinvested in future growth and scale. Therefore it may be more promising and successful to integrate social innovation and social entrepreneurship activities into the strategies and business models of global companies that have enormous influence in societal debates and that can positively shape the governance in a globalized economy. The aim is to specifically target societal problems through fostering cooperation between business and society and through the creation of socially inclusive business models since only such inclusive models allow scalable and sustainable solutions. On the level of the system "economy" this needs to happen under the terms of reference or the binary code of payment/non-payment that is the only relevant "language" in an operatively closed, autopoietic economic system. Otherwise the communication in this system on a specific event doesn't have relevance or cannot even be perceived as relevant. The task for a company then is to reframe societal challenges and problems as economic opportunities and to transfer those into a socio-economic model of cooperation, i.e. a model that is able to simultaneously generate economic benefit for a company as well as to substantially and sustainably solve the targeted societal problem

2 The Globalization Offers Huge Opportunities for Companies-but it Doesn't Come as a Free Lunch

All this is happening against the backdrop of the globalization which has caused a major and highly dynamic upheaval in business and society in recent decades: For the economically developed countries the globalization comes along with huge productivity gains but also with a loss of regulatory power of nation states as well as with an assumed deficit of democratic legitimacy of powerful corporate actors and with an increasing pressure on existing social systems and labor markets. Companies are highly attracted by the globalization as it entails the promise of entering new markets with huge upside potential and the benefit of comparative advantage within their value chain (For a detailed explanation of the reasons behind this phenomenon see Wieland 2004: 14 ff.) However, whether one likes it or not, this opportunity to benefit from global markets and the increase of control and creative leeway of private economic actors doesn't come as a free lunch. It rather comes along with a progression of moral responsibility for the positive development of society and companies are expected to take over some of the responsibilities for which primarily governments should be held accountable against. These problems are rooted, for example, in dysfunctional governance structures and misguided incentives at the political level or an enormous disparity of income and wealth that results in massive social disadvantages for the less privileged people. In principle it is in the enlightened self-interest of companies to accept this increased level of responsibility, but – and this needs to be pointed out very clearly – only if they define their specific spheres of responsibilities as precisely as possible; A process which needs to consider the individual core competencies and resources of each organization. This is important to understand as the change in the tectonics of the governance of modern societies definitively doesn't mean that business can be held accountable for providing solutions to any societal problem or for being the only solution provider. Other stakeholders such as governments, NGOs etc. must accept their specific responsibility in shared dilemmas, based on their resources, skills and expertise and contribute as well to solutions through cooperation and collaborative models. If we accept these developments as facts, it should not be an overly guarded secret anymore that business and entrepreneurship is increasingly seen in terms of social responsibility and that this insight must trigger a more strategic approach how to manage this topic in order to be perceived as a legitimate business and to successfully compete in a changing society. If we try to condense this theoretically we can state that the organizational ability to cooperate becomes an increasingly important factor and the success in such a cooperation driven economy will be dependent on the effective integration of factors such as morality, ethics or culture into the governance of an organization (Wieland 2004; Fürst 2005).

In recent years, a growing number of companies have recognized this trend and put the issue of the organizational responsibility on their agenda. Many of these efforts are rubricated under the terms Corporate Social Responsibility (CSR), Corporate Citizenship (CC) or Corporate Responsibility (CR).[2] However, it seems necessary to stress that most of these efforts didn't really become an integral part of the corporate strategy and the business model as they are conceptualized rather unstrategically, mostly philanthropic in nature and not sufficiently linked with the core business. One of the reasons that can explain this situation is that rather the majority of companies still understand CR quite narrowly as an element of reputation or stakeholder management without making a more systematic

---

[2] We will not dive into detail in regards to explaining (the sometimes subtle) differences of these terms and will just refer to the term Corporate Responsibility in this article.

attempt to create value for business and society at a level of scale with innovative, entrepreneurial means. This is especially unfortunate since it is evident that such more strategic and business related sustainability strategies can promote the competitiveness of companies, beyond potential reputational gains and efficiencies (Nidumolu et al. 2009).

3   Some Remarks on How the Changing Role of Business in Society Impacts the Theory of the Firm

In transactions that have a neoclassical layout in which constitutional and postconstitutional agreements are meant to be complete, where information asymmetries do not exist and where complexity and contingency in highly differentiated societies do not matter, values or ethics are not required to manage risks and to allow cooperation with other stakeholders. The social responsibility assigned to the company is not completely negated in this concept, however it is deliberately excluded from the theoretical core that is oriented towards mathematical modeling, reduction of complexity and social reality. The theory remains focused to conceptualize a company as a mechanism to maximize profits (see Wieland 2009, p. 282).

This theoretical starting position that recently has also been criticized by R. Coase in a comment with the title "Saving Economics from the Economists" (Coase 2012) doesn't seem to adequately reflect business reality and doesn't do sufficient justice to the task to build an empirically relevant economic theory. This is – as we have already outlined – because organizations are increasingly charged with the responsibility to make considerable contributions to the solution of societal problems, i.e. be part of the solution and not the problem. At issue here is the avoidance of potential negative effects of economic transactions that can be described with the dictum of "primum non nocere" as well as the creation of value and benefits for relevant stakeholder groups in society (Drucker 2001a). This is not a normative determination about what a company should specifically deliver or do but simply an empirical statement about the accountability mechanisms of modern societies and the requirements that must be met to obtain the "license to operate and grow".

Since the mid of the 1990s the German business ethicist Josef Wieland is working on his theory of the governance ethics that is trying to integrate these factors through a modern and innovative theory of organizational economics which primary focus is on the collective actor and its formal and informal governance structure and not on the individual actor (see Wieland 2004). The governance ethics is based on the idea that the moral or cultural dimension of each distinct economic transaction can be analyzed at a micro level. It is important to note that the unit to be analyzed is always a clearly distinct economic transaction with its inherent moral dimension. The local, i.e. distinct and specific, application of morals or values is a prerequisite for a successful economic transaction which means increased efficiency of transactions, the reduction of transaction costs and a possible expansion into new partnerships or models of cooperation. The company's purpose, goal and its governance is not about maximizing profits or value but about economizing of relationships with relevant stakeholders[3] whereby this governance needs to be highly adaptive to the requirements of these relationships. Governance is as Oliver Williamson illustrates "[...] the means by which to infuse *order*, thereby to mitigate *conflict* and realize *mutual gains*." (Williamson 2010, p. 5). Any economic cooperation requires social cooperation that can be for example expressed by the acceptance of having responsibility for certain societal problems as this is the only possibility to mobilize the needed resources and to generate societal legitimacy. A firm is therefore a social cooperative project of multiple stakeholders to exploit their resources under the conditions of economic competition. It is a contractually constituted form that enables organized cooperation (Wieland 2009, p. 282). Although this doesn't result in a normative determination regarding the purpose of the company – with good reason –it implies a linkage to the normative social legitimacy of corporate objectives and business activities. This embraces the desirability and necessity of corporate contributions to social welfare beyond the legal requirements. The social character of an enterprise and its nature is thus defined as endogenous (Wieland 2009, p. 282) – for economic reasons -, because otherwise transactions and co-operations cannot be managed successfully. This is a clear difference to the classical economic theory that defines institutional settings in the business environment as an exogenous behavioral restriction and theoretically excludes the role of the company as a responsible collective actor in society.

In the perspective of social theory this definition of the firm conceptualizes a company as a corporate citizen that bears rights and duties as morally proactive citizens (Wieland 2009). This status of a company as a corporate citizen is not to be seen as a legal status, but in the sense as a "concept of citizenship as-a-desirable-activity" (see Wieland 2003, p. 17, referring to Wood/Logsdon 2002, p. 68). Corporate Responsibility in this sense refers back to the values-driven allocation of corporate resources to pursue solutions to social problems (Wieland 2003, p. 18), i.e. the benefits of corporate activities mustn't only be directly allocated to the transaction partners within a legally binding contract, or finance philanthropic contributions as a modus for wealth-redistribution or prevent negative external effects. In fact, one should focus and discuss much more strongly that the allocation of societal legitimacy to specific business activities is especially attached to the question of whether entrepreneurial activity by itself constitutes a contribution to the solution of material societal challenges.

---

[3] See also the work of Peter Drucker: "[...] profitability is not the purpose of, but a limiting factor on business enterprise and business activity. Profit is not the explanation, cause or rationale of business behavior and business decisions, but rather the test of their validity." Drucker 1973, p. 60. See Drucker 2001a; Drucker 2001b, p. 18.

4 Corporate Responsibility Strategy Needs to Operate with a Portfolio of Distinct Intervention Types Though Focusing on Social Business Models

Historically CR had a strong focus on philanthropy that can create social impact if conceptualized with a strategic view but has genuine problems in terms of scalability and replicability. Based on these limitations the genuine entrepreneurial activity understood as a mechanism to create new, sustainable business models through the transformation of societal challenges into innovative and sustainable services needs to be emphasized and specifically addressed in a strategic approach to CR. Therefore, as this is increasingly understood, the trendlines are pointing in the direction of focusing more strongly on activities that are aligning social and commercial ambition and can result in large-scale responsibility engagement, without ignoring philanthropy or zero profit initiatives if such kind of interventions contribute to solving societal challenges. Strategic CR should therefore aim to operate with a portfolio of tailored activities, comprising philanthropic initiatives, zero profit, social business or lower margin business models that are all closely linked to a company's strategy and core competencies and are operated through a strong ethical governance. Figure 1 tries to explore this rationale and typology by showing that specific CR intervention types should be differentiated along the different income segments of the global income pyramid.

Typically, business models of multinational companies are tailored to serve the needs of consumers at the top of the pyramid. Companies can work here with the full spectrum of products and services they have in their portfolio, typically with a focus on highly innovative offerings. CR is here mostly focused on ensuring – through integrity management – that business is done in line with ethical standards (Fürst and Schotter 2013) and that negative external effects are avoided.

The next level of the global income pyramid comprises two different income segments that represent at the upper end emerging markets with a medium level of income and unmet needs and at the lower end subsistence markets with a high level of unmet needs (as an instructive article in regards to segmenting the base of the pyramid we recommend Rangan et al. 2011).

In the segment of emerging markets, companies can approach the existing needs from a CR perspective with market-based solutions that typically should operate with lower prices and therefore offer the possibility to new consumer segments to satisfy their needs with relevant goods and services. The portfolio would here offer solutions such as "low-end disruption" that do not offer the full functionality or "new-market disruptions" that focus unserved needs of potential customers (for the distinction see Christensen et al. 2003).

Further downwards in this segment the preferred intervention type is social business or social enterprise that allows to scale up the model rather quickly as is generates an appropriate level of profitability. Companies need to very carefully differentiate the specific income segments within this income bracket and understand different factors influencing buying decisions such as purchasing power,

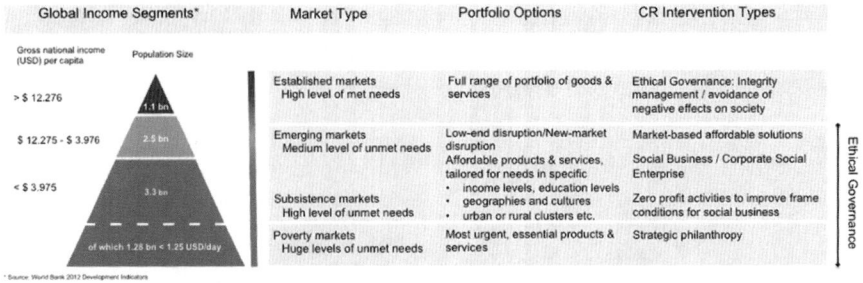

Fig. 1 Differentiating CR intervention types according to income levels

pricing of products and services (very often the price for basic goods is higher in these market segments than in established markets), rural versus urban distribution of population, informal solutions that are provided, cultural determinants, educational and awareness levels, infrastructure gaps, supply chain constraints etc. as only the consideration of these factors will allow to develop a holistic and successful intervention. The consideration and integration of these access-limiting factors into the business model to increase social impact marks the difference to the previous intervention type that primarily focuses on price and affordability. Typically such social business models are built as volume models, i.e. they operate according to the principle of economies of scale, in order to keep the margin levels of the unit low and therefore the product affordable. This is of course problematic during the starting phase of such kinds of models as it needs a certain level of upfront investment in order to quickly achieve scale and pass break-even.

In the next level of the middle of the pyramid – the subsistence market – where significant unmet needs exist and where many people are working in an informal economy, the CR intervention types should be focused on zero-profit models that offer a more limited but still existing potential for scale and can definitely improve the frame conditions for (social) business in the long run.

At the bottom of the income pyramid where people live in total poverty with less than 1.25 USD per day, CR should operate with philanthropic models that have a strong strategic rationale for the company and therefore unfold an appropriate level of impact.

The dictum to avoid unnecessary, negative external effects does apply throughout the whole pyramid and as with the activities in mainstream business all the activities in the middle or the base of the pyramid have to be conducted in accordance with strong ethical standards that a company has committed to as part of its organizational self-governance (Fürst and Schotter 2013). Corporate Responsibility would then refer to the legitimate generation of shared benefit by focusing on trying to solve these societal challenges where a company has competencies and the needed resources. The linkage back to the legitimacy criterion ensures firstly that only those activities are considered that are socially acceptable and desirable and secondly that they are carried out in accordance with ethical principles. Connecting CR with the strategic value drivers and core competencies aims to narrow down the range of activities as only this integration will generate sustainable solutions that are replicable and scalable. If this integration is not done successfully many CR initiatives – however interesting they may be, and independent from a strong moral motivation – are often viewed by decision-takers as primarily cost-generating, not value-adding and up for disposition if an economic crisis will arise or priorities need to be changed.

5 Entreneurship and Corporate Social Entrepreurship: Same Origins – Different Priorities

As already alluded to companies are expected to take a more active role than in the past in regards to solving societal problems – even if they are not directly caused by the company. Companies can reframe these societal expectations as opportunities if they are willing and able to take on this expanded responsibility and fulfill their primary function in the society to satisfy social needs with entrepreneurial means.

To get a better handle on this and also since the term of entrepreneurship is not sufficiently considered in the discussion on the social responsibility of companies though it is such an essential one, it makes sense to investigate and analyze the concept of entrepreneurship a little bit more in detail (for the following see Dees 2001).

It is accredited to Jean Baptiste Say to have given the term entrepreneurship its meaning. For Say entrepreneurship is always generating value by allocating economic resources to the area in which increased productivity and a higher gain can be achieved when compared to an alternative investment. Since the first half of the twentieth Century the term is closely linked with the work of Joseph Schumpeter that describes the dynamic entrepreneur as an innovator that creates progress through entrepreneurial activities and its immanent element of the so called creative destruction (Schumpeter 1912/2006). Entrepreneurs are hereby initiators and catalysts of change without qualifying this change from a normative or ethical perspective. Many authors who have dealt in detail with the concept of entrepreneurship followed the work of Say and Schumpeter and continued in their tradition but also varied the theme. As an example we might look at Peter Drucker and his work, the focus here is less concerned with the entrepreneur as an initiator of change. For him being in an entrepreneur means much more to seize the opportunities that arise because of social change, independently how this change has been brought about. Drucker defines an entrepreneur as follows:

"The entrepreneur always searches for change, responds to it, and exploits it as an opportunity."(Drucker cited by Dees 2001).

The term "opportunity" becomes the focal point of this definition in which Drucker clearly articulates that an entrepreneur perceives societal challenges or change not primarily as a problem but as a chance or opportunity. In line with Say, resources have to be allocated here in order to generate a higher marginal return.

Howard Stevenson made some important further differentiation in the debate on entrepreneurship by adding the element of ingenuity to run a venture or business under the conditions of the limited availability of resources (Stevenson conceptualizes his model of entrepreneurship as opportunity-driven, see Stevenson 2006). This leads Stevenson to the distinction between entrepreneurs and bureaucratic managers whereby an entrepreneur doesn't accept limitations that are determined by a lack of resources. To the contrary, an entrepreneur strives to realize its business objectives even under these conditions while managers consider multiple resource limitation as a reason to restrict their own behavior regarding innovation and to operate with traditional strategies and business models. Stevenson distinguished entrepreneurship from bureaucratic management by defining entrepreneurship as

" (...) the process by which individuals – either in their own or inside organizations – pursue opportunities without regard to the resources they currently control."(Stevenson and Jarillo 1990).

He is very clear that an entrepreneur can operate in a corporate context and is not necessarily managing its own business or enterprise. This concept of the entrepreneur who does not accept a lack of resources as a limitation in his capacity to act but rather tries to mobilize resources through the creation of innovative cooperative models is of great interest specifically in the context of corporate social entrepreneurship. The reason is that many societal problems – however complex they are – can only be solved if entrepreneurs are able to conceptualize solutions as cooperative models in which they combine their specific and often limited resources with complementary resources outside of their own value-chain. It is precisely not about locking down a hermetically closed business model, rather it is the aim to make entrepreneurial skills compatible between stakeholders to enable a cooperation-driven social business model which can create social and simultaneously economic value. Full integration of all competences that are needed doesn't seem to be an option given the size and complexity of many social problems because the integration effort by a single company to manage such multitude of competencies and issues could hardly be absorbed within the corporate boundaries. It is rather the close cooperation of the stakeholders that pursue a social objective and have the ability to couple this with economic value creation that can result in a sustainable solution. This openness to cooperate requires a reciprocal values-based governance in order to align target setting and to protect each partner from exploitation risks caused by e.g. limited or asymmetric information (see e.g. Wieland 2004). However, in business reality, a cooperation between small to mid-scale social enterprises and corporate social enterprises is difficult to structure and to manage because of the inherent issue of scale asymmetry, i.e. social enterprises typically operate in a limited number of countries or regions whereas multinational companies are genuinely looking for scale and operate in a multitude of countries or regions. Scale asymmetry can be prohibitive for multinational companies as it ultimately means that transaction costs increase if the collaboration should be scaled up but the partners are not able to grow and scale up at the same speed or with the same geographical reach.

The aforementioned concepts of entrepreneurship are of specific interest as they have relevance in the area of mainstream business as well as in the social sector. Thus social entrepreneurs are entrepreneurs pursuing a social goal, both in the nonprofit as well as profit sector. In the classical management theory social value occurs as a byproduct of the economic value (see Venkataraman 1997) whereas the distinct feature of social entrepreneurship is to prioritize the creation of social value which is accompanied by creating economic value as this is needed for scalability, replicability and finally sustainability of the solution. Social entrepreneurship entails clear revenue and profit goals as the generation of social innovation and solutions can only be done sustainably if economic benefit is created in the context and under the conditions of an economic system. If we try to amalgate this theoretically we could characterize corporate social entrepreneurship or social business in the following way:

> *Corporate Social Entrepreneurship or Corporate Social Business reframes societal challenges into opportunities with the aim to create social value and the constitutively needed business value through scalable, profitable business models that holistically address access barriers to goods and services. A corporate social entrepreneur doesn't limit himself in the creation of such solutions by current resource constraints but rather mobilizes complimentary resources of different social stakeholders that are all willing to jointly exploit their resources under the conditions of economic competition.*

As we all know, Adam Smith defended the profit motive of individuals because the pursuit of self-interest will lead through the invisible hand to a real increase in social welfare by satisfying the existing demand that is equipped with a certain amount of purchasing power through the supply of needed product or service. The attempt to trade for the public good is not effective and will not result in much good, according to Smith. However, it was obvious to Smith that the pursuit of self-interest should be limited by sympathy or the interest in the fortunes of others that over time builds a system of behavioral, moral rules that limit activities. (see Adam Smith in his work "Inquiry into the Nature and Causes of the Wealth of Nations" from 1776 and Adam Smith "The Theory of moral sentiments" part VII, Sect. II, Chap. 4, 'Of licentious systems').In social business the invisible hand gets less invisible as the allocation of investments does not just happen arbitrarily from an ethical perspective and is not only limited by moral rules but is necessarily and intentionally tied to a specific social and moral interests and investment purpose. Ethics is herewith mutating from a factor that is limiting resource allocation to a factor that is determining the objective and purpose of the resource allocation in order to achieve social value creation.

6 Making Sense of Societal Challenges: Issues and Solutions

In the following section we will describe one reason why corporate social entrepreneurship seems to be underrepresented in corporations of bigger of even global size and briefly explain which governance mechanisms are needed in order to change this situation.

But first let's make a step back: Over many years the thesis of Schumpeter was widely accepted in economic theory that innovation is mainly happening in big corporations and not in ones of smaller size because of asymmetric availability of resources so that big corporations would become the epicenters of innovation (see Schumpeter 1943). The empirical relevance of this hypothesis is not totally clarified and still debated in literature (see Witt 1987, e.g. p. 52 f.). However one could observe that innovation processes especially in large companies are literally stifled despite large resources which results in incremental progress[4] and in a situation that "disruptive innovations" or "catalytic innovations" (Christensen et al. 2006) are frequently originated in small cells of entrepreneurial activities since some time (see Austin and Reficco 2009, referencing the work of Covin and Miles 1999).

Large companies often provide existing customer groups with new products or services that offer improved quality or functionality. Disruptive innovations target the needs of customers that are outside the traditional customer segment of an industry, or companies offer products and services with basic functionality and thereby satisfy or create a demand in a specific, so far untapped consumer segment. Social change occurs here as a byproduct of the business model. Catalytic innovations primarily aim to create social change without ignoring the economic character of business transactions. One reason that disruptive or even more specifically catalytic innovations are underrepresented in large organizations can be identified in exuberant and innovation-averse bureaucracy in business. It is not difficult to imagine that hardly any innovation can be fully developed in an environment where managers are mainly incentivized to administrate or to perpetuate the existing business model and to generate short term success and high profitability rates instead of adjusting or disrupting the business model which typically requires vision, stamina and the strengths to resist short term pressure and profitability expectations.[5] In addition to short term focused incentives that have the potential for being dysfunctional for the capability to innovate, many of the companies also lack the cultural background and the practical experience to

---

[4] As an example may serve the pharmaceutical industry in which biotechs of smaller size are frequently perceived as the core cell of innovation. Similarly the electronics industry where a comparably smaller company such as Apple was driving technological innovation and shaped consumer behavior.

[5] See here the very lucid analysis in Drucker 2001c or Christensen et al. 2006. One can see here very clearly the linkage points to the differentiation between entrepreneur and bureaucratic manager as laid out by Stevenson.

perceive societal challenges that are outside the established business model as a business opportunity. This is then linked to the lack of sufficient skills to develop social business models at a corporate level and as a consequence companies mostly address – almost helplessly – these challenges exclusively with philanthropic means with all the limitations regarding scale and replicability.

If we remind ourselves of the huge scale and urgency of major societal challenges such as access to medicine or to clean water or the impact of climate change it becomes obvious that this perception and skills gap is a problem in terms of generating innovative solutions that are able to sustainably address these problems. It would therefore be of particular interest to understand how the individual and collective mental process of identifying and reframing these challenges to opportunities do occur, that is how such cognitive processes work at the level of an individual but even more importantly of a collective actor. For most of the cases we assume that large companies don't have such set of cognitive and cultural preferences at their disposal that would facilitate an early identification of social challenges as a business opportunity. This is of fundamental importance since we know that the perception and identification of opportunities and risks is always based on cultural patterns, values or identity semantics of social entities (see the work of Weick to the topic of "sensemaking in organizations"; Weick 1995; Weick and Roberts 1993; Weick and Sutcliffe 2003). Perception processes are based on existing values and on cultural patterns that describe social entities and define their identity. Such a constructed perceptual process is indispensably and closely linked to the self-description and self-observation of social systems (see Fürst 2005, p. 140, referencing Japp 1996; Douglas and Wildavsky 1982). A process of "sense-making" happens in organizations which can be described as a method to search, analyze and interpret critical changes and to subsequently enables individual to take "rational" decisions within the given cultural and institutional frame (Weick 1995). "Sense-making" in the context described here is the construction of an organization's identity and institutionalized self-description and entails processes that can enable an organization to perceive societal challenges as a business opportunity and to develop relevant business strategies to address these challenges. The beliefs and the values of a distinct social unit are defined and described in this organizational culture. Karl Weick formulated this as follows:

> "Organizational cultures can codify the organization's understanding of itself and its environment, and thereby clarify the organization's belief and goals for members." (Weick 1995)

Distinct institutions represent cultural determinants from the perspectives of systems theory and institutional economics. These determinants shape specific behavioral dispositions and constitute an environment for individuals that provides them with a certain degree of reliability and safety as they understand what is expected and how they should decide form an ethical, legal or economic point of view. The establishment of distinct institutional settings (values, morals, interpretation patterns, policies, etc.) leads to the emergence of a kind of "collective mind" (see Karl Weick for this term) that represents perception, interpretation, and behavior patterns of individual and collective actors and thus determines the perception and the selection of distinct opportunities.

Belief systems work like glasses, i.e. they allow on the one hand distinct perceptions of issues and sharpen the view one these but on the other hand they literally fade-out issues and realities that are outside of scope. In the context described here, the relevant belief systems are working like sustainability glasses, meaning that an organization is enabled to perceive societal challenges as existing and relevant and to subsequently develop appropriate strategies and social business models to manage these challenges successfully and sustainably.

In conjunction with theoretical considerations in the New Institutional Economics and Organizational Economics, we can say that each company owns an organization-specific "shared mental model" (see Denzau and North 1994) that represents and determines patterns of perception and behavior (see Schlicht with reference to Issac et al. 1991"[. . .] firms and other institutions provide institutional frames which activate certain types of behavior rather than others."). Institutionalization means in this context that a consensus on behavioral expectations exists in each organizational entity. For the negotiated topic in this article this means that both the perception of societal problems as a positive economic opportunity and the related behavior is subject to a specific set of expectations (Japp 1996, p. 121) of a distinct culture. Establishing perception routines as a crystallization of cultural patterns is a central process in the context of enabling corporate social entrepreneurship. A company as a collective actor maintains and conveys perception patterns by its own cultural preferences and institutional settings and the task is to establish organizational governance and learning so that societal challenges can be seen through such sustainability glasses and be interpreted as a business opportunity.

This line of argument is based on the assumption that the world is understood only through a construction process and that this in turn is differentiated and determined through different cultural types. Corporate social entrepreneurship can only succeed if value systems exist in the organization that shape specific perceptual patterns and if such a social entrepreneurial venture is within the "zone of acceptance" (Barnard 1938/1964) of a collective actor. This informal institutional setting is to be strengthened by establishing a distinct formal governance structures that encourage and incentivize the desired behavior, i.e. in this case social entrepreneurship. The informal and formal institutional layers work reciprocal and reinforce the intended pattern of perception in a recursive process. It is known from the risk research done in cultural sociology that the perception bias of an organization is processed by using the distinction of accepting and rejecting information. Information which fits into the frame of values or the cultural setting are perceived as acceptable whereas information that is not compatible with the specific organizational bias and is outside the "zone of acceptance" is typically rejected. The permanent repetition of accepting or rejection information culturally solidifies perceptual episodes into perception patterns of a collective actor (Japp 1996).

To summarize we can say that institutional settings, values and cultural patterns shape the positive or negative perception of information. This means that specific governance structures of an organization should enable its members to perceive a societal problem as an economic opportunity and to deal with it at the through models of corporate social entrepreneurship. Such governance structures are e.g. performance management schemes or career development programs that actively incentivize social entrepreneurship or leadership development programs that provide managers with the skills to develop and manage corporate social enterprises (see. Stolz et al. 2012).It is about the design of the context and about "[. . .] choosing preferences by constructing institutions." (Wildavsky 1987) This kind targeted management should enable corporate social entrepreneurship.

7   How Corporate Social Entrepreneurship 1.0 in Healthcare Can Look Like

Health and healthcare is one of the mega-topics at the beginning of the twenty-first century based on changing demographics in the western or developed countries, the introduction of public health insurance schemes in middle income countries or the huge lack of access to healthcare for poor people in the base of the global income pyramid. With a view on the global burden of disease and specifically mortality, communicable diseases will remain a heavy disease burden specifically in poor countries although the trendline is decreasing. Non communicable diseases will continue to grow and put huge burden in terms of costs, morbidity and mortality on societies in all parts of the world (for data see The Lancet Global Burden of Disease Study 2012). The world population continues to grow and will reach more than 7.7 billion in 2020 (constant fertility variant) which means additional 600 million people will be added to the world population in just 8 years and. Around 2050 the 9 billion mark will be met, i.e. approximately additional 2 billion will live on this planet compared to today – in just 40 years (United Nations Department of Economic and Social Affairs, Population Division: World Population Prospects. The 2010 Revision). Approximately 1.4 billion people of the overall 7 billion live with less than 1.25 USD per day and 2.5 billion live with less 2 USD per day (World Bank 2012 Development Indicators). Although these numbers demonstrate huge levels of inequality and poverty it can also be clearly stated that specific income segments in the middle and at the base of the pyramid bear interesting opportunities as people living in these segments are equipped with a certain purchasing power and are looking for products and services that can serve their basic needs. According to a study conducted by the World Resource Institute and the IFC the pharmaceutical market at the base of the pyramid was an estimated 56.7 billion in 2007 (Hammon et al. 2007. The Base of the Pyramid segment is defined as income less than 3.000 USD per year). All this creates huge challenges but also opportunities for affordable and effective healthcare. Health and healthcare provision have tremendous economic and social impact on societies and specifically on poor people as they are highly vulnerable to stay or to fall back in the poverty spiral.

High morbidity is often the cause for a low or decreasing ability to work, shrinking productivity, loss of income and loss of savings as health expenditures typically have to be paid out of pocket (Niëns et al.). We will not go into further detail in regards to the relation between poverty and health and refer here to an instructive article from Klaus Leisinger (2011). However, it is obvious that the above mentioned facts combined with increasingly vocal debate about the right to health and access to treatment is putting huge expectations on the shoulders of relevant actors in the healthcare sector. In order to provide as many as possible patients with the needed medicine it is sometimes expected from pharmaceutical companies that they should abandon their intellectual property rights or that they should give certain drugs for free or at least lower the prices vary considerably since this would primarily allow access (Attaran 2004, p. 2). However, without neglecting that price and patents have an impact on the level of access to treatments a too narrow view underestimates the complexity of the challenge which is driven by lack of availability of medicine in private and public sector (availability), reduced accessibility of medicine outlets within near distance (accessibility), inadequate quality of health services such as inappropriate prescribing or dispensing practices (quality of care), lack of acceptance of healthcare provision and compliance with the treatment regimen (acceptability) and finally limited affordability as medicine accounts for a high percentage on household expenditure on health(affordability), (For a very instructive analysis of access factors see Obrist et al. 2007; Leisinger 2011). Not to forget factors such as government allocation on health, gender issues, language barriers, low health seeking behavior, ethnic differences, complicated dosage of medicine, lack of cold chain solutions or heat stabile formulations and many more that need to be mentioned to complete the list of factors that are impacting access levels.

All this can in principle be addressed and most actors would agree that all these factors are relevant when looking for solutions. However, it is much more difficult to find an agreement when asking for which of these factors should be prioritized. This disagreement is partially driven by ideology or vested interests but also by the inherent complicated nature of the problem. Therefore solutions that aim to improve access to healthcare are multi-factorial models involving a variety of different stakeholders that are each able to contribute to the solution with their specific skills and competencies that are different but complimentary. Innovative entrepreneurial concepts play an increasingly important role in such considerations since they bear the potential for scale and replicability and factor in collaborative elements as illustrated earlier in this article. For a healthcare or pharmaceutical company the primary purpose and function in society is to develop and produce medicine that is needed by patients and that treat or cure diseases. Assuming they will develop and manufacture innovative medicine or quality generic products that meet the need of patients and focus on relevant disease burden they will be able to sell these products profitably. As already alluded to previously a majority of world's population doesn't have access to appropriate healthcare which urges the question which interventions and collaborative models can contribute to solutions to this huge societal problem. The outcomes of an analysis of the previously mentioned factors that can enable or impede access to healthcare will determine which intervention type such as strategic philanthropy, zero profit, social business or finally traditional business models are most suited to meet the needs of and contribute to a solution. Now, much has been written about social business or shared value (Porter and Kramer 2011) in recent times and one can observe a huge number of highly innovative small social enterprises. However, it is undoubtedly also a fact that the state of affairs in terms of scalable corporate social enterprises doesn't look too rosy, i.e. such initiatives in multinational companies are still rare and limited in terms of scale and impact. An interesting example of a corporate social enterprise that has reached considerable scale in terms of people as well as patients and that is replicated in several countries is the social business from the pharmaceutical company Novartis[6] (reference for the following is FSG "Competing by Saving Lives" 2012).

In 2007, Novartis charged itself with the task to launch an initiative – ArogyaParivar which is Hindi and means healthy families – that is specifically targeting the health needs of villagers in rural India living of 1 USD to 5 USD a day. In India the majority of the population is still living in rural communities, however they account for only a little bit over 20 % of the health spending of which most is out of pocket. Many do not seek formal healthcare and the ones that do typically wait so long until the condition is acute.

When developing the model Novartis very carefully evaluated these factors and the local disease burden in order to be able to offer needed services and the appropriate product portfolio drawn from Novartis originator, generics, over the counter and vaccines business. To insource health related products from other companies that are needed in rural communities is also evaluated in order to address healthcare as holistically as possible. To ensure supply of medicine and to make sales calls, Novartis employs local people as sales force that know the culture, local dialect and understand market situation in these communities. At the same time, in order to improve the low health seeking behavior of many rural villagers, a team of educators is travelling through the villagers organizing health education sessions. To bring quality healthcare services closer to the villagers ArogyaParivar frequently organizes health camps at which physicians are present and available for the rural communities.

Although the challenges described were not fully understood at the beginning of the program it only took 31 months to break even. In 2011 ArogyaParivar covered 42 million people in 33.000 villages across 10 states in India. After the health camps were installed doctors visited these villages 3 times more and has increased from 9 % to 23 % of local populations.

---

[6] Novartis follows the approach of having a targeted CR portfolio where philanthropic activities are mainly driven by the Novartis Foundation for Sustainable Development and comprises zero profit activities such as the Novartis Malaria Initiative which has provided 600 million antimalarial treatments since 2001 (http://malaria.novartis.com/downloads/malaria-initiative/factsheet-malaria-initiative.pdf), April 2013. Novartis has also founded a social business group which is responsible for managing country-specific social business activities such as ArogyaParivar.

As the model has shown to be scalable Novartis is currently rolling out this social business in other countries and is operating this under the umbrella of the new organizational entity Social Business Group. This is an important step as it is obvious that the original model developed in India to bring an innovative solution to rural communities is not directly transferable to other locations and must be adapted because healthcare systems typically vary from country to country. These variances can be characterized by different regulatory environments, cultural dispositions, competitive situation, purchasing power or local disease burden.

Although Novartis has achieved considerable scale and although this is an impressive effort and success we would say that this model is still in its early stage of development and therefore some sort of corporate social enterprise 1.0 which means that refinement and adaptation of the model is needed to unfold its full potential. This is a huge challenge – but a most rewarding one as it is finally about contributing to solutions that are addressing one of the big challenges for humanity: Providing healthcare for people that are most in need!

Literature

Attaran, A. (2004). How do patents and economic policies affect access to essential medicines in developing countries? *Health Affairs, 23*(3), 155–166.
Austin, J., Reficco, E. (2009). Corporate social entrepreneurship. Harvard Business School, Working Paper 09–101.
Barnard, C. I. (1938/1964). *The functions of the executive.* Cambridge: Harvard University Press.
Christensen, C. M., & Raynor, M. E. (2003). *The innovator's solution. Creating and sustaining successful growth.* Boston: Harvard Business School Press.
Christensen, C. M., Baumann, H., Ruggles, R., & Sadtler, T. M. (2006). Disruptive innovation for social change. *Harvard Business Review, 84*(12), 94–101.
Coase, R. (2012, December). Saving economics from the economists. *Harvard Business Review, 90*(12), 36.
Covin, J. G., & Morgan, P. M. (1999). Corporate entrepreneurship and the pursuit of competitive advantage. *Entrepreneurship Theory and Practice, 23*(3), 47–63.
Dees, G. (2001). *The meaning of social entrepreneurship.* http://www.caseatduke.org/documents/dees_sedef.pdf. Accessed 9 June 2013.
Denzau, A. T., & North, D. C. (1994). Shared mental models: Ideologies and institution. *Kyklos, 47,* 3–31. Jg.
Douglas, M., & Wildavsky, A. (1982). *Risk and culture. An essay on the selection of technological and environmental dangers.* Berkeley: University of California Press.
Drucker, P. (1973). *Management: Tasks, responsibilities and practices.* New York: Harper Paperbacks.
Drucker, P. (2001a). Social impacts and social problems. In P. Drucker (Ed.), *The essential Drucker.* New York: Harper Collins.
Drucker, P. (2001b). The purpose and objectives of a business. In P. Drucker (Ed.), *The essential Drucker.* New York: Harper Collins.
Drucker, P. (2001c). The entrepreneurial business. In P. Drucker (Ed.), *The essential Drucker.* New York: Harper Collins.
FSG (2012). *Competing by saving lives.* http://www.fsg.org/Portals/0/Uploads/Documents/PDF/Competing_Saving_Lives.pdf. Accessed 9 June 2013.

Fürst, M. (2005). *Risiko-Governance. Die Wahrnehmung und Steuerung moralökonomischer Risiken*. Marburg: Metropolis.

Fürst, M., & Schotter, A. (2013). Strategic integrity management as a dynamic capability. In: Wilkinson, T. (Ed.), *strategic management in the 21st century*. St. Barbara: Praeger Publishers Forthcoming.

Hammond, A. L., Kramer, W. J., Katz, R. S., Tran, J. T., & Walker, C. (2007). *The next 4 Billion. Market size and business strategy at the base of the pyramid*. Washington, DC: World Resources Institute and International Finance Corporation.

Heracleous, L., & Lan, L. L. (2010, April). Harvard Business Review, *88*(4), 24.

Isaac, R. M., Mathieu, D., & Zajac, Z. (1991). Institutional framing and perceptions of fairness. *Constitutional Political Economy, 2*, 329–370.

Japp, K. P. (1996). *SoziologischeRisikotheorie. Funktionale Differenzierung, Politisierung und Reflexion*. Weinheim/München: Juventa-Verlag.

Leisinger, K. M. (2011). *Poverty, disease and medicines in low and middle-income countries: The roles and responsibilities of pharmaceutical corporations*. http://www.novartisfoundation.org/platform/apps/Publication/getfmfile.asp?id=612&el=4314&se=494451244&doc=251&dse=5. Accessed 9 June 2013.

Nidumolu, R., Prahalad, C. K., & Rangaswami, M.R. (2009, September). Why sustainability is now the key driver of innovation. *Harvard Business Review, 87*(9), 56–64.

Niëns, L. M., Cameron, A., Van de Poel, E., Ewen, M., Brouwer, W. B. F., & Laing, R. (2010). *Quantifying the impoverishing effects of purchasing medicines: A cross-country comparison of the affordability of medicines in the developing world*. http://www.plosmedicine.org/article/info%3Adoi%2F10.1371%2Fjournal.pmed.1000333. Accessed 9 June 2013.

Obrist, B., Iteba, N., Lengeler, C., Makemba, A., Mshana, C., Nathan, R., et al. (2007). Access to health care in contexts of livelihood insecurity: A framework for analysis and action. *PLoS Medicine, 4*, 1584–1588.

Porter, M., & Kramer, M. (2011). Creating shared value. *Harvard Business Review, 89*(1/2), 62–77.

Prahalad, C. K. (2004). *Fortune at the bottom of the pyramid: Eradicating poverty through profits*. Upper Saddle River: Wharton School Publishing.

Prahalad, C. K., & Hammond, A. (2002). Serving the world's poor, profitably. *Harvard Business Review, 80*(9), 48–58.

Rangan, V. K., Chu, M., & Petkoski, D. (2011). Segmenting the base of the pyramid. *Harvard Business Review, 89*(6), 113–117.

Schlicht, E. (2008). Consistency in organizations. *Journal of Institutional and Theoretical Economics JITE, 164*(4), 612–623.

Schumpeter, J. A. (1912/2006). Theorie der wirtschaftlichen Entwicklung. Berlin: Dunker & Humblot.

Schumpeter, J. A. (1943). *Capitalism, socialism and democracy*. London: Unwin University Books.

Smith, A. (1759).In S. M. Soares (Ed.), *The theory of moral semtiments*. MetaLibri, 2005, v1.0p. http://www.ibiblio.org/ml/libri/s/SmithA_MoralSentiments_p.pdf. Accessed 9 June 2013.

Smith, A. (1776). In M. S. Sálvio (Ed.) (2007), *An Inquiry into the nature and causes of the wealth of nations*. MetaLibri, v.1.0p. http://www.ibiblio.org/ml/libri/s/SmithA_WealthNations_p.pdf. Accessed 9 June 2013.

Stevenson, H. H. (2006). A perspective on entrepreneurship. *Harvard Business School*, Case Study 9-384-13 (revised from original 1983 version).

Stevenson, H. H., & Jarillo, J. C. (1990). A perspective of entrepreneurship: entrepreneurial management. *Strategic Management Journal, 11*, 17–27.

Stolz, I., Fürst, M., & Mundle, D. (2012). Beyond volunteerism! Real world learning to foster responsible leadership, serve patients and develop innovative solutions to healthcare challenges. In: T. Wehner, & G. C. Gentile (Eds.), *Corporate Volunteering: Unternehmen im Spannungsfeld von Gemeinschaft und Gesellschaft (AT)*. GablerVerlag.

The Lancet Global Burden of Disease Study (2012). http://www.thelancet.com/themed/global-burden-of-disease. Accessed 28 Jan 2013.

United Nations Department of Economic and Social Affairs, Population Division: World Population Prospects. The 2010 Revision. http://esa.un.org/wpp/Documentation/publications.htm. Accessed 28 Jan 2012.

Vekataraman, S. (1997). The distinctive domain of entrepreneurship research. In J. Katz (Ed.), *Advances in entrepreneurship, firm emergence and growth* (Vol. III, pp. 119–138). Greenwich: JAI Press.

Weick, K. E. (1995). *Sensemaking in organizations.* Thousand Oaks: Sage.

Weick, K. E., & Roberts, C. (1993). Collective mind in organizations: Heedful interrelating on flight decks. *Administrative Science Quarterly,* 357–381.

Weick, K. E., & Sutcliffe, K. M. (2003). *Das Unerwartete managen. Wie Unternehmen aus Extremsituationen lernen.* Stuttgart: Klett-Cotta.

Wieland, J. (1996). *Ökonomische Organisation, Allokation und Status.* Tübingen: Mohr.

Wieland, J. (1998). Kooperationsökonomie. Die Ökonomie der Diversität, Abhängigkeit und Atmosphäre. In G. Wegner & J. Wieland (Eds.), *Formelle und informelle Institutionen. Genese, Interaktion und Wandel.* Marburg: Metropolis.

Wieland, J. (2003). Corporate citizenship und strategische Unternehmenskommunikation in der Praxis. In J. Wieland, & M. Behrent (Hrsg.): Corporate citizenship und strategische Unternehmenskommunikation in der Praxis. München/Mering: Hampp Verlag.

Wieland, J. (2004). *Die Ethik der Governance.* Marburg: Metropolis.

Wieland, J. (2009). Die Firma als Kooperationsprojekt der Gesellschaft. In J. Wieland (Ed.), *CSR als Netzwerkgovernance – Theoretische Herausforderungen und praktische Antworten.* Marburg: Metropolis.

Wildavsky, A. (1987). Choosing preferences by constructing institution: A cultural theory of preference formation. *American Political Science Review, 81,* 3–21.

Williamson, O. E. (2010). Corporate governance: A contractual and organizational perspective. In L. Sacconi, M. Blair, E. Freeman, & A. Vercelli (Ed.), Corporate Social Responsibility and Corporate Governance. Basingstoke: Palgrave Macmillan.

Witt, U. (1987). *Individualistische Grundlagen der evolutorischen Ökonomik.* Tübingen: Siebeck.

Wood, D. J., & Logsdon, J. M. (2002). Business citizenship. From individual's to organizations. In: *Business Ethics Quarterly, 3,* pp. 59–94.

World Bank 2012 Development Indicators (2012).

Part III
Implementation and Instruments

Embedding Sustainable Entrepreneurship in Companies: The Eternal Internal Challenge

Aileen Ionescu-Somers

1 Introduction

Since the late 1980s, a major focus of executives responsible for sustainability strategy rollout in companies has been on "finding the business case" for sustainability and on convincing mainstream managers to exploit that case as much as possible internally and externally. Research has proven that the stronger the business case, and the more value drivers it builds upon, the stronger the internal and external appeal of associated projects, and the more robust the internal alignment within corporate organizations (Steger 2004). However, research has also shown that the "right" values, and corporate mind- and skill sets go a long way to supporting this process through the recruitment and creation of sustainability entrepreneurs that change behavioral patterns within firms.

Today, as sustainability increasingly enters the mainstream of business thought and action, we observe that discussion around the strength of the business case for sustainability still persists. Simply because managers still struggle to find that robust and watertight business case for even the most economically relevant sustainability projects. In global companies at least, managers are reaching the limits of exploiting the easier wins or "lower hanging fruits". What is now needed is more managers that effectively act as sustainability entrepreneurs pushing the frontiers of innovation within the firm.

One of the reasons for the challenges in moving forward with sustainability agendas in firms is the tension that exists between short term pressure and long term benefits. Some sustainability issues, although extremely economically relevant in the medium to long term, have not proven to be "make it or break it" issues in the shorter term when applying traditional business logic. This is because the business logic for sustainability has often focused on a backward focused risk-averse logic.

A. Ionescu-Somers (✉)
CSM Platform IMD, Chemin de Bellerive 23, 915, Lausanne 1001, Switzerland
e-mail: aileen.somers@imd.ch

C. Weidinger et al. (eds.), *Sustainable Entrepreneurship*, CSR, Sustainability, Ethics & Governance, DOI 10.1007/978-3-642-38753-1_12, © Springer-Verlag Berlin Heidelberg 2014

The opportunity-focused, more forward looking argument has so far not been center stage except for rare exceptions.

As we write, even the highly risk intensive issue of "food safety" which has led to very many significant scandals – most recently and ominously around transparency in the European meat food chains – has not led companies to reinvent food systems aggressively enough. The fact is that when it comes to managing sustainability risks, many companies have contented themselves with taking incremental steps. More daring radical innovation for sustainability is much less prevalent.

So, this being the case, where is the potential for a giant step around some of the most urgent sustainability issues out there? Changing the mindset and in-company enabling mechanisms around strategic innovation for sustainability to promote entrepreneurship within companies will be a major prerequisite. This has a great deal to do with how companies of the future will address innovation strategically. It also has a great deal to do with how managers perceive sustainability, the leeway they have to be "sustainable entrepreneurs" within their own companies, and how they network formally and informally to make sure that sustainable entrepreneurship gets embedded.

2 Drivers and Types of Strategic Innovation for Sustainability

Joseph A Schumpeter's theories are often referred to in discussions related to drivers of innovation in the field of entrepreneurship (Baregheh et al. 2009). Schumpeter makes a distinction between continuous change occurring within existing systems and the circular flow of the existing economy, and more discontinuous change, which disturbs the flow (Schumpeter 1934). Building on Marxist thought and the now well-known concept of "creative destruction", he reflected on how a new product, process, type of organization or market can change the economy, sometimes pushing traditional organizations to eventual extinction (Schumpeter 1942). Over time, we have observed multiple forces of creative destruction operating in the business environment; the internet, for example, is breaking the boundaries as we write, rendering – for example – the printed press and a host of other technologies increasingly obsolete. The introduction of the Apple iPhone and latterly the iPad is another good example, which has led to long-standing robust and reputable companies such as Nokia to feel the intense pain and pressure of Apple's disruptive technology.

Schumpeter took a rather wide view of "innovation". He defined it as encompassing change in products (both goods and services), processes, operations, management systems, business models or external relations (Schumpeter 1934; OECD 2005; Baregheh et al. 2009). He defined the process of technological change as consisting of different stages from invention to innovation to diffusion, leading to differing degrees of change on a spectrum from small adjustments to existing technologies and processes to fundamentally new technologies and processes (Dewar and Dutton 1985; Jaffe et al. 2003).

To go further with this idea, there is a spectrum of types of innovation for sustainability. The first type is characterized by the more traditional step-by-step, incremental change referred to above, an approach currently adopted by very many companies mainly owing to their risk averse nature and particularly during these last few, crisis driven years. The second – radical innovation – means the creation of new business models but still within a traditional business system. However, the third relates to game-changing systemic transformation which means reinventing business systems that are essentially faulty since unsustainable by their very nature. As we look out at the business environment and the urgency around specific sustainability issues and impacts, it is apparent that too few companies are in the two latter spaces (Steger 2004).

Nowadays, we are increasingly hearing the words "market transformation" bandied about; by this we mean the changes that are necessary to make production and consumption within markets sustainable worldwide. Take the example of commodity value chains around the world. They are increasingly stressed, and the predictions are in some cases dire and severe, owing to a myriad of complex social environmental and economic challenges linked to the fact that there are finite resources on a finite planet that are being exploited by increasingly more people. The debate around market transformation indicates that what is needed is a funda-mental rethink of existing structures, business models and markets. This implies discontinuous transformation (Schlegelmilch et al. 2003).

Schumpeter pointed out that discontinuous change is generally instigated from within organizations. This means that consumer demand is not always the main driver of innovation. His view, rather, was that business innovation itself that mostly drives changes in consumer demand (Schumpeter 1934). Taking the Apple iPhone example again; it was not consumer demand that initially drove this innovation. Go back further to Edison; it was not consumer demand that asked for the electric light bulb. Rather these successful and game changing innovations were propelled by entrepreneurship and ingenuity of visionary thinkers.

In the rest of this chapter, we examine the case of Unilever's Lipton tea brand in terms of its contribution to a discussion on market transformation and discontinuous change, applying a filter of organizational renewal to create enabling factors to drive the change. The production, processing and consumption of tea is a commod-ity value chain amongst many others. We describe how Unilever created a breeding ground for in-company entrepreneurship that allowed the company, not only to make its own tea value chain sustainable and to gradually roll the sustainable tea strategy out around the world, but also to create a "domino" effect within the entire tea industry leading to a tipping point toward market transformation (Braga and Ionescu-Somers 2011; Ionescu-Somers et al. 2011). Although one can think of it as an example of "creative destruction" in the Schumpeter sense, perhaps "creative construction" is a better set of words given the persistent existence of a viable tea value chain – but an enhanced and more sustainable one – as a result of Unilever's entrepreneurial and innovative approach.

3 Leadership Role in Stimulating Sustainable Entrepreneurship

Most global companies nowadays have initiated a strategy of sorts around their most business relevant sustainability issues, but the level and quality of strategic integration is still highly – and for some industries such as food & beverage, dangerously – variable (Ionescu-Somers and Steger 2008). Social and environmental reports abound. New internal functions dedicated to sustainability have sprung up all over the place in the last decade. Corporate purpose statements increasingly encompass social and environmental, as well as economic dimensions. Given the hype around sustainability issues, their omnipresence on the media radar screen, and the increasing armies of NGOs expressing the critical public voice more and more insistently, one could get complacent in thinking that companies feel under pressure to urgently minimize negative externalities and rapidly neutralize their worst impacts on the planet. Unfortunately, many sustainability strategies still remain sidelined mostly owing to a lack of organizational will to adopt a "Full Monty" approach to strategic roll out. This means using all the tools available to the strategic roll out process.

Unilever is pressing for a "Full Monty" approach in its company; firstly, because it sees sustainability as pushing core business strategy. In 2009, Unilever launched a new corporate vision to double the size of the business while reducing overall environmental and social impact across the entire value chain. It calls this the Unilever Sustainable Living Plan (USLP). This is not just the sustainability strategy of the organization, it is the core strategy. In other words, there is no difference between the two. Unilever CEO Pol Polman writes in the 2011 annual report:

> Unilever's future success depends upon being able to decouple our growth from our environmental footprint, while at the same time increasing our positive social impacts. These are the central objectives of the Unilever Sustainable Living Plan which we launched in November, 2010.

The thinking embedded in the USLP – balancing risk with opportunity – has laid a strong foundation for sustainable entrepreneurship within Unilever as an organization. In the specific case of Lipton tea, it would lead to large-scale social and environmental improvements on tea production plantations supplying Unilever, and at the same time a reversal of the commoditization trend of the tea market.

4 Understanding the Business Relevance of Social and Environmental Impacts

Unilever is one of the world's largest tea companies, purchasing around 12 % of global tea production. The company is also vertically integrated in the value chain and is present from production to commercialization (this means that it also owns plantations as well as sourcing from outside the firm).

Tea plantations are located in tropical forest areas in about a dozen counties. Tea grows year round and is a labor intensive crop employing over 13 million workers of which around 9 million are smallholders. At the time of the beginning of our story in 2006, numerous and diverse sustainability issues around tea predominated; mainly because of low margins and underinvestment at the farm level:

- Contamination of soil, surface water and final product by pesticide residues;
- Soil erosion and degradation and sedimentation of rivers;
- Land conversion and logging for firewood, often leading to deforestation and loss of biodiversity;
- Low wages and poor working conditions and/or housing on plantations, particularly for seasonal workers;
- Uneven value distribution:
- Safety issues leading to work related injuries and agrochemical contamination problems;
- Pollution and energy inefficiency;
- Gender issues and ethnic discrimination.

At Unilever, the above-mentioned issues were identified as risks that were compounded by a persistent state of oversupply, keeping a downward pressure on prices. Although falling prices might have seemed like an optimum scenario for consumers, they were a threat to the long-term economic health of the tea industry since they were a barrier to many of the issues listed above being addressed.

The company had established its first sustainable agriculture program in the 1990s, developing guidelines for sound agricultural practices for key crops. The guidelines were developed in consultation with key stakeholders and extended to suppliers. However, in 2005 came a particularly groundbreaking move; social and environmental issues were brought into product brand innovation and development plans. The process for this at Unilever was called: Brand Imprint. This was a proprietary planning tool developed by Unilever to fuel brand innovation by integrating social economic and environmental considerations. It was a move that would open the doors for a concept that was previously considered "niche" by many managers to rapidly go mainstream within the organization. This move led to a "tipping pont" which got the organization thinking of sustainable sourcing as opportunity, and not only risk. Note what Unilever's corporate report said of sustainable sourcing at that time:

"Sustainable sourcing not only helps us to manage a key business risk, it also presents an opportunity for growth, allowing our brands to differentiate themselves to the growing number of consumers who choose products based on their sustainability credentials".

5 Creating a Breeding Ground for In-firm Sustainable Entrepreneurship

5.1 Focusing on Customer Needs and the Business Context

Companies in the food & beverage industry worldwide are experiencing serious pressures in sourcing agricultural commodities. The reasons are known and recognized by global industry players. World demography is rocketing (we will be nine billion on this planet by the year 2050). In key geographical areas where large populations are getting richer, people are switching to animal protein-based diets, with ensuing ever increasing pressure on water and other resources. Food production has greater negative impact than any other human activity, being responsible for most deforestations and habitat loss worldwide. Agriculture uses twice as much water as all other water uses combined; 70 % of total freshwater drawdown worldwide. It is also the single largest source of pollution of rivers and streams, produces more carbon dioxide than any other activity, and has managed over time to reduce topsoil globally by half.

A simple equation "no resource (ergo no raw materials) = no business" applies. Yet, industry moves to ensuring sustainable sourcing strategies in supply chains are precariously slow, with most collaborative initiatives on a precompetitive level often slanting toward a relatively low common denominator. Moreover, sourcing managers are locked in behaviors that pertain to an earlier, less resource-threatened business context.

Unilever market research showed that sustainability was of growing concern in key emerging markets and that it could potentially be turned into an effective differentiating factor when effectively communicated to consumers. It was recognized that communication about sustainability had so far been low key and with lack of visibility for consumers, harvesting impacts on brand value and therefore the bottom line were hardly being realized.

5.2 Leveraging Relevant Applied Technology and Sharing Knowledge

In 2006, Lipton was the very first Unilever brand to carry out a Brand Imprint exercise. The fact that Michiel Leijnse – who had just joined the brand from Ben & Jerry's – was the one to instigate this was significant. Ben & Jerry's ice-cream was a socially responsible brand taken over by Unilever in the late 1990s. At the time of the acquisition, it was agreed that Ben & Jerry's would continue to pursue its triple bottom line (environmental, social and economic) mission and would operate independently from Unilever's US ice cream business by having an independent board of directors to provide leadership for the company's social mission and brand integrity. In other words, the DNA of Ben & Jerry's was allowed to remain intact. In fact, Unilever managers willingly admitted that the acquisition was a good opportunity to learn about socially responsible brand imprint.

Lipton already had solid experience in integrating sustainability into tea production. Lipton's teas estates in Kericho (Kenya) and Mufindi (Tanzania) fully complied with Unilever's own standards of sustainable agriculture. The company was already working to align supplier's practices with those standards and had partnered with the Kenya Tea Development Agency (KTDA) to promote sustainable practices among smallholders.

Unilever saw that it was strategically well placed to be a catalyst for transformation of the entire tea industry. This recognition did not emerge out of nowhere; for years Unilever had developed a strong corporate reputation in sustainability, recognized by its consistent performance as the industry leader on the Dow Jones Sustainability indices. Unilever had been a pioneer in addressing sustainability issues, including sustainable agriculture. Its work with WWF in creating the Marine Stewardship Council (Steger and Raedler 2002) or with other companies in creating the Sustainable Agriculture Initiative (SAI) Platform for example, has been well documented and it has been a pioneer in instigating several impressive base of the pyramid models; the Unilever Shakti project (Ionescu-Somers et al. 2002) is one particularly successful example.

Therefore, Unilever had a record in building a strong track record in keeping its finger on the pulse of the market, not only through market research, but also through extensive consultation with key stakeholders by actively participating in the sustainability platforms referred to above, sometimes playing a key leadership role.

As leader of the Lipton tea initiative, Leijnse "cross fertilized" the corporate DNA from Ben & Jerry's . In a further move to consult with stakeholders and share knowledge, he assembled an assessment group of brand developers, supply managers, corporate responsibility executives, outside consultants and Unilever managers from different functions to run the Brand Imprint exercise.

5.3   Developing and Understanding Possible Scenarios

During the 4 month brand imprint exercise, the group carried out integrated analyses of:

- The footprint (social, economic and environmental) of the Lipton Brand across the value chain and,

- The influence of consumers, market forces, key opinion formers (customers, suppliers, NGOs and governments)

Based on the findings, the group designed an integrated brand strategy. The group identified significant business opportunities that would render the brand "shiny and new" by linking brand preference to sustainability and enabling the company to engage in positive dialogue with consumers and enhance brand value.

The group reached the conclusion that consumers would not necessarily accept Lipton's self-declared sustainability excellence as credible. To overcome this risk, they felt they should get the support and endorsement of a third party. They looked at a number of third party certification schemes such as Fairtrade, UTZ certified and the Rainforest Alliance Certified. Criteria for selection of a partner were:

- Does the shadow brand, or seal, have brand recognition?
- Could the certifier's message overshadow that of the Lipton brand?
- Did the organizations have adequate scale to certify an extensive supply base such as the tea required by Unilever for its Lipton brand?
- Did the organization have the flexibility and capacity to certify large estates as well as smallholders?

Having built various scenarios, Unilever concluded that a partnership with the Rainforest Alliance was the most feasible option for an organization of the size and breadth of Unilever, and with a wide reaching global brand such as Lipton Tea.

5.4   Hardwiring the Business Case

Seeking the green light from Unilever's top management was the next step. The team's proposal was to convert the whole brand to certified sustainable tea, meaning that all Lipton products would eventually be Rainforest Alliance certified. This also meant that sustainability in tea production would be mainstreamed at Unilever, a contrast to the customary approach of introducing a variant brand onto a niche market. The team wanted to change *mainstream* brand positioning and create long-term value by building a very strong link between sustainability as an attribute and the brand.

Unilever decision-makers asked highly business relevant questions: How costly would this be? How rapidly could the conversion happen? What return would Lipton get on the income side? This latter question was the most difficult to answer. Converting to sustainable tea meant paying a premium to growers while keeping the retail price unchanged. In such circumstances, how could the brand remain profitable? Their consumer research had indicated that in spite of best intentions, mainstream consumers did not tend to follow words with action and pay more for sustainable products. Did this mean that the additional cost would need to be absorbed in the margin, thus reducing profitability for the firm?

Because of careful scenario planning, the group had the answer ready. They had predicted substantial growth in market share. They were confident that additional supply chain costs would be recovered through that growth.

But what if other major tea brands also switched to certified tea? Where then was the competitive advantage for Unilever? Again, the answer had been carefully thought out. If a significant share of both tea producers and buyers around the world switched to certifying their tea products, prices would also inevitably increase across the board, reversing the trend to commoditization of tea, and allowing retail prices to increase gradually across the board. But because Unilever held the largest market share, a large portion of the income growth would come to the company. Predicting an end to the downward spiral of prices and quality variance on the global tea market was also part of the strong business case.

As we can see, the scenario and forward thinking planning of Unilever allowed for questions that would address identified external barriers; lack of push from investors and customers. The discussion with Unilever's key decision-makers took 5 months. It was not an easy win, but when it came, it came with an added bonus; it was recommended that PG Tips and Lyons tea – another two well-known Unilever brands – would also move to certified sustainable tea.

5.5 Embedding Sustainability Performance Requirements in Key Processes

The USLP has set significant objectives to be achieved by 2020 such as:

- Help more than a billion people to improve their health and well-being;
- Halve the environmental footprint of Unilever products;
- Source 100 % of agricultural raw materials sustainably.

To underpin these broad goals, Unilever has introduced around 60 time-bound targets spanning social, economic and environmental performance across the value chain, from sourcing of raw materials through to the use of products in homes.

On the level of Lipton Tea, Leijnse's team worked on related cascaded objectives that would ultimately contribute to the ambitious overall USLP goals. His team decided to first convert all Lipton tea sold in Western Europe to Rainforest Alliance certified tea by 2010 before certifying all Lipton tea globally by 2015. This was guided by a predicted limited availability of certified tea in the first years, and by differing levels of consumer awareness and interest in sustainability in specific markets.

To secure first mover benefits, Unilever took the daring decision of "leapfrogging to mainstream" rather than testing through pilot projects. This meant that the team would have to work concurrently on supply chain and market rollout.

In the food industry, it is critical not to change the "mouth feel" of a product since this can rapidly lead to consumer rejection of a previously loved brand. The certified tea had to be available not only at the right time, but also from the customary suppliers since each product uses a complex and unique blend made of different origins and qualities of tea. In other words, the blend could not be altered.

Significant brand risks were at stake if the rollout was not well timed with marketing deployment. Once targets were made public, conversion became a one-way street. There would be no going back.

5.6   Creating a "Domino Effect": A Value Chain Transformation

After over a decade of extensive research at IMD on industry-specific business logic for sustainability and their associated roll out strategies, it is interesting to note that key leading companies are struggling with a number of persistent barriers to strategic integration of their sustainability strategies. Some of these are external barriers: such as the absence of a "level playing field" or regulatory framework enabling companies to take more aggressive steps (Steger et al. 2009). Note that the steps taken by the Lipton tea team described above represents a "quasi-regulatory" solution that effectively led to what is called a "value chain conversion" in tea.

From the beginning of its journey, Unilever had foreseen that its own value chain conversion would help the integration of sustainable practices in the entire tea industry. In 2006, Unilever had dropped out of the Ethical Tea Partnership (ETP) an initiative created by the tea industry in 1997 and focusing on supply chain issues on a pre-competitive basis. Unilever's decision was motivated by resistance of other players to adopting broader and more far reaching sustainability criteria going beyond self-assessment.

But when eventually Lipton, PG Tips and Lyons certified tea was available on shop shelves in Western Europe, Japan, North America and Australia, a virtual domino effect took hold (see Fig. 1). In August 2009, the ETP announced a new collaboration with the Rainforest Alliance certification program to build capacity within the tea industry and streamline the certification process for tea producers. Subsequently, Yorkshire tea, Twining's and Tetley went public with commitments to obtaining Rainforest Alliance certification for their brands globally. To give an idea of the scale of impact, by the end of 2011, some 70 % of tea volume sold by UK retailers would be Rainforest Alliance Certified. It was also predicted that by 2016, Unilever and the other mainstream brands would convert some 20 % of the world's tea producers to certified sustainable tea. In other words, a robust market transformation process took hold following Unilever's pioneering move.

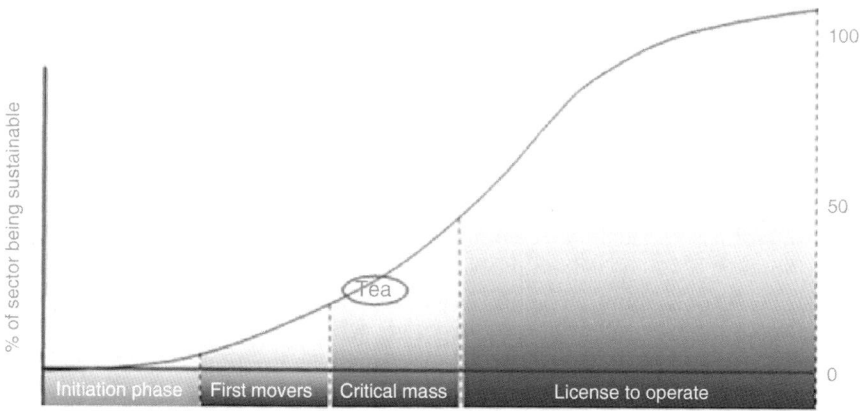

Fig. 1 The Unilever driven "domino effect" taking hold in the tea value chains (IDH Monitoring 2010)

5.7 Going the "Full Monty" with Strategic Alignment

Sustainability experts in specialist areas such as sustainable sourcing and in energy and climate change, often report that there are often vast knowledge gaps within firms on the innovation opportunities, efficiency benefits and risk reduction potential that attending to key sustainability issues present. Relative ignorance about external effects and their business relevance, even amongst highly educated, competent managers, can prevent the building and rolling out of project- and unit-specific business cases such as the Unilever Lipton tea transformations described above.

Key functions that dominate policies within firms, such as finance, sales and marketing, can be locked into specific unsustainable short-term focused behaviors that are consistently rewarded and thus perpetuated. Incentive systems are rarely modified to encompass the longer term objectives of a sustainability strategy. They remain doggedly fixed on the short term dictated by the business models within which firms operate, creating a tension with longer term goal achievement. In a daring move in 2011, Unilever simply stopped reporting to "Wall Street" on a quarterly basis because, as CEO Paul Polman boldly states, he wants stakeholders (including shareholders) to look at Unilever's success long-term not simply quarter to quarter. In this way, Unilever aims to be an ongoing model of sustainability for other corporations, particularly those looking to expand to emerging markets such as Brazil, Russia, Indonesia, China and South Africa.

The SAI Platform mentioned earlier is an industry partnership of over 40 companies focused on rolling out sustainable agriculture sourcing guidelines and standards. It identified similar internal barriers in food and beverage organizations as those identified by IMD empirical research more generally. IMD and SAI Platform have thus co-developed a master class experience to help empower managers to embed sustainable agriculture strategies. Moreover, in a bid to close these same knowledge gaps, recently some key organizations have collaborated together to produce the world's first practitioners guide to the sustainable sourcing of agricultural raw materials (De Man and Ionescu-Somers 2013).

Firms engaging in moves towards embedded sustainability strategies must seek the "Full Monty" with creating sustainable entrepreneurs, meaning that they need to use the full panoply of tools available to them to promote a culture of strategic innovation for sustainability in the firm. This means recruiting managers with the right mindset to start with, rebooting internal value systems, and rewarding managers for changing behaviors and achieving longer term sustainability objectives. It means having executive development policies and programs that ensure managers are kept abreast of rapidly evolving changes in the external environment that are leading to both business risks and opportunities. It means empowering sustainability entrepreneurs to effect bottom up changes in organizations, and creating networks of "sustainability champions" for whom sustainability is not only a "nice to have" but a business imperative.

Literature

Baregheh, A., Rowley, J., & Sambrook, S. (2009). Towards a multidisciplinary definition of innovation. *Management Decision, 47*(8), 1323–1339.
Baregheh et al. (2009). Oslo manual, third edition. Towards a multidisciplinary definition of innovation, working party of national experts on scientific and technology indicators. Paris: OECD.
Braga, T., Ionescu-Somers, A., & Seifert, R.V. (2011) Unilever sustainable tea; Part 1 Leapfrogging to Mainstream, IMD/IDH – Dutch Sustainable Trade Initiative.
De Man, R., & Ionescu-Somers, A. (Forthcoming in April, 2013). *Sustainable Sourcing of Agricultural Raw Materials: A Practitioner's Guide.* Produced by partners: SAI Platform, IMD Global Center for Sustainability Leadership, the International Trade Center and the Sustainable Trade initiative. Also supported by Business for Social Responsibility, Sedex and the Sustainable Food Laboratory.
Dewar, R. D., & Dutton, J. E. (1985). The adoption of radical and incremental innovations: An empirical analysis. *Management Science, 32*(11), 1422–1433.
IDH Monitoring Protocol (2010–2015). Mainstreaming global supply chain sustainability. Dutch Sustainable Trade Initiative, 2009, p. 9.
Ionescu-Somers, A., & Steger, U. (2008). *Business logic for sustainability: A food and beverage industry perspective.* New York: Palgrave Macmillan.
Ionescu-Somers, A., Steger, U., & Amann, W. (2002). *Hindustan Lever – Leaping a Millenium*, IMD Case Study: 3–1073. Lausanne.
Ionescu-Somers, A., Braga, T., & Seifert, R. (2011) Unilever tea (A): Revitalizing Lipton's supply chain, IMD case study 6–0327.
Jaffe, A. B., Newell, R. G., & Stavins, R. N. (2003). Technological change and the environment. In K.-G. Mäler & J. R. Vincent (Eds.), *Handbook of environmental economics: Environmental degradation and institutional responses. Vol 1* (Vol. 3). Amsterdam: Elsevier Science. 2003–2005.

Organization of Economic Cooperation and Development (OECD). (2005). The measurement of scientific and technological activities: Guidelines for collecting and interpreting innovation data

Schlegelmilch, B. B., Diamantopoulos, A., & Kreuz, P. (2003). Strategic innovation: The construct, its drivers and its strategic outcomes. *Journal of Strategic Marketing, 11*(2003), 117–132.

Schumpeter, J. A. (1934). *Theory of economic development.* Cambridge: Harvard University Press.

Schumpeter, J. A. (1942). *Capitalism, socialism and democracy.* London: Routledge.

Steger, U. (Ed.). (2004). *The business of sustainability: Building industry cases for corporate sustainability.* New York: Palgrave Macmillan.

Steger, U., & Raedler G. (1999). Marine stewardship council (A): Is a joint venture possible between "Suits and Sandals"? and departing in uncharted waters. IMD case studies: 2–0080 and 2–0081. Lausanne.

Steger, U., Ionescu-Somers, A., Salzmann, O., & Mansourian, S. (2009). *Sustainability partnerships: The manager's handbook.* New York: Palgrave Macmillan.

Fostering Sustainable Innovation Within Organizations

Peter Vogel and Ursula Fischler-Strasak

1 Introduction

Over the past years many corporate business leaders have started to shift their strategy from a pure profit seeking one towards a balance in simultaneously striving to achieve economic, environmental and social goals (Elkington 1998; Preuss 2007; Roth 2009). As a result, challenges on the sustainability agenda have emerged as a new source of opportunities for innovation and competitive advantage (Fichter 2006; Hockerts 2008; Hansen et al. 2009). Research has shown that entrepreneurs are the main drivers of innovation, economic growth and social change (Audretsch 2002); hence, organizations try to adopt entrepreneurial approaches in order to spur their own innovativeness (Hamel 1999; Ireland et al. 2009). However, as recent publications have discussed, the promotion of entrepreneurship is a difficult and multifaceted issue requiring the consideration of dynamic processes describing the interplay of multiple external factors, local conditions and the individual innovators (Isenberg 2010; Krueger 2012; Vogel 2013). These difficulties are particularly distinct when discussing sustainable innovation, as risk-related reluctance in instigating this kind of innovation can still be observed among corporate leaders (Hall 2002). If established companies plan to take part in creating tomorrow's economy, it will be necessary for them to challenge prevailing assumptions about innovation processes (Hamel 1999). The purpose of this chapter is to investigate the main success factors of entrepreneurial ecosystems and discuss ways how to assimilate these in an organizational context.

P. Vogel (✉)
Ecole Polytechnique Fédérale de Lausanne (EPFL), CDM-ENTC, Station 5, Lausanne 1015, Switzerland
e-mail: peter.vogel@epfl.ch

U. Fischler-Strasak
National Center for Engineering Pathways to Innovation (EPICENTER), Stanford University, USA
e-mail: ursula.fischler@gmail.com

C. Weidinger et al. (eds.), *Sustainable Entrepreneurship*, CSR, Sustainability, Ethics & Governance, DOI 10.1007/978-3-642-38753-1_13,
© Springer-Verlag Berlin Heidelberg 2014

2 Entrepreneurial Ecosystems as a Basis for Innovation

The economic and societal perspectives on entrepreneurship have drastically changed over the last half-century. While today there exists the widely accepted view that entrepreneurship is one of the major drivers of the global economy, social well-being, job creation, economic competitiveness and innovation (Audretsch 2002; Thurik and Wennekers 2004; ManpowerGroup 2012), this was not as clear in the past when it was the common belief that the large corporations and not startups were the sole creators of economic progress (Schumpeter 1942, p. 106).

Historically, the individual entrepreneur was the focus of attention of scholars and practitioners. In recent years, however, external factors as well as the interdependencies with the entrepreneurs and the outcomes have received increasing attention (Van de Ven 1993; Isenberg 2010; Krueger 2012; Vogel 2013). Entrepreneurial Ecosystems constitute "an interactive community that is composed of varied and interdependent actors and factors that evolves over time and promotes new venture creation" (Vogel 2013, p. 5). Figure 1 depicts the major components of an entrepreneurial ecosystem including the external environment (also referred to as "habitat"), the local entrepreneurship-specific factors and the individual entrepreneurs that create the new companies.

While this framework visualizes the fundamental elements of an entrepreneurial ecosystem, some other important framework conditions that influence the successful implementation of an entrepreneurial ecosystem should be mentioned.

- *Each ecosystem is unique!* Silicon Valley is consistently referred to as one of the prototype entrepreneurial ecosystem with regards to innovations (Herrmann et al. 2012). It is a unique combination of different factors such as outstanding talents (partly driven by academic institutions such as Stanford), the ample availability of capital and a truly entrepreneurial culture that are "allowed to circulate freely [...] and meld into whatever combinations are most likely to generate innovation and wealth" (Hamel 1999, p. 73). In environments like this, work is more than just a job – it is a lifestyle! People fully identify with what they do and they are incentivized to innovate. However, when developing a new entrepreneurial ecosystem it is not advisable to try and merely duplicate ecosystems such as the Silicon Valley as many of the underlying factors are quite different across the globe and cannot be altered easily (e.g., the culture or a country's political system). It is important to first understand a community's strengths and weaknesses in order to develop a strategic roadmap for the successful creation of a truly unique entrepreneurial ecosystem (Vogel 2013).
- *Holistic and supervised implementation!* It is advantageous to focus on the implementation of multiple local ecosystem factors in parallel. Setting up single initiatives (e.g. a training program to foster entrepreneurship) without the other critical elements being in place will most likely not lead to the desired outcome. Furthermore, it is recommended to execute the implementation plan in a coordinated and supervised manner. Ideally, that would mean to have an independent team dedicated to setting up and executing on this strategy (Vogel 2013).

Fig. 1 The entrepreneurial
ecosystem (graphic adopted
from Vogel 2013)

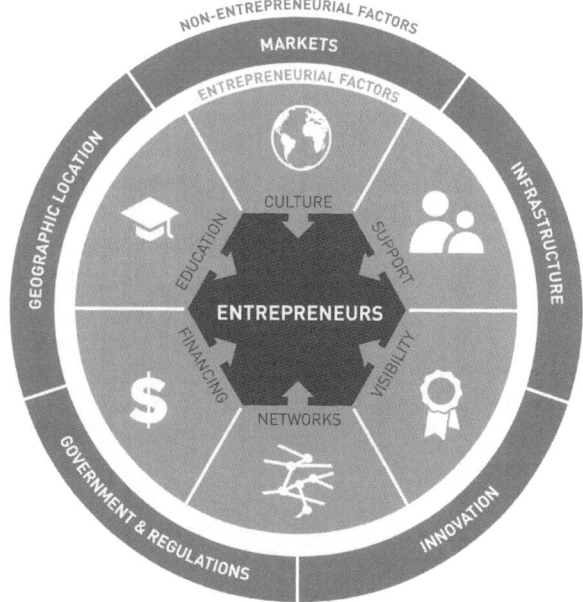

- *Dynamic bureaucracy!* Bureaucratic processes should not cause stagnation
  during the creation of an ecosystem. Building an entrepreneurial ecosystem as
  well as starting up and running a company are equally dynamic and therefore
  require dynamic and flexible processes (Krueger 2012).
- *Building an entrepreneurial culture!* An entrepreneurial culture is essential to
  building a successful entrepreneurial ecosystem. Only if entrepreneurship is seen
  as an attractive career option and entrepreneurs are seen as responsible and
  respectable individuals, will individuals dare to take the step and become
  innovators.

In most large organizations, building structures and a culture of innovation is a
challenging but not impossible undertaking that requires fundamental conceptual
rethinking. Based on the core dimensions of entrepreneurial ecosystems, the
following section will discuss recommendations for building environments of
innovation in a corporate context.

3 Adopting Entrepreneurial Ecosystems to Foster Innovation Within Existing Organizations

Corporate entrepreneurship is seen as a major mechanism for revitalizing
organizations and enhancing firm performance (Antoncic and Hisrich 2001).
Hence, organizations are trying to learn how to build a work environment that

Fig. 2 Intrapreneurial
innovation ecosystem

fosters innovation and nurtures the entrepreneurial passion of their employees. Inspired by the previously presented entrepreneurial ecosystem, Fig. 2 depicts the proposed intrapreneurial innovation ecosystem; a dynamic inter-relatedness between the internal- and the external context with the center containing the innovation process to spur organizational innovativeness. This section will discuss the five core dimensions of innovation – namely culture, architecture, communication and collaboration, talent management and financing – and then provide some insights into how corporate leaders can adopt them within their organization to drive innovation.

3.1 A Culture of Innovation

"Most organizations fail at unleashing one of their most valuable resources: human creativity, imagination, and original thinking. They lack a systematic approach to building a culture of innovation" (Linkner 2011). When talking about a culture of innovation, it is important to consider the factors that play a critical role, among them the organizations' value system, visions and norms on the one hand and the employees' mindset, passion and tolerance of failure on the other hand. Three central approaches to building a culture of innovation are discussed: (1) building an honest value system with innovation at the core, (2) furthering a leadership style which involves all employees in the innovation process, and (3) establishing a business culture where failure is broadly accepted and is a central element leading to new product or service development.

3.1.1   Live Honest Company Values

The majority of corporate value systems sound something like "Integrity – Communication – Respect – Excellence". These, however, seldom reflect the true values of the company. For an effective and efficient workplace to evolve, it is important that employees share and identify with the values of the company and feel like their own values are reflected in the corporate culture. Reed Hasting, founder and CEO of Netflix, defined the corporate values of his organization in his globally recognized document called "Netflix Culture: Freedom & Responsibility" (Shontell 2013) where he touches on a large variety of critical elements which comprise a culture of innovation. These include the cultural fit of individual employees, selflessness, rewarding performance as opposed to effort as well as freedom and responsibility for every employee at the core.

> When building a value system that ought to nurture innovativeness, it is important to consider that values should reflect the true behavior and skills of the employees and not simply impose corporate values upon the company (Netflix 2009).

3.1.2   Involve All Employees in Idea Generation Processes

Without proactively developing new ideas, the ability to respond to dynamic market pressures, or to envision new products or services, organizations are at risk of losing their competitive position and becoming slow and unresponsive to rapidly changing market demands. One reason why entrepreneurial ecosystems have more vibrant markets for ideas than larger corporations is the absence of "prejudices about who is or is not capable of inventing [new products, services or] new business models" (Hamel 1999, p. 78). Companies often live hierarchies where experience counts far more than imagination. Leadership needs to overcome these obstacles by breaking down barriers and eliminating "arbitrary distinctions between 'suits' and 'creatives'" (Florida and Goodnight 2005, p. 126). It is not rare to see higher success rates of ideas that have been executed in the ranks without support from above (Amabile and Khaire 2008; Hamel 1999).

> Involving all employees in the idea generation process will increase the odds of finding the best solutions to existing problems. It also sends the important signal that everyone is on the same team and strives towards the same goal.

3.1.3 Embrace a Culture Which Accepts Failure

During the process of generating new ideas, there is neither place for judgment nor the fear of failing (Brown 2009) as failure is a necessary element of an innovation culture (Hurley and Hult 1998), and "from failure comes learning, iteration, adaptation, and the building of new conceptual and physical models through an iterative learning cycle" (Hess 2012). New ventures typically enter the market with an early version of a product or service (beta version) and adapt it according to customer feedback; this is also true for some disrupting technology-based innovators such as Google, Facebook or Apple. Innovators within larger organizations, however, often need to go through a series of tedious internal approval processes, which aim at eliminating failure. But in the end, it is exactly the dynamic, iterative cycle of shaping and refining ideas that allows for innovative products, services or business models to emerge and shake up industries.

> If organizations want to become more innovative from the inside, they need to build the right culture – a culture that embraces a mentality of "failing fast and failing cheap".

3.2 Architecture for Innovation

"Organizations are designed to promote order and routine; they are inhospitable environments for innovation. Those who don't understand organizational realities are doomed to see their ideas go unrealized" (Levitt 2002, p. 137). Creativity requires an environment that stimulates and encourages new ideas to flourish, organizational processes that are aligned with functional responsibilities (Andriopoulos and Dawson 2009) and an overall organizational strategy in order to systematically take these new ideas to the market. The following thoughts show how an 'architecture for innovation' can facilitate the generation of creative ideas within an organization.

3.2.1 Establish Creative and Fun Workspaces

In the past, a workplace was viewed exclusively as a physical environment for work. However, in today's highly competitive environment this old fashioned perspective is becoming "increasingly unsuited to emerging patterns of work" (GSA 2006, p. 3). It requires exciting workplaces that promote an atmosphere for continuous innovation and creativity. While many organizations have embraced home-office practices over the past years to accommodate their employees' desire for flexibility, they now start to realize that new and innovative "ideas spring from spontaneous chats between employees in the cafeteria or at the gym. That is [one] reason [why] they provide on-campus perks" (Cain Miller 2013). However, an innovative workplace is not just about a gym, day-care services, free espresso and food, health benefits and nice colorful offices with lots of sticky notes; it is also about meaningful tasks, a collective mindset, a positive mood, honesty and trust, loyalty, effectiveness as the benchmark for success, stunning people as well as a creative, fun and inspirational environment (Vogel 2012b).

> "A new breed of workers (the knowledge workers) is emerging to provide the required creativity and innovation" (Schriefer 2005). It requires stunning and fun workplaces to attract these innovators, satisfy their needs and provide incentives to stay with the organization to ensure continuous innovation.

3.2.2 Launch Idea Labs and Flexible Commercialization Processes

Developing new products, services and business models is done best outside of existing business units through dedicated, independent creativity labs, which are strategically located at the edge of the organization, having one foot inside the organization and one foot outside. This is exactly where great ideas are born through the collision with other ideas (Johnson 2010). However, it is essential that these models are aligned with the overall strategy, facilitating the formation of alliances with internal stakeholders, such as the R&D units (Blank 2012). If organizations want to ensure that their idea lab does not fail to bring creative ideas to the market[1], a non-bureaucratic but efficiency-driven development and evaluation process needs to follow the idea lab sessions (Amabile and Khaire 2008).

> By leveraging the fact that each individual has a set of ideas at any point in time (Hill and Birkinshaw 2010) organizations can systematically build creative spaces for collaboration that facilitate the recombination of these different 'idea sets' to form novel and innovative ideas.

3.3 Talent Management for Innovation

Managing "true innovators, people who are totally sold out to new ideas and new concepts" (Philipps 2007), is difficult but not impossible. Startups, mostly

---

[1] (such as the failure of the Qualcomm "Venture Fest" (Dos Santos 2013))

unconsciously and out of necessity, typically have an innovator-friendly approach to recruiting rewarding and retaining their talents. They need to offer them exhilarating work, which gives them intrinsic motivation to innovate, alongside an upside potential, because if they don't, they will most likely turn in their badges and leave (Hamel 1999). In corporate environments, talents are hardly ever utilized to their full potential. In fact, many talents that bring up creative ideas inside larger corporations face managerial responses such as "this is how things are done around here". The following section describes some talent management approaches to spur each employee's innovativeness.

3.3.1 Educate for Creativity and Innovation

"Creativity and innovation are becoming increasingly important for the development of the 21st century knowledge society" (Ferrari et al. 2009, p. iii). While creativity has long been seen as a characteristic of eminent people, a more differentiated perspective has emerged in recent years (Craft et al. 2001) describing certain elements of creativity – those related to attitudes towards finding effective solutions to everyday problems – as being accessible to everyone. With the right training, tools and techniques, every employee, team and organization can learn how to challenge the *status quo* and generate creative ideas (Seelig 2012).

> Individuals can be taught to develop the ability to generate and recombine ideas to address identified challenges.

3.3.2 Facilitate Talent Mobility

Talent drives economic growth and innovativeness. Talent mobility – both inside and outside the organization – is an enabler for organizations to close skill gaps, retain talent, remedy talent shortages as well as move more people to employability and employment (WEF 2012). In fact, if companies do not work on internal talent mobility programs, "highly creative and ambitious people who feel trapped in moribund businesses are going to leave" and never come back (Hamel 1999, p. 83). Corporate leaders need to ask themselves whether they want their employees to leave and create wealth for themselves by creating a spinout; for others by joining a competitor; or if they want them to stay and create wealth for the company by innovating within. Participating in talent mobility programs can be an exciting alternative to switching employers (Vogel 2012a); however, organizations need to "provide incentives for employees to abandon the familiar for the unconventional" (Hamel 1999, p. 83).

Innovation happens at the boundary of organizations. By allowing talents to freely move across departments as well as outside of the organization, business leaders can facilitate the generation of innovative ideas.

3.3.3 Customize Rewards to the Individual

If organizations want to retain their innovative talents, it is critical that these remain motivated at any point in time. A central part of this motivation and pro-entrepreneurial behavior is a well-structured reward system that encourages risk-taking and intrapreneurship (Ireland et al. 2009). However, due to fundamentally different identities that shape motivations and key decisions of entrepreneurs (Fauchart and Gruber 2011), the most effective incentives for intrapreneurs will most likely vary across employees. While some individuals are driven primarily by monetary incentives, others are driven by the purpose they follow or recognition by others. Therefore, some individuals might be incentivized best by receiving shares in the organization or large bonuses whereas other employees are best rewarded by being put on an even more challenging project (Florida and Goodnight 2005; Amabile and Khaire 2008).

If organizations want to maximize the innovativeness of their employees, they need to tailor the reward system to the employee's identity.

3.4 Communication and Collaboration for Innovation

"Open communication of information, ideas and feelings is the lifeblood of innovation" (Wycoff 2004). Yet, many large organizations are stuck in secrecy-driven R&D processes that seldom involve external stakeholders. In recent years, there have been two major trends in the entrepreneurial domain that have changed the way innovation is done: "pivoting" and "open innovation / crowdsourcing". The applicability of these two methods in an organizational context are briefly discussed below.

3.4.1 Pivot Ideas with Stakeholders

Organizations often start off with an idea that they think people want. They spend lots of time and money building a "perfect" product only to fail when they reach out to prospective clients learning about their indifference (Ries 2011). Driven by the fear of harming the brand, corporations have institutionalized rather rigid processes when preparing a new product or service for entering the market. Many startups, on the contrary, have engaged in "pivoting", a radically new approach to innovation. They generate new ideas and rudimentary products or services, test them quickly and cheaply on the market, shed them if they do not cause the anticipated traction and move on to the next idea.

> Business leaders should encourage the pivoting of ideas with their network of stakeholders in order to effectively select winning ideas without threatening the brand reputation.

3.4.2 Leverage the Network Through Open Innovation

As opposed to the traditional early twentieth century paradigm of closed innovation – meaning that an organization has the entire innovation process under internal control – "open innovation is a paradigm that assumes that firms can and should use external ideas as well as internal ideas" to advance their innovativeness (Chesbrough 2003). In a world where inconceivable amounts of knowledge are widely dispersed, it is impossible for an organization to solely rely on their internal R&D; otherwise, they will lag behind or be overtaken by some small, agile competitor. With regards to the accumulation of innovative ideas, organizations can leverage their existing networks (e.g., clients, research organizations and universities) towards engaging in joint R&D projects and additionally involving the broad public through open innovation competitions (e.g., the Cisco I-Prize). These help to fill knowledge-gaps, solve problems or come up with the next big thing.

> In order to become or remain an innovative trendsetter, organizations could embrace "open innovation" and "crowdsourcing" as modern, inclusive processes of idea generation.

3.5 Marketplace for Capital

While both venture capitalists and CFOs are interested in funding successful projects, they surely do not follow the same approach. While the CFO's goal is to never make an investment that fails to deliver an adequate yield, a venture capitalist's goal is to have at least one big winner amongst the wide range of projects. Along the same lines, an innovator inside an organization typically has to go through a line of hierarchical decision processes, whereas an external innovator can pitch to multiple investors, ideally have them compete with each other and then select the best offer. The following section will briefly describe how organizations can create an internal innovation-friendly marketplace for capital.

3.5.1 Build Entrepreneurial Processes to Select Winning Opportunities

Unlike in Silicon Valley, where it's rare to find successful start-ups that do not have to pitch their ideas multiple times, organizations are often designed in a way that an innovator has only one opportunity to pitch a new idea and a "no" (possibly due to an incompatibility with balance-sheet-driven KPIs) immediately means the end to it (Hamel 1999). In the case of an entrepreneur this normally looks fundamentally different with a first-stage risk-free evaluation from the 3Fs (family, friends or fools) and subsequent incremental feedback from the market.

Organizations should embrace both, monetary and non-monetary evaluation criteria to evaluate ideas and simultaneously make multi-channel approaches available for protagonists that seek funding to present their concepts (Hamel 1999).

3.5.2 Balance of M&A Activities with Intrapreneurship

While many companies buy their innovations from external entrepreneurial ventures through expensive M&As (Hess 2012), creating an internal marketplace for capital that is paired with a proper intrapreneurial innovation process would allow tapping into a largely unutilized source of innovation. Yet the internal marketplace for capital needs to be separated from standard budgeting processes in order to offer flexibility at relatively low bureaucratic levels.

Corporate leaders who want their organization to be innovation leaders should follow a simultaneous approach of M&A as well as funding of intrapreneurial projects.

3.5.3   Create an Internal Crowdfunding Portal

It is common knowledge that both entrepreneurs and intrapreneurs face difficulties in attracting early-stage funding in order to take their ideas to the market (Cassar 2004). In recent years, crowdfunding has rapidly evolved as a viable alternative to more traditional sources of funding such as banks or equity capital. Crowdfunding is a collective monetary effort of individuals, better known as the "crowd", to support projects and ideas initiated by other people or organizations (Belleflamme et al. 2011). Despite its prominence in startup funding, larger corporations are not making use of this novel concept. Innovators within organizations typically have one source of funding for their ideas and if their superiors reject it, their only chances are to either drop the idea, leave the company and join the competition, or build their own business.

> By creating an internal crowdfunding portal that facilitates the inter-departmental funding of ideas, corporations can capture a significant share of the value that is currently being lost.

4   Conclusion

Innovation is a crucial element of sustainable growth (Global Innovation Index 2012) and appears to be largely driven by entrepreneurs (Audretsch 2002), as big companies "are designed to be bad at innovation" (Wessel 2012). In most large organizations, building structures and a culture of innovation is a challenging but not impossible undertaking, requiring fundamental changes. The purpose of this chapter was to share insights into entrepreneurial ecosystems as major sources of innovation and offer some thoughts as to how certain elements from the entrepreneurial world could be translated into a corporate context. It should not be regarded as a case-proven formula for how business leaders could enhance their organization's innovativeness, but rather as a thought stimulus to critically reflect upon interal innovation processes and to courageously adopt one or more ideas from entrepreneurial ecosystems. To stay competitive, organizations must nurture innovativeness in a variety of areas and particularly embrace sustainability as a new source of opportunities. The question that remains is whether organizational leaders want to be part of defining the future or whether they capitulate to dauntless entrepreneurs.

Literature

Amabile, T. M., & Khaire, M. (2008). Creativity and the role of the leader. *Journal of the Management Training Institute, SAIL, Ranchi, 36*, 48–51.

Andriopoulos, C., & Dawson, P. (2009). *Managing change, creativity and innovation.* London: Sage.

Antoncic, B., & Hisrich, R. D. (2001). Intrapreneurship: Construct refinement and cross-cultural validation. *Journal of Business Venturing, 16*(5), 495–527.

Audretsch, (2002). The dynamic role of small firms: evidence from the US. Small Business Economics. *18*(1), 13–40.

Belleflamme, P., Lambert, T., & Schwienbacher, A. (2011). Crowdfunding: Tapping the right crowd. *CORE Discussion Paper No. 2011/32.*

Blank, St. (2012). *The future of corporate innovation and entrepreneurship.* Blog post. http://steveblank.com/2012/12/03/the-future-of-corporate-innovation-and-entrepreneurship/. Accessed Feb 2013.

Brown, T. (2009). *Change by design: How design thinking transforms organizations and inspires innovation.* New York: HarperCollins.

Cain-Miller, C. (2013). *Will Yahoo increase productivity by banning people from working at Home?* The New York Times online: http://bits.blogs.nytimes.com/2013/02/25/will-yahoo-increase-productivity-by-banning-people-from-working-at-home/. Accessed Feb 2013.

Cassar, G. (2004). The financing of business start-ups. *Journal of Business Venturing, 19*(2), 261–283.

Chesbrough, H. W. (2003). *Open innovation: The new imperative for creating and profiting from technology.* Boston (MA): Harvard Business Press.

Craft, A., Jeffrey, B., & Leibling, M. (2001). *Creativity in education.* London: Continuum International Publishing Group.

Dos Santos, R. (2013). *Qualcomms corporate entrepreneurship program – Lessons learned Part1 & Part 2.* Blog post: http://steveblank.com/2013/01/30/qualcomms-corporate-entrepreneurship-program-lesson-learned-part-2/. Accessed Feb 2013.

Elkington, J. (1998). *Cannibals with forks: The triple bottom line of 21st century business.* Stony Creek: New Society Publishers.

Fauchart, E., & Gruber, M. (2011). Darwinians, communitarians, and missionaries: The role of founder identity in entrepreneurship. *Academy of Management Journal, 54*(5), 935–957.

Ferrari, A., Cachia, R., & Punie, Y. (2009). Innovation and creativity in education and training in the EU member states: Fostering creative learning and supporting innovative teaching: Literature review on innovation and creativity in E&T in the EU Member States (ICEAC). JRC Technical Note 52374.

Fichter, K. (2006). *Nachhaltigkeitskonzepte für Innovationsprozesse.* Wiesbaden: Fraunhofer IRB Verlag.

Florida, R., & Goodnight, J. (2005). Managing for creativity. *Harvard Business Review, 83*(7), 124–131.

Global Innovation Index. (2012). *Stronger innovation linkages for global growth. INSEAD.* http://www.wipo.int/econ_stat/en/economics/gii/index.html. Accessed Feb 2013.

GSA Office of Government Wide Policy. (2006). *Innovative workplaces: Benefits and best practices*: http://www.gsa.gov/graphics/pbs/Innovative_Workplaces-508_R2OD26_0Z5RDZ-i34K-pR.pdf. Accessed Feb 2013.

Hall, J. (2002). Sustainable development innovation: A research agenda for the next 10 years. *Journal of Cleaner Production, 10*, 195–196.

Hamel, G. (1999). Bringing silicon valley inside. *Harvard Business Review, 77*(5), 70–84.

Hansen, E. G., Grosse-Dunker, F., & Reichwald, R. (2009). Sustainability innovation cube – a framework to evaluate sustainability-oriented innovations. *International Journal of Innovation Management, 13*(4), 683–713.

Herrmann, B. L., Marmer, M., Dogrultan, E., & Hotschke, D. (2012). Startup Ecosystem Report. The Startup Genome Project.

Hess, E. D. (2012). *Creating an innovation culture: Accepting failure is necessary*. Forbes Magazine: http://www.forbes.com/sites/darden/2012/06/20/creating-an-innovation-culture-accepting-failure-is-necessary/. Accessed Feb 2013.

Hill, S. A., & Birkinshaw, J. M. (2010). Idea sets. *Organizational research methods, 13*(1), 85–113.

Hockerts, K. (2008). *Managerial perceptions of the business case for corporate social responsibility* (CBS working paper series, Vol. (03.2007)). Frederiksberg: Copenhagen Business School.

Hurley, R. F., & Hult, G. T. M. (1998). Innovation, market orientation, and organizational learning: An integration and empirical examination. *The Journal of Marketing, 62*(3), 42–54.

Ireland, R. D., Covin, J. G., & Kuratko, D. F. (2009). Conceptualizing corporate entrepreneurship strategy. *Entrepreneurship Theory and Practice, 33*(1), 19–46.

Isenberg, D. J. (2010). How to start an entrepreneurial revolution. *Harvard Business Review, 88* (6), 41–49.

Johnson, S. (2010). *Where good ideas come from: The natural history of innovation*. New York: Penguin Group.

Krueger, N. F. (2012). *Candidates guide to growing a more entrepreneurial economy*. http://papers.ssrn.com/sol3/papers.cfm?abstract_id=2098094. Accessed Feb 2013.

Levitt, T. (2002). Creativity is not enough. *Harvard Business Review, 80*, 137–144.

Linkner, J. (2011). *Seven steps to a culture of innovation*. http://www.inc.com/articles/201106/josh-linkner-7-steps-to-a-culture-of-innovation.html. Accessed Feb 2013.

ManpowerGroup, (2012). *Youth unemployment challenge and solutions*. http://www3.weforum.org/docs/Manpower_YouthEmploymentChallengeSolutions_2012.pdf. Accessed Feb 2013.

Netflix. (2009). *Netflix culture: Freedom & responsibility*. www.slideshare.net/reed2001/culture-1798664. Accessed Feb 2013.

Philipps, J. (2007). *How to manage an innovator. Innovate on purpose*: http://innovateonpurpose.blogspot.ch/2007/11/how-to-manage-innovator.html. Accessed Feb 2013.

Preuss, L. (2007). Contribution of purchasing and supply management to ecological innovation. *International Journal of Innovation Management, 11*(4), 515–537.

Ries, E. (2011). *The lean startup: How today's entrepreneurs use continuous innovation to create radically successful businesses*. New York: Crown Business.

Roth, S. (2009). *Non-technological and non-economic innovations: Contributions to a theory of robust innovation*. Switzerland: Peter Lang.

Schriefer, A. E. (2005). Workplace strategy: What it is and why you should care. *Journal of Corporate Real Estate, 7*(3), 222–233.

Schumpeter, J. A. (1942). *Capitalism, socialism and democracy*. New York: Harper and Row.

Seelig, T. (2012). *inGenius: A crash course on creativity*. London: HarperCollins.

Shontell, A. (2013). *Sheryl Sandberg: 'The most important document ever to come out of the valley'*. Business insider online: http://www.businessinsider.com/netflixs-management-and-culture-presentation-2013-2?op=1. Accessed Feb 2013.

Thurik, R., & Wennekers, S. (2004). Entrepreneurship, small business and economic growth. *Journal of Small Business and Enterprise Development, 11*(1), 140–149.

Van de Ven, H. (1993). The development of an infrastructure for entrepreneurship. *Journal of Business Venturing, 8*(3), 211–230.

Vogel, P. (2012a). *Die Generation Y ist weniger loyal zum Arbeitgeber – aber loyal zum Deal*. HR Today: https://www.jobzippers.com/media/press/201206-HRToday.pdf. Accessed Feb 2013.

Vogel, P. (2012b). *Unleashing the talent of the NEXT generation*. As part of the futurework forum seminar series "The workplace of the future" at the Lorange Institute of Business, Zurich.

Vogel, P. (2013). The employment outlook for youth: Building entrepreneurial ecosystems as a way forward. *Conference Paper for the G20 Youth Forum 2013*, St. Petersburg.

Wessel, M. (2012). Why big companies can't innovate. *Harvard Business Review Blog*: http://blogs.hbr.org/cs/2012/09/why_big_companies_cant_innovate.html. Accessed Feb 2013.

World Economic Forum. (2012). *Talent mobility good practices: Collaboration at the core of driving economic grow. Global agenda council.* http://www.weforum.org/reports/talent-mobility-good-practices-collaboration-core-driving-economic-growth. Accessed Feb 2013.

Wycoff, J. (2004). *The big ten innovation kills and how to keep your innovation system alive and well. Innovation Network*: http://www.innovationnetwork.biz/library/BigTenInnovationKillers.htm. Accessed Feb 2013.

Corporate Capability Management: Collective Intelligence in Use for Improvement on a Company's Sustainability, Innovativeness and Competiveness

Daniel Velásquez Norrman, Martin Riester, and Wilfried Sihn

1 Introduction

Despite of proven immense impact on short-term profitability, short payback periods, serving as a multiplier for performance enhancements or by annual cost-savings and being recognized for its significance on the innovativeness and competiveness of a company, successful continuous improvement (CI) as defined in the paper is rare. An approach with prerequisites of a successful exception is the Fraunhofer Austria Corporate Capability Management (CCM) concept. CCM is defined as the systematic and holistic approach to ongoing improvements on organization's capabilities in order to efficiently enhance a company's **sustainability**, **innovativeness** and **competitiveness**. The concept comprehends discrepancies between research findings on critical success factors and contemporary industrial practices. The paper demonstrates that a gap between best practices and the actual implementation in companies is present. It concludes that the CCM concept addresses potentials for cost-savings, increased innovativeness and sustainability even left out by advanced CI practices.

2 Demand for Enhanced CI-Concepts

The significance of CI for a company's innovativeness and competiveness has already been recognized (Bessant and Caffyn 1997; Shingo 1988; Caffyn 1999). In fact, CI of work processes was estimated second most important to short-term

D. Velásquez Norrman (✉) • M. Riester
Division of Production and Logistics Management, Fraunhofer Austria Research GmbH,
Theresianumgasse 27, Vienna 1040, Austria
e-mail: daniel.norrman@fraunhofer.at; martin.riester@fraunhofer.at

W. Sihn
Fraunhofer Austria Research GmbH, Theresianumgasse 27, Vienna 1040, Austria
e-mail: wilfried.sihn@fraunhofer.at

C. Weidinger et al. (eds.), *Sustainable Entrepreneurship*, CSR, Sustainability,
Ethics & Governance, DOI 10.1007/978-3-642-38753-1_14,
© Springer-Verlag Berlin Heidelberg 2014

profitability at the same time as internal quality improvement groups were seen as the most important source of innovation in work processes and procedures (Soderquist and Chanaron 1997). Nevertheless, as our online study, industrial projects conducted by Fraunhofer Austria and former research point out – the context in which a **successful** CI process takes place still needs to be reformed. In addition to contemporary practices, three explicit aspects overlooked by organizations must be considered by efficient CI. Hence discrepancies between contemporary practices and demands on processes for ongoing improvements are herein elucidated and the Fraunhofer Austria Corporate Capability Management concept (CCM), a systematic and holistic approach to successful improvements on organization's capabilities, is explained.

2.1 Research Focus

The research and development of the CCM concept is based on results in the area of continuous improvement. Over the last decade, the importance of managing ongoing improvements has increased and spread to new fields. Many companies still lack appropriate approaches and methods to effectively address CI in their organization. The research aims on overcoming these gaps and providing a concept for successful CI.

In order to conduct the work different methodological approaches has been relevant for this paper. A state-of-the-art analysis, an online survey, and continuous improvement projects at industrial companies were conducted to obtain up-to-date information on relevant trends and challenges. Whereas the state-of-the-art analysis and the CCM-concept are presented, only a few results from the study has been chosen for presentation in this paper

2.2 Brief History Description on CI

CI as defined by Bessant et al. (1994) in "a company-wide process of enabling a continuing stream of focused incremental innovation" or as herein understood as "an approach that continuously seeks to identify, evaluate and implement sustainable enhancements targeting the elimination of waste in all systems and processes of an organization, products and services", goes back to the eighteenth century during which initiatives such as management encouraged employee-driven improvements were undertaken (Schroeder and Robinson). Over the decades, the need to continuously improve on a larger scale within the organization became essential (Bhuiyan and Baghel 2005) and the approach stretched geographically. The scope of CI, initially used in the manufacturing process, evolved into a much broader term, constituting a management tool for ongoing improvement involving everyone in an organization (Bhuiyan and Baghel 2005; Kossoff 1994; Imai 1986).

CI has as of today established itself as one of the core strategies for manufacturing and an imperative as to meet challenges posed by the contemporary competitive environment. One of these challenges is the continuously changing environment which puts attention on companies to incorporate **flexibility** into its system if to be able to change and match market needs. Hence the most important thing stays the ability to change and to do it quickly enough (Yamashina 1995).

Generally the minimal costs and cost reduction motives involved in the implementation and maintenance of CI has been one of the main reasons to its expansion in Europe (Boer et al. 1999). During the boost CI has had, also in association with the introduction of the TQM movement or CI methodologies such as e.g. Lean Manufacturing, Six Sigma and Kaizen (Bhuiyan and Baghel 2005), the phrase has become increasingly popular. Even if CI is present, and to some seem to have the characteristics of another worn-out buzz-word, considerable **potentials** are left unexploited by even the early adopters of CI.

2.3 Successful CI

A web-based online study on CI and CCM carried out by Fraunhofer Austria in 2012 had the respondents divided into three categories dependent on their score with regard to the three criteria "employee participation rate", "idea for improvement implementation rate" as well as "target range of CI process". The top 15% of the respondents with the highest score were categorized as High-performers whereas the 15% of the respondents with the lowest score were labeled Low-performers. In matter of economic benefits the results were evident – the study showed on cost savings being up to **three times** higher per employee and year amongst the High-performers when compared to the Low-Performers.

The difference between basic forms of instituted programs in order to apply CI and successful CI, i.e. being efficient in taking advantage of available potential, can be immense. The Critical success factor for CI has amongst others been studied and identified by Gibb and Davies (1990) in research on Australian small to medium enterprises (SMEs), been described by Bessant and Francis (1999) as practices within the behavioral model describing the evolution of CI capability as well as listed by Caffyn (1999) as the core organizational abilities and key behaviors for CI. Even if the emphasis of the study on CI and CCM was not on the identification of critical success factors, the study results were used together with former research results and industrial project experiences in the derivation of **explicit aspects** that must be stressed and covered by the CCM concept.

Hence CCM has been developed in order to exploit the potentials for capability enhancements left unutilized. New **sources** of intelligence and new **fields** of improvement with **sustainability** being one in particular are addressed. Former flaws have been recognized due to the lack of a systematic approach and methods to implementation of processes for ongoing improvements regard to: (1) **corporate approach**, i.e. comprehension of external stakeholders, (2) **operation specific**, i.e. transfer of CI methodologies to indirect and company specific operations, and (3) **empowered controlling**, i.e. the communication and monitoring for a sustainable CI.

2.4 Corporate Approach

The enforcement of CI as a management tool for ongoing improvement involving everyone in an organization does not consider the relation to external stakeholders. A similar inadvertence of external stakeholders is seen with the types of CI based on the organizational designs presented by Berger (1997). Focus is foremost on the benefits of multifunctional work groups whereas external stakeholders are left unmentioned.

Caffyn (1999) addresses the ability to move CI across organizational boundaries, i.e. effective working across internal and external boundaries at all levels is defined as one of the core organizational abilities for CI. The view of CI for continuous and incremental improvements or the one of intermittent and not incremental, i.e. **innovation** (Imai 1986), comes into play when the contribution of external stakeholders to CI are to discus. Whereas CI is strongly linked with continuous incremental improvements, also understood under Kaizen, it does not exclude incremental innovation. In the ongoing process for improvements targeting the elimination of waste in all systems, innovations are important and must hence include external stakeholders as possible source for ideas. Close working relationship with key customer was seen as the second most important source of innovation in work processes and procedures in a study amongst French SMEs, close working relationship with key supplier qualified as the eighth most important source (Soderquist and Chanaron 1997). Singh and Singh (2011), comes to similar results in their investigation having customer relationship rated as most important in carrying out continuous improvement activities in the manufacturing organizations. The comprehension of **external stakeholders** considers the potential of these stakeholders in their feasible contribution to incremental improvements as well as innovation.

Results from the Fraunhofer Austria study on CI and CCM showed on an average participation rate, i.e. number of involved employees to total number of employees on-site, of around 30 % over the last 2 years amongst the respondents. The comparison between the best and the worst in class showed on a more than **eight times** higher participation rate amongst the High-performers than amongst the Low-performers. Integration of external stakeholders in the process of ongoing improvements increases the potential for a higher participation rate that then theoretically even may exceed 100%.

The fundamental process by which firms gain the benefits of internal and external knowledge, create competitive advantage and develop capability can be summarized in the term Knowledge Integration (KI). The characteristic of KI, i.e. integration of knowledge, is a vital part of CI and an important driver for innovation and productivity performance (OECD 2004). Comprehension of stakeholders in the concept of successful CI does not lead to differentiation but integrates knowledge indispensable for competitive survival and creates **firm-specific** innovation (Mohannak 2012). Innovation being firm-specific is also more valuable inside the organization than in the market, less subjected to imitation, and contributes to the ongoing improvements of a company's capabilities.

The contemporary practices of collaboration with stakeholders are mainly limited to explicit stakeholder such as customers within the new product development or suppliers for the supply chain management. Mohannak (2012) discusses current relevant frameworks and proposes an own conceptual framework for KI in R&D firms and with emphasis on the new product development. Described are e.g. critical success factors such as strategic communities (SC), company specific knowledge integration system dependent by the type of knowledge the company wishes to integrate (goals), team building capability and knowledge integration through communication networks within and outside the organization. The importance to integrate external stakeholders is hence evident but seldom addressed by concepts on capability improvements.

The process of ongoing improvements is neither limited to incremental improvements nor internal ideas for improvement, but improves on corporate capability when external stakeholders are comprehended with efficient processes for transfer and integration of knowledge in the work with continuous improvements.

2.5 Operation Specific

Whereas the enforcement of CI methodologies within non-manufacturing processes, has reached a certain stage of maturity in regards to implementation, it can still not be considered as mastered amongst others in matters of employee participation rate. The differences between CI maturity level between manufacturing and non-manufacturing operations are also seen in the previously described study on CI and CCM where the participation rate amongst manufacturing companies was almost **three times** higher than amongst trading companies.

Reasons to why CI has not reached the same status in non-manufacturing operations are foremost seen in the history of CI as previously described. The philosophy of incremental or continuous improvements was originally used for enhancing manufacturing processes and first more recently gained popularity in indirect operations (Yamashina 1995). Another eligible reason is that business processes in various senses differ from manufacturing processes. Wiegand and Nutz (2007) separate the operations of a company into direct and indirect operations, where direct operations work with goods and materials whereas the indirect operations are mainly concerned with information.

The classification of operations into direct and indirect operations has the operations of Manufacturing, Assembly, In- and Outbound Logistics and Maintenance arranged to direct operations and the operations of Accounting, Controlling, Purchasing, Sales, R&D, IT, Procurement and Human Resources to the indirect operations. Indirect operations are then characterized with task mostly not being well-defined, consolidated with a high degree of creativity and employee individual design as well as primarily made up of overhead costs. Hence the approaches and methods so successfully used on manufacturing processes must not necessary imply the same results when applied on indirect business processes. The difference in characteristics between manufacturing and business processes is in fact seen as such major ones that methods must be adapted (Laqua 2012). The differentiation of Lean Production and Lean Administration, both with the target of sustainable elimination of waste through continuous improvements programs but for direct respectively indirect operations, is just one example. The evolution of criteria of manufacturing paradigms from cost over quality, variety, responsiveness and to sustainability (Koren 2010) as new area for improvements is a second example on how successful CI has to be operation specific.

Methods and approaches that have successfully been used in the direct operations cannot directly be transferred onto indirect operations, but must first be adapted in regards to their characteristics before they can be applied in the process of ongoing improvements.

2.6 Empowered Controlling

Controlling of CI is distinguished as a critical success factor in the Fraunhofer Austria study on CI and CCM. The share of companies measuring the number of submitted ideas was twice as high amongst High- to Low-performers. When the controlling of savings through continuous improvements is considered, the share is almost **three times** as high amongst the High-performers. Monitoring and measurement of CI is also described as a practice in the higher levels of CI evolution (Bessant and Francis 1999) at the same time as Singh and Singh (2012) stress that efforts in e.g. measuring and reporting CI productivity and costs as long overdue.

Slightly more than the half of over 200 companies questioned in a study on Management Tools reported that they have established a standardized CI-Process in their organization. The share is reduced to almost a third when it comes to taking advantage of the CI-process as to achieve ongoing operational process improvement (Stegner 2010). Bessant (2000) comes to similar results in a survey conducted by CI research advantage (CIRCA) at UK firms, around 50% have instituted some form of systematic program to apply CI and 19% claims to have a wide spread and sustained process of CI in operation.

More widespread knowledge and enthusiastic ideas, demand organizations to be able to integrate them through mechanisms such as directions and organizational routines (Grant 1996). Task of the management is to support this process and anchor it in the management system. The lack of support from management level is second to the unwillingness to change amongst employees the biggest reason to why increase in productivity projects are stalled (Schneider et al. 2011) and the presence of support as a core ability for continuous improvements accordingly to Caffyn (1999). The management role is highlighted in CI systems using control charts suggested by MacKay (1988), requiring a management team to decide which processes to attack, to establish teams to work on the project, to allocate resources and to review progress.

Goals and results must continuously be measured and when necessary adapt actions or goals as to secure the sustainability of the ongoing improvement process. In this manner two major obstacles within controlling and CI arises, defining and measuring CI goals as well as coordination individual employee motivation with e.g. target agreement to maximize in terms of incentive structure (Maras 2009). Complementary to the explicit aspects addressed by a concept for successful CCM, added value of a monitored and measured CI process is further seen in **market valuation** of a company. Measuring intellectual capital is becoming more important for companies in matters of stock market valuation, as to attract venture capital or build a partnership.

The sustainability and hence the success of the process of ongoing improvements is dependent on the incorporation of management support and employee motivation. Management support and employee motivation is in turn empowered with controlling of incentive structures coupled to CI targets.

2.7 Requirements on Successful CI-Concept: CCM

Substantial unexploited potentials arise as a result from contemporary practices not being:

- **Corporate**, i.e. comprehend external stakeholder and new source for improvement (i.e. collective intelligence) through efficient processes for transfer and integration of knowledge
- **Specific**, i.e. adapted to company and process specific characteristics, as well as novel areas for improvement, and
- **Sustainable**, i.e. implemented mechanism for individual motivation and link to CI targets

These potentials must be addressed by concepts aiming at a systematic and holistic approach to ongoing improvements on organization's capabilities in order to efficiently improve on company's sustainability, innovativeness and competitiveness. Hence the Fraunhofer Austria CCM concept has been developed and practiced.

3 Concept of Corporate Capability Management (CCM)

Fraunhofer Austria's CCM concept focuses on afore mentioned and described requirements, necessary for a successful implementation of a CI-concept, means: (1) corporate, (2) specific and (3) sustainable.

3.1 CCM Main Idea

CCM considers not only current employees as potential **sources** for ideas and not only direct operations as **fields** to improve a company's performance. CCM represents an enhanced concept which includes several CCM-**stakeholder** groups and novel fields such as **sustainability** and **indirect operations**. Furthermore, the improvement of a company's performance is not the only objective the CCM is taking into account. CCM explicitly considers the improvement of an organizational culture as a major additional objective. For realizing these objectives, the CCM-concept pursuits two ways of gaining ideas, (1) individually initialized by stakeholder groups (bottom-up) and (2) specifically set activities by responsible persons within a company (top-down). Additionally, the CCM-concept includes an approach for a sustainable controlling, adaptable to a company's specific structure and processes as well as applicable by each company, regardless of its industry classification. Besides this, the CCM-concept provides two adaptable generic core modules as support for setting up a specific CCM, aligned to a company's needs, as well as a generic roadmap for a stepwise and sustainable implementation of CCM within a company. Therefore the CCM-concept represents a holistic and sustainable approach for accessing and utilizing a company's corporate capabilities (Picture 1).

3.2 CCM Core Module #1

The core module #1 basically consists of three circles which are named "**stakeholder**", "**objectives**" and "**methods** (for collecting, evaluating and implementing ideas)". Forming those three generic circles, starting with the inner one, according to a company's specific structure and processes, is the first step for the implementation of a sustainable CCM-concept.

3.2.1 Stakeholder

Contrary to classical approaches of CI, the CCM-concept does not see ideas for improving a company's performance just in mind of a company's current staff. Ideas for small or significant improvements of a company's performance are also

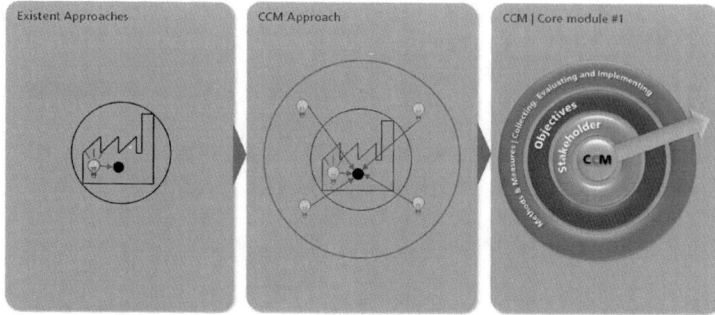

Picture 1   CCM approach and core module #1

seen in mind of several **external people**, who are or were directly or indirectly in contact with a company. These people are all considered by the CCM-concept as possible sources for ideas and are defined as "CCM-stakeholders". Examples for such stakeholder groups can be: employees, customers, suppliers, research organizations, retired persons (former employees), inter-trade organizations, etc.

In practice, out of these stakeholder groups, relevant ones have to be defined by a company. Generally, just two or three are selected in a first step and the company's specific CCM-concept is designed for this selection first. Additional stakeholder can be added later at any time. It's not necessary to include all possible stakeholders from beginning on.

3.2.2   Objectives

As already mentioned, improving a **company performance**, i.e. process quality, service quality and product quality, represents just a part of the objectives the CCM-concept focuses on. Another objective, the CCM-concept is targeting, is the **organizations culture**. Organizations culture is considered as a combination of the factors "working environment", "corporate social responsibility" as well as "sustainability". The reason behind this setting of these two major objectives is based on the assumption that there are interdependencies between an organizations culture and its performance as described above. The possibility to place ideas regarding the improvement of an organizations culture and see those ideas being realized influences the stakeholder groups and their willingness to generate and share ideas for a company's performance in a positive way. The same assumption is applicable vice versa, means a good company performance influences the loyalty of stakeholder groups and their willingness to generate and share ideas to improve respectively to support an organizations culture in a positive way.

Depending on the company and its corporate strategy the relevance of particular components of the two major objectives are different respectively needs to be individually adapted. E.g. "product quality" is not necessarily relevant for the service industry. However, important is that the basic structure of the official

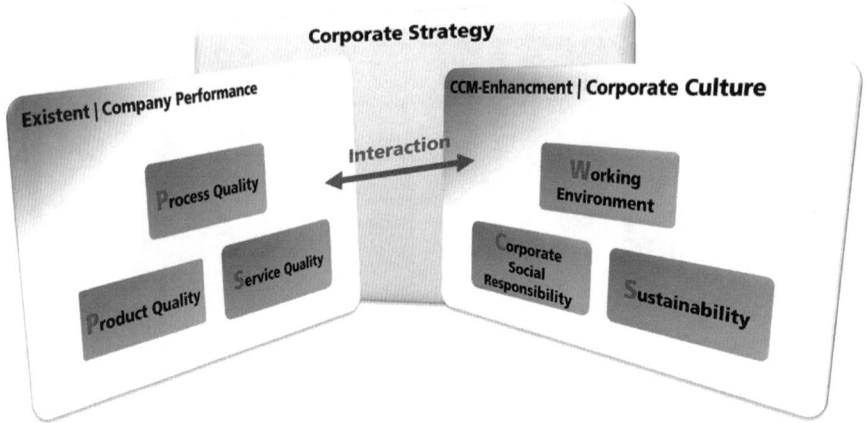

Picture 2 Enhanced objectives of CCM

targeted objectives always consist of a company's performance as well as of its organizational culture (Picture 2).

After setting the basic structure of objectives, they have to be matched with the selected stakeholder groups, i.e. the following question needs to be answered: Are there stakeholder groups, who just should deal with certain objectives? Does a company want to treat all set objectives with all stakeholder groups? E.g. the stakeholder group "suppliers" could be matched with the objectives "process quality" and "sustainability" but not with "working environment" because there is no meaningful link. Basically, defined stakeholder groups and objectives can be matched with each without any restrictions as long as it makes sense.

All defined pairs are basis for designing the third circle (methods and measures) of the core module #1.

3.2.3 Methods and Measures

For gaining a maximum output (max. number of ideas/improvement potential), it is mandatory to **define a suitable set of methods and measures** individually aligned to the stakeholder groups a company is dealing with and the objectives the company is focusing on with each stakeholder group. E.g. for gaining ideas of employees for production process optimization a company can organize a weekly meeting (measure) and apply "value stream mapping" as method. For gaining ideas of customers for new products likely it is more productive to organize a quarterly meeting (measure) and apply the method "brain-storming".

However, considering and matching possible methods and measures only on the level of "gaining" respectively "collecting" ideas is not sufficient enough for ensuring a holistic and sustainable approach of CCM. Further methods and measures for "evaluating" and "implementing" ideas are mandatory. Therefore

Picture 3 Core module #2

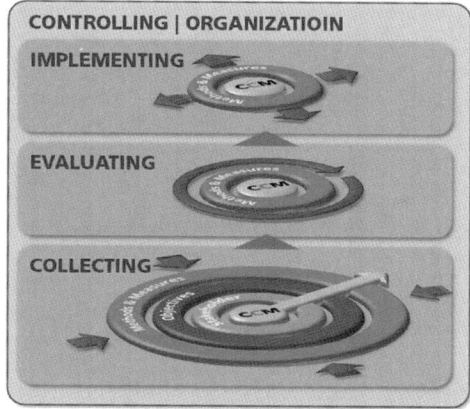

the CCM-concept provides a second core module which is explained in the following.

3.3   Core Module #2

The illustration of core module #2 points out the necessity of applying methods and measures on the level of "collecting", "evaluating" and "implementing" ideas (Picture 3).

3.3.1   Levels of Methods and Measures

As explained, for **collecting** ideas each stakeholder group needs a special set of methods and measures.

Also for **evaluating** the ideas which the stakeholder groups generated, transparent measures and methods need to be developed and applied. E.g. reviewing all submitted ideas in a weekly session (measure) by applying standardized evaluation forms and independent evaluators (method).

In a similar way, methods and measures for **implementing** the collected and positively evaluated ideas need to be defined. These measures and methods are essential for realizing a measurable benefit for a company. E.g. realizing employee ideas categorized as "easy realizable" within the next 2 weeks after evaluation (measure) under participation of the employee who submitted the idea (method).

Like already indicated within the description of the methods and measures in core module #1, it is mandatory to link all defined methods on each level either to a continuous or a periodic recurring cycle to ensure an ongoing utilization. E.g. weekly employee meetings, quarterly meetings with suppliers or monthly reviews of achieved CCM results.

3.3.2 Controlling and Organization

After "levels of methods and measures", the part "controlling and organization" represents the second major part of the core module #2.

This part is essential for ensuring the long-term success of the CCM-concept. According to a company's structure an appropriate CCM **organization** needs to be designed and set up. This includes three major points: (1) designing processes, e.g. how will collected ideas be forwarded to the evaluators? (2) Naming of other responsible persons, e.g. who is responsible for coordinating CCM in total or evaluating ideas? (3) Design communication processes. E.g. how are results communicated to participating stakeholder groups?

In addition, based on afore mentioned points, infrastructural requirements will be derived, i.e. info boards on shop-floor level, IT exchange platforms or the like.

Besides setting-up appropriate organization structures, the implementation of a suitable **controlling** is needed for ensuring an enduring transparency of ongoing activities and their performance. This includes basically two essential points. (1) The definition of key figures which are used for expressing the success of the concept. Depending on a company and the branch it's assigned to, determined figures can be different. But independent of these, the CCM-concept intends to **measure** the **output** (quantitatively and qualitatively) and the **input** (quantitatively and qualitatively) which is caused by all defined measures and methods. Means e.g., cost savings, increasing motivation of employees or number of collected ideas on the one hand, as well as invested time (personnel costs) and material costs on the other hand.

The second point intends to (2) anchorage (formal or informal) CCM-objectives within objective agreements of departments or responsible persons.

A further point could be the definition of a bonus system. In contrast to the points mentioned before, this third point is considered as optional, depending on the philosophy a company is pursuing.

3.4 CCM-Roadmap

The design of the core modules #1 and #2 as well as further steps for setting up and utilizing CCM in a company are summarized in the CCM-roadmap. It represents a guideline regarding the major steps for designing, implementing and operating CCM (Picture 4).

3.4.1 Development Phase

The development is structured into the five steps already described above. (1) Define relevant stakeholder groups, (2) set objectives for each stakeholder group

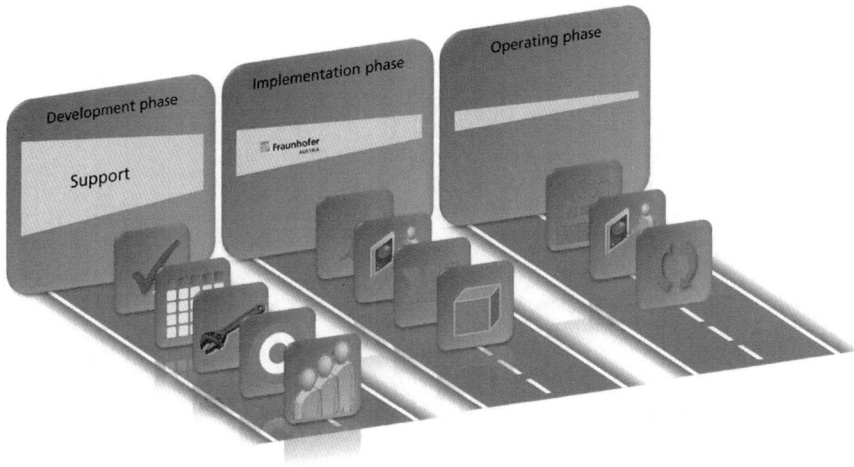

Picture 4 CCM-Roadmap

(3) define methods for collecting, evaluating and implementing ideas, (4) define a periodic or ongoing cycle for the utilization of each method, (5) define a appropriate organization as well as a formal controlling for the CCM in total.

3.4.2 Implementation Phase

The implementation phase is divided into four major steps. (1) Implement required infrastructure, means e.g. the built-up boards on shop-floor level for communicating CCM-objectives and offering the possibility to turn in ideas. (2) Implement essential processes and organizational measures like defined in step 5 within the development phase. (3) Instruct stakeholder groups and train CCM responsible people e.g. head of departments. (4) Operational start of CCM in a determined pilot sector of a company.

3.4.3 Operating Phase

The operating phase is built up of the three major steps. (1) Real time operation i.e. extending the CCM-concept to other areas within a company and applying the defined methods and measures according to the agreed periodic or ongoing cycles. (2) Instruction and training of additional stakeholder groups respectively people and CCM responsible people. (3) Ongoing adaption, improvements and enhancements of methods to new or changed circumstances.

3.5 Findings

Based on a state-of-the-art analysis, the Fraunhofer Austria study on CI & CCM as well as industrial project experiences, **discrepancies** between **critical success factors** for a company's innovativeness and **contemporary** industrial **practices** were identified. A gap between best practices and contemporary concepts as well as actual implementation in companies is present. In particular three explicit aspects in relation to CI were discussed as overdue.

The result was that existing CI concepts leave out on critical potentials in unutilized sources and fields. Findings further showed that the difference between best practices and under-performers in the sense of High- and Low-performers is striking. High-performers showed cost savings being more than three times higher per employee and year when compared to the Low-performers.

Subsequent the term **successful CI** was defined as the concept exploiting these identified potentials. Three explicit aspects discussed had to be transformed into requirements on a successful CI concept:

- **Corporate**, i.e. comprehend external stakeholder and new source for improvement through efficient processes for transfer and integration of knowledge
- **Specific**, i.e. adapted to company and process specific characteristics, as well as novel areas for improvement, and
- **Sustainable**, i.e. implemented mechanism for individual motivation and link to CI targets

Based on these findings the concept of Corporate Capability Management (CCM), defined as the systematic and holistic approach to ongoing improvements on organization's capabilities in order to efficiently enhance a company's **sustainability**, **innovativeness** and **competitiveness**, was developed.

The explanations given in this paper, point out the holistic approach of the CCM-concept. CCM enables companies to (1) access a broader field of possible idea sources, (2) be innovative in novel fields and (3) generate measureable benefits out of them. Furthermore it provides a proceeding for a stepwise designing, implementing and operating process of the system for continuously improvements on capabilities. The CCM concept was built up in two modules and a roadmap. Module #1 shows objectives, method and measurements whereas Module #2 provides a detailed view on methods and measures relating to the three levels of "collecting", "evaluating" and "implementing" ideas. The CCM-roadmap was designed with regards to the different phases when establishing a CI-system.

Finally, the CCM-concept, in a long term perspective, tends to institutionalize CCM as a function within an organization.

Literature

Journal Articles

Berger, A. (1997). Continuous improvement and: Standardization and organizational designs. *Integrated Manufacturing Systems, 8*(2), 110–117.

Bessant, J. (2000). Developing and sustaining employee involvement in continuous improvement. *IEE Seminar on KAIZEN: from understanding to action*, London, Vol. 2, No. 1, pp. 1–18.

Bessant, J., & Caffyn, S. (1997). High-involvement innovation through continuous improvement. *International Journal of Operations & Production Management, 14*(1), 7–28.

Bessant, J., & Francis, D. (1999). Developing strategic continuous improvement capability. *International Journal of Operations & Production Management, 19*(11), 1106–1119.

Bessant, J., Caffyn, S., Gilbert, J., Harding, R., & Webb, S. (1994). Rediscovering continuous improvement. *Technovation, 14*(1), 17–29.

Bhuiyan, N., & Baghel, A. (2005). An overview of continuous improvement: From the past to the present. *Management Decision, 43*(5), 761–771.

Caffyn, S. (1999). Development of a continuous improvement self-assessment tool. *International Journal of Operations & Production Management, 19*(11), 1138–1153.

Gibb, A., & Davies, L. (1990). In pursuit of frameworks for the development of growth models of the small business. *International Small Business Journal, 9*(1), 15–31.

Grant, R. M. (1996). Prospering in dynamically-competitive environments: Organizational capability as knowledge integration. *Organization Science, 7*(4), 375–387.

Singh, J., & Singh, H. (2011). Assessment of the importance level of continuous improvement strategies in manufacturing industry of Northern India. *International Journal of Management and Business Studies, 1*(1), 8–13.

Singh, J., & Singh, H. (2012). Continuous improvement approach: State-of-art review and future implications. *International Journal of Lean Six Sigma, 3*(2), 88–111.

Yamashina, H. (1995). Japanese manufacturing strategy and the role of total productive maintenance. *Journal of Quality in Maintenance Engineering, 1*, 27–38.

Books

Boer, H., Berger, A., Chapman, R., & Gertsen, F. (1999). *CI changes: From suggestion box to organizational learning, continuous improvement in Europe and Australia*. Aldershot: Ashgate.

Imai, M. (1986). *Kaizen: The key to Japan's competitive success*. New York: Random House.

Koren, Y. (2010). *Globalization and manufacturing paradigms, the global manufacturing revolution: Product-process-business integration and reconfigurable systems*. Hoboken: Wiley.

Kossoff, L. L. (1994). *Closing the gap: The handbook for total quality implementation*. Knoxville: Spc Press.

Laqua, I. (2012). *Lean Administration: Das Ergebnis zählt*. Ludwigsburg: LOG_X Verlag GmbH.

Schneider, R., Schöllhammer, O., Meizer, F., Lingitz, L., Westkämpfer, E., & Sihn, W. (Eds.). (2011). *Lean office 2010 – Wie schlank sind Unternehmen in der Administration wirklich?* Stuttgart: Fraunhofer IPA, Fraunhofer Verlag.

Shingo, S. (1988). *Non-stock production: The Shingo system for continuous improvement*. Cambridge: Productivity Press.

Soderquist, K., & Chanaron, J. J. (1997). Managing innovation in French small and medium-sized enterprises: An empirical study. *Benchmarking for Quality Management & Technology, 4*(4), 259–272.
Wiegand, B., & Nutz, K. (2007). *Prozessorganisation*. München: Oldenbourg Wissenschaftsverlag.

Book Chapters

Mohannak, K. (2012). Organisational knowledge integration: Towards a conceptual framework. In L. J. Uden, F. Herrera, J. B. Perez, & J. M. C. Rodriguea (Eds.), *7th international conference on knowledge management in organizations: Service and cloud computing* (pp. 81–92). Salamanca: Springer Verlag.

Online Documents

MacKay, R. J. (1988). Continuous improvement with control charts. Publication, University of Waterloo. http://www.bisrg.uwaterloo.ca/archive/RR-88-04.pdf
Maras, D. (2009). *Corporate performance management, Wie effektiv ist Ihre Unternehmenssteuerung?*, Resource document. PriceWaterhouseCoopers. http://www.pwc.ch/user_content/editor/files/publ_adv/pwc_cpm.pdf
OECD. (2004). *The significance of knowledge management in the business sector, policy brief*. Paris: OECD. http://www.oecd.org/innovation/researchandknowledgemanagement/33641372.pdf
Stegner, T. (2010). *Studie "Einsatz von Management-Tools in der Unternehmenssteuerung"*, Resource document. Michel-Institut. http://www.mi-gmbh.de/pdf/auswertung_studie2010.pdf

Greening the Bottom Line

Marc R. Pacheco

1 Introduction

The issue of climate change is quite obviously an environmental concern. However, far too often, we ignore the myriad other burdens associated with global warming. The grave problems caused by climate change are not solely environmental in nature; rather, they are predicaments of public health, national security, public safety, and economic development, to name only a few. Climate change and global warming touch upon practically every aspect of our society. Those who continue to ignore, deny, or downplay the critical developing issues caused by climate change place our nations and our world at risk. Those who embrace these charges understand that battling global warming will lead not only to an enhancement in our environment and quality of life, but to economic benefits and job growth as well. To put it simply, embracing the climate challenge allows us to avoid the worst effects of climate change and to green the bottom line.

2 My Philosophy

As Chair of the Massachusetts Senate Committee of Global Warming and Climate Change and Senate Chair of the Joint Committee Environment, Natural Resources, and Agriculture, I've often been asked my philosophy about environmental protection. Many of the questions center around the dilemma of environment versus the economy. Namely, does one focus on the environment at the expense of the financial state of our nations or does one advance an economic agenda to the detriment of our ecosystem? However, to me that is a false choice. As mentioned

M.R. Pacheco (✉)
Office of Senator Pacheco, State House: Room 312B, Boston, MA 02133, USA
e-mail: Kyle.Murray@masenate.gov

C. Weidinger et al. (eds.), *Sustainable Entrepreneurship*, CSR, Sustainability, Ethics & Governance, DOI 10.1007/978-3-642-38753-1_15,
© Springer-Verlag Berlin Heidelberg 2014

earlier, I do not believe the two are mutually exclusive. With some effort, we can simultaneously grow our economy and protect our world. It requires some creative thinking and the political courage to stand up to some powerful interests, as well as a private sector that demonstrates leadership and vision. By falling victim to the black-or-white thinking that we must choose either the economy or the environment, we significantly hamper our ability to progress in both.

Unfortunately, within the market, special interests are often resistant, and in many cases openly defiant, to change. Therefore, to truly achieve any real progress for sustainability and to embrace the challenges of climate change, our governments must set the framework within which the private sector can thrive and be sustainable. As it has become clear that the world is rapidly progressing on the path to embracing sustainability and green technology, it makes no sense for government incentives to lag behind. Ten years ago, no one would have dreamed China would lead the way in manufacturing green technologies, but it is currently doing just that. They are manufacturing clean technology at a truly impressive pace, and much of the developed world is purchasing green technology from them, at least in part. If China, which has actually begun to consider a carbon tax, can offer this much support for green manufacturing, then certainly other developed nations, including the United States, must do so as well.

Truthfully, the government really can only establish the incentives and the framework within which a free market can expand and develop sustainable growth. To truly achieve market expansion, we need entrepreneurs who have the vision and commitment to embrace a sustainable future and government leaders that are truly willing to embrace a private-public partnership. In doing so, it is important for the state to work to find the appropriate balance between incentives and regulation for the market. This can be achieved through a variety of methods, including grants and bonds for climate-friendly energy generation, financial disincentives for carbon-spewing fossil fuels, and other nonmonetary incentives. However, these are just a few of a vast landscape of creative ideas. The government cannot do it all on its own. Rather, the government must set the framework and work with sustainable entrepreneurs in the market to embrace sustainability.

3   The Expansion of United States Climate Legislation

We in the United States are aware of our somewhat less than sterling reputation among developed nations when it comes to providing national leadership in solving the climate crisis. Though that notoriety is apt for some of the less progressive parts of our nation, other states in the United States are doing some truly remarkable things. Massachusetts is currently leading the way in terms of energy efficiency, and we are at the forefront of clean technology, green jobs, and environmentally-friendly legislation. In 2007, I was able to get the support of then-Senate President Robert Travaglini to form the Senate Committee on Global Warming and Climate Change, the first such committee in the nation. In recent years, the state of Massachusetts has enacted a suite of legislation that has vaulted the Commonwealth into a position as one of the leading green states in the nation. In 2008, Senate and House environmental legislators were able to win the support of Senate President Therese Murray and most of the other leaders, and we passed the Oceans Management Act, the Green Communities Act, the Biofuels Act, the Global Warming Solutions Act, and the Green Jobs Act. Any one of these pieces of legislation alone would have been impressive. However, the fact that the Legislature passed all five in the same session is remarkable and demonstrates Massachusetts' continued dedication to the environment. The Biofuels Act has helped encourage the growth of an advanced biofuels industry, while the Oceans Management Act has required the state to develop a groundbreaking comprehensive plan to zone our oceans from the coast to the three-mile limit. The Green Jobs Act, as its name suggests, has helped to create environmental jobs in the state. Next, the Green Communities Act has helped provide cities and towns with means to improve energy efficiency and reduce energy consumption and has provided incentives to promote widespread renewable development in the state. Finally, the Global Warming Solutions Act, of which I was the chief sponsor, has aided in battling climate change by requiring the state to develop a comprehensive regulatory program to fight global warming and has provided the executive branch with the ability to utilize market-based measures to reduce carbon emissions and set aggressive, but obtainable goals for emission reduction. These aims were up to a 25 percent reduction of the 1990 emission numbers by 2020 and an 80 percent reduction by 2050 (Mass. Gen. Laws ch. 298, 2008).

The state of Massachusetts is not alone in our quest to pursue sustainability. For example, the majority of states in New England are involved in a Regional Greenhouse Gas Initiative, which puts a cap-and-trade system for CO_2 emissions from power plants into place. California, Oregon, and Washington have been involved in the West Coast Governors' Global Warming Initiative since 2004. This initiative utilizes a broad variety of techniques aimed at reducing global warming pollution. Even some states that one would not necessarily associate with progressive environmental actions, like Kansas, Arizona, Wisconsin, and New Mexico, are participating in greenhouse gas programs. Arizona and New Mexico are involved in the Southwest Climate Change Initiative, while Kansas and Wisconsin, along with other states such as Minnesota, Iowa, and Illinois, are participating in the Midwestern Greenhouse Gas Reduction Accord.

4 The Climate Commitment of United States Political Leaders

Since his reelection, President Obama has continued and strengthened his stated commitment that energy and climate change are a top priority on his agenda for his second term. (Broder 2013). He also created a Climate Change Adaptation Task Force in 2009, which has a goal of recommending methods that the federal government can use to fight climate change (Climate Change Adaptation Task Force n.d.). Additionally, the President has frequently pointed out the climate change issue in major speeches, most recently in his inaugural address and state of the union speech (Rabe and Borick 2013). Former Secretary of State Hillary Clinton, acting for the administration, used her platform to engage the climate debate. Clinton also has underscored the dire need for advancement in this area, and she has worked with the Environmental Protection Agency and other environmental organizations, both domestic and international, in order to realize this goal.

Further, several of the United States' leading political figures have used their leadership positions and various foundations to embrace and try to solve the climate crisis. For example, former President Bill Clinton's Clinton Foundation has started the Clinton Climate Change Initiative. The goal of the Clinton Climate Change Initiative is to "implement[] programs that create and advance solutions to the root causes of climate change" (Clinton Climate Inititive n.d). The foundation has implemented such initiatives such as a large-scale solar program and a building retrofit program to help curb the ever-worsening climate situation. The work that former President Clinton is doing in this area and his engagement of the world community in his work are truly inspirational. Former Vice President Al Gore, one of the first political leaders in America, if not the world, to speak out on climate change when he was a young member of Congress, also has been heavily involved in grappling with the climate crisis. Though many know of Gore's climate work on the film *An Inconvenient Truth*, his initiative has had and continues to have a greater impact. Vice President Gore's Climate Reality Project has similar goals to the Clinton Climate Change Initiative. However, whereas the Clinton Initiative focuses on developing programs for solving the climate change issue, the Climate Reality Project is dedicated to "unleashing a global cultural movement demanding action on the climate crisis" (Climate Reality Project n.d.). As someone who has taken part in Gore's climate messenger training, I can tell you first-hand of the invaluable work that the former Vice President is doing. He has taken the fairly complicated issues associated with climate change and boiled them down to a message that can be easily understood. Finally, the Secretary of State, John Kerry, has long been a student and leader on this issue. Not only has he written a book with his wife Theresa Heinz Kerry about the subject, but he also sponsored an impressive climate bill in the Senate in 2009 (Florenz 2013). Secretary Kerry understands the potentially dire consequences of inaction, and also realizes that an investment in the green market has the potential to drive a struggling economy forward (Leber 2013). Therefore, though pockets of the United States remain defiant to the green revolution, many states and a number of notable, powerful legislators have affirmed the importance of investment in a better way for both economic and environmental reasons. On both accounts, the cost of inaction is simply too high.

5 The Effect of Sound Policy on Jobs

The marriage of sound policy and extensive outreach toward the market has allowed the clean technology field to flourish in recent years. Since 2008, Massachusetts has seen remarkable growth in green jobs. The clean energy field alone has experienced impressive expansion. From 2007 to 2010, employment in the clean energy sector rose by close to 65 percent (Executive Office of Energy and Environmental Affairs 2010). According to the Massachusetts Clean Energy Center's 2011 Industry Report, in 2011, there were an estimated 64,310 clean energy workers in the commonwealth (Massachusetts Clean Energy Center 2011). The 2012 report stated that by 2012, that figure had increased to around 71,523 employees, an increase of 11.2 percent (Massachusetts Clean Energy Center 2012). One noteworthy takeaway from the 2012 report is that expansion in the industry was expected to continue unabated. The 2012 report estimated that the employment in the clean energy sector would grow by another 12.4 percent to around 80,405 jobs in 2013 (Massachusetts Clean Energy Center 2012). These figures are especially notable when compared to the fact that employment in Massachusetts was expected to rise by only around 1.4 percent in the 12 months after the report was published (Massachusetts Clean Energy Center 2012).

However, the clean energy field in Massachusetts is but a small part of the overall green economy in the United States. According to a report by the Brookings Institution in 2011, the clean economy sector employed about 2.7 million United States workers (Brookings Institution 2011). The Brookings Institution report defined a clean economy as "economic activity-measured in terms of establishments and the jobs associated with them-that produces goods and services with an environmental benefit or adds value to such products using skills or technologies that are uniquely applied to those products" (Brookings 2011). The Bureau of Labor Statistics released its own figures in 2012, using a rather different definition. It defines green jobs as either "[j]obs in businesses that produce goods or provide services that benefit the environment or conserve natural resources" or "[j] obs in which workers' duties involve making their establishment's production processes more environmentally friendly or use fewer natural resources" (Green Jobs Overview 2013). Utilizing this description, the bureau found around 3.1 million green jobs, which represents about 2.4 percent of the total employment in the United States (Green Jobs Overview 2013). This number may seem somewhat small; in actuality it symbolizes a significant and ever-growing piece of the overall economy.

In fact, all of these figures could actually be seen as underestimations. The definition which the bureau utilized to generate the greater sum has several fairly significant weaknesses which could cause their comprehensive model to be low. The statistics used by the bureau do not include those areas where they did not expect to find many clean jobs (Pollack 2012). Additionally, the bureau's methods only include those businesses which provide green goods and services (Pollack 2012). It fails to mention "jobs involving processes that make a business greener"

(Pollack 2012). Finally, for this survey, the bureau applied information from the North American Industry Classification System. Unfortunately, the data regarding green jobs have not been updated since 2007 (Pollack 2012). Therefore, some of the fields which should have been defined as a green job may have fallen into other categories (Pollack 2012). For example, the installation of solar photovoltaics is covered by existing construction sectors (Pollack 2012). A figure that did not have these limitations would have the potential to raise the green employment measures to a significantly greater quantity.

Regardless of which interpretation one uses, the job numbers remain noteworthy. Though the employment statistics themselves are quite impressive, it is the trends that many of the numbers suggest that are truly important. For example, newer clean technology growth has been "explosive," and growth in the industry was actually stronger than the overall economy during the recession (Brookings Institution 2011). On average, annual employment growth was 0.034 percent higher for every 1 percent increase in the share of an industry's employment in green jobs in the United States (Pollack 2012). On that same note, states that had higher shares of green jobs tended to fare better during the recession (Pollack 2012). The green economy is also not solely for high-skilled or college-educated individuals. Rather, a strong investment in clean technology provides a number of opportunities for low- and middle-skilled workers. Interestingly, a green economy actually provides jobs at a higher rate for those levels of skilled worker than the general economy (Brookings Institution 2011). Along those same lines, a heavy percentage of the jobs in a clean economy are in the manufacturing and export realms. While manufacturing represents only around 9 percent of the United States general economy, it makes up around 26 percent of the clean economy (Brookings Institution 2011). Further, median wages in the clean economy are roughly 13 percent higher than the median job in the United States (Brookings Institution 2011). At a time when many low- and middle-skilled workers are still struggling to find work, investing in the green economy can stimulate a whole new clean economy. Many critics of green technology also simultaneously cite the need to spur American manufacturing. They do not seem to understand that an investment in a clean economy would do exactly that. With the correct policies in place to encourage investment in green technology, we can help get many of our low- and middle-skilled unemployed citizens who are desperate for a job back to work.

Though it is quite clear that green technology can spur job growth, a valid question remains: at what cost? More specifically, if the funds provided for clean technology were also applied to traditional fossil fuels, would the fossil fuel sector see greater job growth? The answer is resoundingly no. In September 2008, the University of Massachusetts at Amherst released a report which examined this very question. Their results demonstrated that an investment of $100 billion in a green economy could result in the creation of two million jobs (Political Economy Research Institute 2008). A similar investment in traditional fossil fuels would bring about only around 540,000 new jobs (Political Economy Research Institute 2008). While still a significant figure, it strikingly pales in comparison to the job-driving force that is the green economy.

Unfortunately, the answer to the job growth query only leads to another question: if green technology can stimulate job growth better than fossil fuels, why are the costs associated with a clean economy generally higher than the traditional fuel source? There are a number of factors which come into play here, including the fact that the fossil fuel industry has been established for more than 100 years while the green industry is relatively new, but one of the main reasons is that the oil and natural gas industry receives significantly more in subsidies than clean technologies. Though opponents of a clean economy and deniers of climate change will frequently deride subsidies that go to green technologies, they are strangely silent on those that go to oil and natural gas. The simple fact remains that the price for oil and natural gas remains lower due to enormous historical tax breaks and other subsidies. Between 2002 and 2008, the oil and natural gas industry received about $72 billion in subsidies and were actually trending upward during the period (Environmental Law Institute 2009). During that same time frame renewables received only around $29 billion (Environmental Law Institute 2009). Further, close to half of that $29 billion went to corn-based ethanol, a fuel source that some say is of questionable climate benefit (Environmental Law Institute 2009). Additionally, in its first 15 years of development, the oil and natural gas industry was subsidized at a rate five times that of renewable technologies during the same development period (DBL Investors 2011). I believe if renewables were subsidized at a level similar to oil and natural gas, they would have an opportunity to become truly competitive in an open market.

6   The Effect of Sound Policy on the Economy

It is very easy to just talk about the merits of a clean economy, but a legitimate question remains as to what impact green policies have had in driving revenue creation. Thankfully, and not all that surprisingly, the green market has proven to be an excellent force in generating economic growth. For example, the global market for solar and wind technology has increased from $6.5 billion in 2000 to over $131 billion in 2010 (Clean Edge 2011). This figure ignores the additional $56 billion for biofuels (Clean Edge 2011). Overall global revenue from biofuels, solar, and wind totaled around $248.7 billion in 2012 (Clean Edge 2013). While already a fairly remarkable sum, it is expected to rise to $426.1 billion by 2022 (Clean Edge 2013). In the United States, revenue from clean technologies continues to grow at a remarkable pace. Revenue from clean energy installations in the United States for the period of 2012–2018 is expected to be nearly $269 billion (The Pew Charitable Trusts 2012). This $269 billion represents 14.5 percent of the expected global total (The Pew Charitable Trusts 2012). During this period, revenue is also expected to grow at a compound annual rate of around 14 percent (The Pew Charitable Trusts 2012).

Progressive policies, like those utilized in Massachusetts, can have a significant return on investment. One of the main components of the Green Communities Act was a $1.1 billion investment in energy efficiency and renewable technologies

(Cater Communications 2012). This investment is expected to bring back close to $2.5 billion, all while significantly reducing the state's carbon footprint and increasing energy independence (Cater Communications 2012). Energy independence is an important virtue for Massachusetts, as according to 2010 numbers, the state spends around $28 billion on energy annually, but only about $6 billion in the Commonwealth (Executive Office of Energy and Environmental Affairs 2010). This desire for energy independence tied into the state's decision to participate in the RGGI program. Participation in the RGGI program has brought more than $1.6 billion to its ten participating states and allowed them to keep more than $765 million in the local economy due to decreased fossil fuel demand (The Analysis Group 2011).

Venture capitalists and angel investors have seen these promising trends and have been flocking to invest in green technology. In 2004, venture capital and private equity for clean technology totaled only around $0.6 billion (Bloomberg New Energy Finance 2013). However, since 2007, that figure has not dropped below $4.0 billion and actually topped out at $7.0 billion in 2008 (Bloomberg New Energy Finance 2013). Venture capitalism and private equity has totaled an investment of around $36 billion in the green markets since 2004 (Bloomberg New Energy Finance 2013). Though hurt by the recession like all industries, investment in a clean economy has remained strong. According to Clean Edge, a leading clean-technology research and consulting firm, venture capital investments for United States clean technology companies totaled around $5 billion in 2012 (Clean Edge 2013). Though this is down from the $6.6 billion invested in 2011, it still amounted to around one-fifth of the total venture capital investment for 2012 (Clean Edge 2013). Additionally, the declines in green venture capitalism were in line with the overall declines in venture capital investment during that same time period.

7 The People Who Make It Happen

As stated earlier, I believe that the government must set the rules, but it is the private sector that must drive the market forward. However, when one focuses on the big picture and the overall state of the green market, it can be too simple to overlook the individuals who help push such growth. It is when creative individuals cooperate and work toward a common goal that true innovation can shine and sound policy has the opportunity to be successful. Many of the major environmental groups, such as the Sierra Club, the Conservation Law Foundation, the Nature Conservancy, the Audubon Society, the Worldwatch Institute, and the Sustainable Future Campaign, to name just a few of the countless impactful organizations, have played major roles in developing policy. However, the process is dynamic when you combine those who have a keen business sense with those who also have a powerful passion for the environment. Environmental Entrepreneurs, or E2, is a collection of such like-minded individuals. E2 is not solely an environmental group. Rather, they are a business group comprised of individuals who are focused on environmental issues and with environmental credentials (Environmental Entrepreneurs 2012). The goal of E2 is to "create a platform for independent business leaders to promote environmentally sustainable economic growth" (Environmental Entrepreneurs n.d.). E2 is not a clean energy representation group, nor are they paid lobbyists. Rather, they are business leaders who desire a platform to "leverage their professional experience and networks to influence policy and shape the debate around environmental and sustainability issues" (Environmental Entrepreneurs n.d.). E2 represents the perfect melding of green and business that allows clean growth to proceed in an environmentally-safe and economically-beneficial manner. They are a national community of over 850 independent business leaders who have founded, funded, or developed more than 1,400 companies (Environmental Entrepreneurs n.d.). These 1,400 companies have resulted in more than 500,000 clean jobs (Environmental Entrepreneurs n.d.).

E2's influence can be seen in a number of rather high-profile environmental policy efforts. Their first major campaign with which they were involved was the so-called Pavley bill in California (Environmental Entrepreneurs 2012). This bill was groundbreaking, as it defined carbon dioxide as a pollutant (Environmental Entrepreneurs 2012). It was thanks in no small part to E2's advocacy that the Pavley bill was able to go through. E2 also was crucial to the suite of legislation that passed in 2008 in Massachusetts that was previously mentioned. They brought a sharp business view to the table to compliment the environmental aspects. Again, thanks to the combined efforts E2 and other business advocates, the legislature was able to pass the Massachusetts Oceans Act, the Global Warming Solutions Act, the Green Jobs Act, the Green Communities Act, and the Biofuels Act.

I recently spoke with David Miller, Executive Managing Director of the Clean Energy Venture Group, or CEVG, and one of the directors for E2's New England chapter. Miller was one of the co-founders of EPrime, a networking forum for clean energy entrepreneurs, and he also is a founding member of the MIT Enterprise Forum's Energy Special Interest Group. His company, Clean Energy Venture Group, is an angel investment group which focuses on innovative clean energy companies. Miller actually began his career in telecommunications software. However, a deep concern about the impacts of climate change led him into to the clean energy field. Clean Energy Venture Group has invested in 18 companies and are adding more each year. According to Miller, they did not originally start out planning to invest in that many companies. However, since 2008 there has been such an influx of higher-quality companies into the community that they did not wish to pass up all the opportunities.

One of the companies in which the Clean Energy Venture Group has invested, Next Step Living, is a perfect example of what can happen when sound policy and entrepreneurship work in harmony. Next Step Living is located in Massachusetts and is dedicated to residential energy efficiency. According to Miller, the partners at his company became aware of Next Step Living through some of their connections in the field. Generally, CEVG did not invest in this type of company. Instead, they focused more on high-growth technological inventions and intellectual property. However, because Miller felt the idea's time had come and that Next Step Living had an innovative business model, he and another partner in the group decided to invest. They took a position on the board, and within six months, the company was able to raise even more capital. This second round of investment helped catalyze the business and allowed them to evolve their strategy over time. At that time, venture capitalists were extremely hesitant to take on that sort of project. Without deep intellectual property or technology, Next Step Living had difficulty drawing in venture capital funds. Therefore, they had to do another round of investment, drawing upon family offices and other non-venture fund sources. The company continued to progress solidly from that point, until they finally got to a level of scale where they drew in a significant investment from VantagePoint Capital Partners, a venture capital group. VantagePoint, along with existing investors Black Coral Capital and Mass Green Energy Fund, recently contributed around $18.2 million to help advance Next Step Living (Next Step press release). Today the company has nearly 500 employees, representative of the company's rapid rise. According to Miller, without the solid policy framework in place, this sort of success story would not be possible (D. Miller, personal communication, March 14, 2013).

8 Conclusion

It is clear that through the marriage of sound policy and aggressive, Sustainable Entrepreneurship, everyone wins. Not only can this harmony improve the quality of our lives, but the economy as well. As has been pointed out by a number of prominent individuals, including former President John F. Kennedy, President Barack Obama, Al Gore and others, the Chinese word for "crisis" is represented by the characters for "danger" and "opportunity." By dealing with the climate crisis in a smart and sustainable manner, we have the ability to eliminate the danger posed by global warming and have the opportunity to truly green the bottom line.

Literature

Bloomberg New Energy Finance. (2013). *Sustainable energy in America 2013 factbook*. Retrieved from http://bnef.com/WhitePapers/download/266
Broder, M. (2013, March 8). *Energy and climate on the White House Agenda*. Retrieved from http://thecaucus.blogs.nytimes.com/2013/03/08/energy-and-climate-on-the-white-house-agenda/
Brookings Institution. (2011). *Sizing the clean economy*. Retrieved from http://www.brookings.edu/~/media/Series/resources/0713_clean_economy.pdf
Cater Communications. (2012). *The truth about Massachusetts' Green Communities Act*. Retrieved from http://www.clf.org/wp-content/uploads/2012/08/GCA-Myth-vs-Fact_02-20-12.pdf
Clean Edge. (2011). *Clean energy trends 2011*. Retrieved from http://www.cleanedge.com/sites/default/files/Trends2011.pdf

Clean Edge. (2013). *Clean energy trends 2013*. Retrieved from http://www.cleanedge.com/sites/default/files/CETrends2013_Final_Web.pdf

Climate Change Adaptation Task Force. (n.d.) In *White House, Council on Environmental Quality*. Retrieved from http://www.whitehouse.gov/administration/eop/ceq/initiatives/adaptation

Climate Reality Project. (n.d.) Retrieved from http://climaterealityproject.org/about-us/

Clinton Climate Initiative. (n.d.) In *Clinton climate initiative*. Retrieved from http://www.clintonfoundation.org/main/our-work/by-initiative/clinton-climate-initiative/about.html

DBL Investors. (2011). *What would Jefferson do?* Retrieved from http://www.dblinvestors.com/documents/What-Would-Jefferson-Do-Final-Version.pdf

Environmental Entrepreneurs (Producer). (2012). About E2 [Video file]. Retrieved from http://vimeo.com/47672612

Environmental Entrepreneurs. (n.d.). Retrieved from http://e2.org/jsp/controller?docId=14673

Environmental Law Institute. (2009). *Eliminating U.S. Government subsidies to energy sources: 2002–2008*. Retrieved from http://www.elistore.org/Data/products/d19_07.pdf

Executive Office of Energy and Environmental Affairs. (2010). *Massachusetts clean energy and climate plan for 2020*. Retrieved from http://www.mass.gov/eea/docs/eea/energy/2020-clean-energy-plan.pdf

Florenz, K. (2013, February 4). *John Kerry U.S. Secretary of State: Good Omen for Climate?* Retrieved from http://www.huffingtonpost.com/karlheinz-florenz/john-kerry-us-secretary-o_b_2615555.html

Green Jobs Overview. (2013). In *Bureau of labor statistics*. Retrieved from http://www.bls.gov/green/overview.htm#Overall

Leber, R. (2013, January 24). *Kerry pledges to confront climate change: 'I Will Be A Passionate Advocate' of action*. Retrieved from http://thinkprogress.org/climate/2013/01/24/1492061/kerry-pledges-to-confront-climate-change-advocate-of-action/?mobile=nc

Mass. Gen. Laws ch. 298.

Massachusetts Clean Energy Center. (2011). *Massachusetts clean energy industry report*. Retrieved from http://www.masscec.com/masscec/file/MassCEC%20Industry-Rept_DesignFinal%281%29.pdf

Massachusetts Clean Energy Center. (2012). *Massachusetts clean energy industry report*. Retrieved from http://www.masscec.com/index.cfm/pid/11151/cdid/13909

Political Economy Research Institute . (2008). *Green recovery: A program to create good jobs and start building a low-carbon economy*. Retrieved from http://www.peri.umass.edu/fileadmin/pdf/other_publication_types/peri_report.pdf

Pollack, E. (2012, October 10). *Counting up to green*. Retrieved from http://www.epi.org/publication/bp349-assessing-the-green-economy/

Rabe, B., & Borick, C. (2013, March 13). *The climate change rebound*. Retrieved from http://www.realclearpolitics.com/articles/2013/03/05/the_climate_change_rebound_117283.html

The Analysis Group. (2011). *The economic impact of the regional greenhouse gas initiative on Ten Northeast and Mid-Atlantic States*. Retrieved from http://www.analysisgroup.com/uploadedfiles/publishing/articles/economic_impact_rggi_report.pdf

The Pew Charitable Trusts. (2012). *Innovate, manufactures, compete*. Retrieved from http://www.pewenvironment.org/uploadedFiles/PEG/Newsroom/Press_Release/Innovate,%20Manufacture,%20Compete.pdf

Sustainability Reporting: A Challenge Worthwhile

Matthias S. Fifka

1 Introduction

Sustainability reporting (SR) is increasingly becoming a standard, especially among large companies, and the publication of a respective report can be considered a common business practice nowadays among multinational corporations. In a recent study, KPMG (2011) found that 95 % of the world's largest 250 corporations issued such a report. 10 years earlier, only 45 % had done so (KPMG 2002). Moreover, SR has not only become a standard, it has also gradually been standardized with regard to the content disclosed. In 2011, already 80 % of the world's largest 250 corporations applied the guidelines provided by the Global Reporting Initiative (GRI) for determining the contents of their reports.

Nevertheless, there is a significant regional gap, as many countries in the Americas, Africa, Asia, and Eastern Europe are falling behind Western European countries with regard to reporting. This gap, however, has been diminished recently, and emerging markets are catching up, as can be demonstrated by the examples of Hungary and Mexico. While in 2008 only 26 of the 100 largest companies in Hungary had issued a sustainability report, the number went up to 70 in 2011. A similar change could be observed for Mexico, where the number rose from 17 to 66 (KPMG 2011).

In addition to the regional gap, there is also a "size gap". Numerous studies on various geographic regions have shown that small and medium-sized enterprises (SME) are still lacking significantly behind with regard to reporting in comparison to larger companies (e.g., Brammer and Pavelin 2008; Da Silva Monteiro and Aibar-Guzmán 2010; Morhardt 2010; Stanny and Ely 2008; for an overview see Fifka 2011b). Especially SME often regard SR as a costly and complicated process – an

M.S. Fifka (✉)
Dr. Juergen Meyer Endowed Chair for International Business Ethics and Sustainability, Cologne Business School, Hardefuststr. 1, Köln 50677, Germany

University of Dallas, 1845 E. Northgate Drive, Irving, Texas 75062, USA
e-mail: m.fifka@cbs-edu.de

C. Weidinger et al. (eds.), *Sustainable Entrepreneurship*, CSR, Sustainability, Ethics & Governance, DOI 10.1007/978-3-642-38753-1_16, © Springer-Verlag Berlin Heidelberg 2014

impression that is not unjustified – and do not see sufficient advantages to compensate for the perceived burden of reporting. As a consequence, most SME and their associations are heavily opposed against legislation making SR mandatory.

Against this background, the purpose of the following chapter is manifold. First of all, the development of SR shall shortly be examined, as it has undergone a significant change over time. Based on that, a definition that matches the current understanding of what SR consists of will be provided. Then difficulties and benefits of SR – for the individual company as well as for society – will be discussed. Based on these findings, recommendations for companies on SR will be provided, ere a conclusion will be drawn.

2 The Development and Status Quo of Sustainability Reporting

2.1 The Historical Evolvement of Sustainability Reporting

Non-financial reporting has its origins in the 1970s, when companies began to disclose information on social issues like employee treatment, equal opportunities, and benefits provided for workers, aside from the regular publication of financial information in annual reports. Also issues like product quality and safety were addressed. This mostly reflected the concerns that companies were confronted with by the public. As Gray et al. (1990, p. 598) have observed, companies reacted to increasing calls "for the disclosure of information", and especially multinational corporations came under scrutiny because of their power "to control and move resources internationally". Companies from Western Europe were leading this development in reporting and began to publish a separate "social balance sheet", a "bilan social", or a "Sozialbilanz". As the names indicate, social issues stood at the forefront.

In the 1980s, more attention was gradually given to the environmental dimension of business activities and their impact on the natural environment. Nuclear disasters like in Harrisburg in the United States (1979) or Chernobyl in the Ukraine (1986), chemical disasters like in Bhopal, India, (1984), and oil spills, e.g. the Exxon Valdez accident in Alaska (1989), drove that development. Nevertheless, the focus of non-financial reporting still remained on the social dimensions throughout this decade.

This profoundly changed in the 1990s, when environmental reporting was shifted to the center of non-financial disclosure. As Gray (2002, p. 691) observed, the environment became the "talisman of worth", and environmental issues were given "the prime focus of attention" (Owen 2008, p. 243). The "talisman of worth" refers to the increasing realization of businesses that environmentally friendly practices and products could create significant comparative advantages and help to improve image and reputation (Welford and Gouldson 1993). Thus, the environmental report became the standard of the 1990s with regard to non-financial disclosure.

After the turn of the millennium, a "merger" of non-financial reporting occurred. Social and environmental information was combined in companies' reports. Later on, following the Triple-Bottom-Line approach developed by Elkington (1997), companies added financial information to provide information on "people, planet, and profit". Though only the most important financial information was given, while social and environmental issues took up considerably more space (cp. Fifka and Drabble 2012), the financial information was inevitably published twice – in the annual and the separate non-financial report. This led companies to increasingly integrate social and environmental information into the traditional annual report, a practice that has become known as "integrated reporting". While in 2008, only 4 % of the world's largest 250 companies had included social and environmental information into their annual report, already 27 % did so in 2011 (KPMG 2011). However, this information is often condensed, and thus many companies still publish a more extensive separate report.

Another development that has to be discussed here is assurance. Since its introduction in the 1970s, readers have questioned the validity and reliability of social and environmental information provided by companies, which – unlike financial data – was not certified by independent auditors. To address this credibility problem, companies have begun to hire auditors to assure their non-financial reporting. In 2011, 46 % of the world's largest 250 companies conducted assurance activities on their reports. More than 70 % of those which did so did a major auditing firm. Therefore, it comes as no surprise that large auditing firms have created departments specialized on the assurance of social and environmental data.

Closely connected to assurance is increasing standardization of the reports provided. Social, environmental, and financial information – even if provided correctly and assured by a third party – is mostly meaningless if it is not comparable to the performance of other companies. What, e.g., does it tell if a company states that it emits a certain amount of carbon dioxide per unit produced if this number cannot be compared to a competitor's performance? The competitor in turn might disclose the total amount of carbon dioxide emitted, and again a comparison and a meaningful interpretation of the numbers – at least to non-experts – would not be possible. Over the course of time, many reporting guidelines and standards have evolved – some industry-specific, some broad – but it can safely be said that the GRI has become the prevalent standard, at least for large corporations. Also following the Triple Bottom Line, it contains 7 economic, 17 environmental, and 31 social core indicators on which companies should provide information.

2.2   Coming to a Definition of Sustainability Reporting

In historic perspective, non-financial reporting has occurred under many names. Throughout the development process just described, the terminology has always followed the respective business practices. In the 1970s, non-financial reporting was mostly referred to as "social reporting", "social disclosure", or "social accounting". Despite the fact that "reporting/disclosing" is not necessarily the same as "accounting", because they describe different procedural steps (Yongvanich and Guthrie 2006), the terms are mostly used synonymously until today (Spence 2009).

Due to the increasing importance of environmental issues in the 1980s and 1990s, "environmental reporting" soon became a prominent term. The following combination of social and environmental information paved the way for "social and environmental reporting" or "social and environmental accounting". After the turn of the millennium, these terminologies were gradually replaced, and the era of "corporate (social) responsibility reporting", "corporate citizenship reporting", and "sustainability reporting" began. Although corporate (social) responsibility, corporate citizenship, and sustainability are not the same and describe different concepts with different backgrounds, there is hardly any differentiation between the three terms in the business world, especially when it comes to the title of non-financial reports (Fifka and Drabble 2012).

This terminological evolution is well reflected by the titles of the surveys carried out by KPMG on corporate reporting. The first three studies of the years 1993, 1997, and 1999 were published under the title "International Survey of Environmental Reporting". The 2002 study referred to "Sustainability Reporting", and the three most recent studies (KPMG 2005, 2008, 2011) were titled "International Survey of Corporate Responsibility Reporting".

As just pointed out, despite potential differentiations between the terms and underlying concepts, it can be stated that today to most people "corporate (social) responsibility reporting", "corporate citizenship reporting", and "sustainability reporting" mean the same thing. However, even if there is agreement on the synonymous character of the terms, still substantial disagreement exists on what "sustainability reporting", the term chosen for this chapter and also by the GRI as the prevalent reporting standard, does actually contain. There is dispute on whether SR only refers to voluntary disclosure or if it also contains information provided because of legal requirements (Gray et al. 1997; Kolk 2008). On top of that, there is the controversial discussion on whether SR *should* contain mandatory elements, as it already is the case in some European countries, e.g., France and Denmark.

Thus, it can be said that SR has remained an ambiguous term with many facets. In order to address this heterogeneous character, a broad definition of SR is chosen here: SR is the voluntary or mandatory practice of measuring and publicly disclosing information on the economic, social, and environmental performance of a firm.

3   Difficulties and Benefits of Sustainability Reporting

As pointed out in the introduction, meaningful SR is connected to many hurdles, but at the same time it holds significant advantages in store, for companies and society alike. The nature of these difficulties will be discussed in the next subchapter, ere benefits of SR will be presented.

3.1 Difficulties

The provision of information on a company's social, environmental, and economic performance is often subject to psychological, technical, and financial barriers.

Psychological barriers mostly consist of a reluctance to provide in-depth information because of underlying fears that the public will only view reporting as a marketing initiative, that competitors might obtain important information, and that the company could become vulnerable to attacks by the public and the media if it discloses unfavorable information on itself (Dando and Swift 2003). All of these fears are not without substance. Indeed, scholars and the public alike cultivate a substantial mistrust when it comes to CSR, sustainability, and the related reporting activities. Often they are perceived as mere attempts to polish up one's reputation, while a sincere motivation to generate a social benefit is presumably lacking. Ulrich (2008), e.g., has remarked that CSR, sustainability, and related terms are solely catchwords used by Public Relations strategists, who report on them in order to somehow convey what the respective company does to contribute to society's well-being. Especially, in continental Western Europe, such skepticism is driven by a "latently critical attitude towards business in society" (Backhaus-Maul 2008, p. 492).

Against this background, companies are especially reluctant to publish unfavorable information with regard to their social and environmental performance. Inevitably, they are caught in a dilemma here that is not to be underestimated. When only publishing positive information, the readers might very likely reach the conclusion that the information provided is biased, since it is unlikely that the performance has exclusively been positive. The publication of unfavorable information might help to defy the impression of biased reporting, but it carries the danger that the readers' attention will focus on the negative aspects, which then will come to dominate the overall impression. Moreover, such information also provides a basis for attacks, especially by the media and antagonistic non-governmental organizations (NGOs).

Finally, companies, especially if they disclose information according to far-reaching standards such as the GRI, fear that competitors might get an insight into important information connected to their products or production methods. Standards-based reporting does indeed require the disclosure of in-depth information, e.g., with regard to emissions, resources used, work accidents, or trainings provided. However, if such information is provided by competitors as well – due to the increasing use of standards, this development can already be observed – than this fear is marginalized, as there is mutual disclosure.

Far greater barriers to SR than skepticism are technical and financial hurdles. Meaningful SR that exceeds the provision of superficial information gathered for marketing purposes will inevitably be connected to a significant technical and financial effort. To measure a large number of social and environmental indicators requires technical expertise first of all. Many companies do not have this expertise because they lack the necessary engineering and cost accounting staff. In these cases, they will have to hire external specialists, which is a costly undertaking. Even if the expertise is existent in-house, there is substantial cost for the man-hours invested.

This assessment, however, is only the first step in SR. A second and usually costly step may consist of external assurance of the data provided through an independent auditor. As pointed out above, there is an increasing tendency to have such external verification conducted in order to demonstrate the credibility of the information provided to the potential readers, who might otherwise doubt its correctness. Considering the large number of data that is to be audited if reporting is done according to standards such as the GRI – there are 55 indicators, as mentioned previously – extensive work by the auditors is necessary, which requires significant financial resources.

The final step consists of actually disclosing the information, irrespective of a previous external audit. Such disclosure can occur internally and externally, whereas the latter is usually more costly since the information has to be circulated widely. The media for doing so are manifold: printed reports, online-reports (mostly in pdf-format), the company homepage, press releases, or newsletters. The provision of a stand-alone report as an electronic file on the company homepage has become a certain standard. However, many companies will also provide a printed version upon request, though this is connected to substantially higher costs for production. Most companies will also issue press releases on respective activities as they can reach a wider audience for relatively low cost (Fifka 2011a). Newsletters, either electronic or in print, are a more difficult medium since potential addressees for these mailings have to be identified previously.

The technical and financial effort that is required by meaningful SR inevitably leads to the question what the benefits of SR are – for a company as well as for society.

3.2 Benefits of Sustainability Reporting

SR is subject to a dilemma: while the costs – be it for the collection, auditing, or publication of the information – are measurable, the potential benefits can hardly be determined in financial numbers. Thus, a quantitative cost-benefit-analysis is not possible or very difficult at best. Nevertheless, there are significant benefits that result from SR.

3.2.1 Benefits of Reporting for Companies

The most considered benefit of SR is an improvement of reputation and image. SR essentially demonstrates that a company is willing to provide information – usually on a voluntary basis – not only on its financial, but also on its social and environmental performance (Hooghiemstra 2000). This willingness indicates that the company seeks transparency on its operations and products, and has nothing to hide.

Such transparency creates significant goodwill. First of all, it is increasingly becoming a requirement for receiving the so-called "license to operate", which is not an administrative license, but a social license consisting of the company's acceptance by its stakeholders (Schaltegger and Burritt 2010). Consumers, clients, employees, civil society as a whole, and NGOs in specific as well as the government are expecting companies to give an insight into their way of doing business and the economic, social, and environmental impacts resulting from it. Though it must be said, as described above, that SR is at times received with criticism because it is assumed to be a mere marketing effort, it can safely be stated that not disclosing information at all will be perceived with even more skepticism. It will be judged as either ignorance or the attempt to hold back unfavorable information.

The license to operate and a favorable reputation are not only helpful with regard to marketing and sales. They are also vital for attracting and retaining qualified employees and for maintaining their work satisfaction. Especially in a world with a toughening "war for talent", this benefit is not to be underestimated. Moreover, the voluntary disclosure of information can prevent tighter governmental legislation, as political decision makers might not deem it necessary to make reporting mandatory if companies come out with the respective information on their own initiative.

Finally, SR allows companies to take a closer look at their operations and products. By doing so, potential risks and possibilities to reduce costs can be identified. A company, e.g., might become aware of the fact that it is heavily dependent on fossil fuels or on materials obtained from countries subject to political instability, which poses a threat to the company's supply chain and its ability to operate. This might be the incentive to search for replacements, which can be more environmentally friendly or obtained without problems in the long run. Another prime example is more efficient operations. SR requires companies to examine their waste production, their freshwater withdrawal, and their emissions, e.g. In an effort to reduce these, the attempt to use less materials and to design production methods more efficiently will be undertaken, which can lead to lower costs (Aras and Cowther 2009; Schaltegger and Burritt 2006). Moreover, the introduction of more environmentally friendly production methods and products can contribute to the health of employees and consumers, which can also be considered as a benefit for society.

3.2.2 Benefits of Reporting for Society

The essential advantage of SR for society is access to information that it would most likely not obtain otherwise (Reynolds and Yuthas 2008). Certainly, the risk exists that this information is not fully correct or at least has been selected in a way favorable for business. However, it is reduced substantially through external verification and the application of standards. Moreover, companies run a high risk if they provide false or strongly biased information, because they must always consider the possibility that NGOs or the media through investigative practices will check on information that seems unrealistic or suspicious. In short, credibility is usually not the problem.

However, what is still not uncommon with regard to SR is the practice of only providing qualitative information, especially when no standards are used. Such information usually is more of a "narrative" of some selected activities and does not allow for any measurement. Despite the lack of "hard" facts, even such qualitative reporting might be a first careful step towards a more profound disclosure, especially when done by SME that lack resources or knowledge to provide more meaningful data.

In any case, society gets an insight on how a business affects its environment. Especially when quantifiable information is disclosed by a business and its competitors, SR enables stakeholders to compare their economic, social, and ecological performance, and to react. The companies that perform better might attract more customers and potential employees. It is these mechanisms that put pressure on business to disclose and to improve, because after all it is a competitive situation. Seeking improvement – be it financial, social, or ecological – will in turn be beneficial for a company's stakeholders.

One might very well argue that a consumer does not consider a sustainability report before purchasing a product and a job seeker does not use it for deciding on where to apply. Nevertheless, simply the possibility to do so must be considered an advantage for society, because there is a broader base of information which can be used in decision-making processes of all kinds. Moreover, investors and shareholders are increasingly demanding information on social and environmental performance, because they have come to the awareness that businesses which neglect these factors are endangering their financial performance because of reputational risks.

Furthermore, it is often argued that NGOs and journalists are the only readers of sustainability reports, aside from investors, and indeed we have little information on who actually reads reports (Spence 2009). However, even if activists and journalists were the only readers, SR would be beneficial for society as a whole, because they act as catalysts and distillers that make the public aware of crucial information to which they got access through SR.

Finally, it is exactly this information that gives stakeholders the possibility to hold companies accountable for what they do. Accountability in this context means that companies will have to stand in and assume the responsibility for the economic, social, and environmental impact of their operations. Such accountability cannot exist without the proper accounting and reporting practices. From an ethics and governance point of view, such accountability is justified, because companies profit from society in numerous ways – e.g., they make use of public infrastructure, and rely on qualified employees provided through the educational system – and, thus, they should be accountable to society in return.

Overall, there is considerable business and social pressure for companies to engage in SR, which leads to the question how successful reporting should be conducted.

4 Recommendations for Sustainability Reporting

There are numerous factors that companies should consider when they provide information on their sustainability efforts.

First of all, companies need to overcome a reluctance towards reporting because they expect that it will only be perceived as marketing. The potential damage from non-reporting is far greater than being confronted with accusations of public relation efforts. Furthermore, most companies already undertake some social or environmental initiative, and thus they should not hesitate to report on it.

It should be borne in mind, however, that simply talking about charitable activities, usually enriched with some nice pictures showing happy people, will not be sufficient in the long run. Though it might be a start, it is exactly this "glossy" style of reporting that creates the impression of being a pure marketing initiative. The sustainability communication strategy should be tightened to the core business. Reporting on issues that cannot be connected to the actual business by the audience will hardly be credibly. Even large corporations with significant reporting experience do not always adhere to this premise. There is the famous example of a multinational oil corporation that reported extensively on donating bicycles for Africa, which inevitably seemed more ironic than coherent with the company's business.

As this example illustrates, designing the appropriate social and environmental policies and programs is essential in the first place. When the sustainability strategy as such is flawed because it is not comprehensive or not bound to the core business activity, then it will be very hard to undertake meaningful SR. What companies should avoid in any case is to provide altered or incorrect information, since the danger of such manipulation being uncovered is always present, due to the work of journalists and activists. The resulting loss in credibility and reputation can be significant.

Concerning the question on what should be reported, an exchange with the stakeholders can be helpful. Companies should identify through round tables, questionnaires, surveys, or conferences what their stakeholders would like to see reported (Azzone et al. 1997). While it may be impossible to take all of their interests or desires into account, it can at least be made sure that no essential issues are neglected.

As the points just discussed show, the provision of accurate quantifiable data is inevitable sooner or later, not only because it will be desired by stakeholders, but also because it demonstrates a company's commitment to report and to undertake considerable effort (Perrini and Tencati 2006). Such a commitment is also shown by reporting on a regular basis. Providing some information unregularly at will is not convincing, because it creates the impression that the company is only disclosing information when it is convenient. When quantifiable data is provided, clear goals should be articulated, e.g., it should be stated until when a certain amount of waste is to be reduced. Numbers alone do not mean much if they do not serve to measure progress. This is in the interest of business and the addressees of reporting alike. Consequently, a company should not hesitate to state when it has not reached a certain goal or refrain from disclosing unfavorable information. If no failures or drawbacks are reported, then the positive aspects lose credibility as well.

Concerning qualitative and quantitative information, the right "mixture" of the two is important. Though quantitative data is essential and also inevitable, when a reporting standard is applied, companies should not forget that some readers will not seek to go through long columns of numbers, especially the ones who do not have the necessary expertise to comprehend their meaning. A report thus should also articulate in a clear-cut and understandable manner what the company does and seeks to do in terms of sustainability.

Applying a standard can be favorable in many ways, especially for larger enterprises that have the technical and financial means to do so. First of all, following a standard answers the initial question on what content should be reported at all, as it makes demands on what a report must contain. Moreover, using a standard counters the claim that companies would only be providing discretionary data, and it enables a comparison to competitors. In this context, not following a standard when competitors do might easily be discredited as the "easy way out", which will be detrimental to a company's reputation.

Though it is the nature of standards to reduce the flexibility of the ones who follow them, many standards – also the GRI as the most prominent one – are designed in a way that leaves companies with room to maneuver. It is not expected, e.g., that all of the 55 core indicators are measured and reported right away. It is also possible to file a sustainability report according to the GRI by disclosing fewer indicators. Overall, it is important to understand SR as a learning process that should be approached step by step. This is especially the case for SME. They should not attempt to undertake full-scale reporting right from the start, but implement it gradually.

5 Conclusion

SR provides significant advantages for companies and societies alike. Businesses can improve their reputation and market position with it, communicate their economic, social, and environmental efforts internally and externally, attract and retain employees, and get an insight into their operations to reach greater cost efficiency. Moreover, SR is increasingly becoming an important element in obtaining and maintaining the license to operate. From society's point of view, SR provides access to information for a variety of stakeholders on how a business impacts its economic, social, and ecological environment. This information would otherwise be very difficult or impossible to obtain.

This "social" benefit of reporting and the possibility to hold companies account-able for what they do have sparked a widespread political debate – primarily in Europe – on whether SR should be made mandatory. Reporting required by the law has considerable advantages and disadvantages. The most obvious advantage is guaranteed access to information for stakeholders. Moreover, legal requirements would most likely determine a specific set of indicators that have to be disclosed. Thus, the playing field for companies would be made even and comparability of different reports would be ensured.

However, as it already is the problem with existing standards, the requirement to report on indicators determined by law might be difficult across different industries and even neglect questions of relevance. While obtaining detailed environmental information about the operations of an electricity company or a logistics provider will be highly relevant, e.g., it might be less so for a bank or an insurance company. Issues of data protection in turn will be more important for a provider of financial services than for an electricity company. The resulting possibilities create a dilemma: Regulatory determination of different reporting indicators for different industries and the following enforcement are a huge administrative burden on the one side. On the other, a compromise on a set of indicators that is applicable to all industries will inevitably mean that important industry-specific information might not be disclosed by the respective companies.

Another challenge for mandatory reporting is the consideration of different company sizes. As pointed out previously, SME often lack the expertise and the financial means to undertake significant SR. In order not to overwhelm those companies with unbearable legal requirements, a differentiation with regard to what must be reported has to be made. Again, such a differentiation is a tremendous administrative burden.

Overall, legally mandatory reporting is bound to many problems, especially with regard to the content that has to be disclosed. Therefore, voluntary reporting is preferable, and there is considerable initiative, as the increasing application of the GRI and widespread reporting among large corporations show. However, it has to be attested that such initiative is still mostly limited to large MNC that are constant subject to public scrutiny. As Fifka (2011a) has demonstrated, only 44 % of the 100 largest German companies provide a sustainability report, and reporting clearly decreases with company size. Moreover, some industries – such as banking and insurance – also fall behind (Fifka 2011a). Though it would not be justified to generally "accuse" smaller corporations and SME as well as financial service providers of not reporting, as some have made considerable effort, across the board many of the respective companies significantly lack behind and even resist reporting.

These observations should have the following implications for politics. Business should be given a precisely determined transition phase to organize and carry out meaningful SR, based on internationally recognized standards. The standards, e.g., could be introduced by industry associations, because they can take industry or size-specific factors into account. If business does not act accordingly and the necessary steps are not taken, then the introduction of SR required by the law is inevitable, as societies have a right to be informed about how business impacts their economic, social, and ecological environment.

Pending mandatory reporting should be even more of an incentive for companies to introduce SR for two reasons. Firstly, widespread and meaningful reporting might prevent legislation, which will give business itself more possibilities to decide on content to report and standards. Secondly, assuming that mandatory reporting will be introduced, the companies that have engaged in reporting already will be prepared for legal requirements and enjoy a competitive advantage versus those which are unprepared.

Literature

Aras, G., & Cowther, D. (2009). Corporate sustainability reporting: A study in disingenuity? *Journal of Business Ethics, 87,* 279–288.
Azzone, G., Brophy, M., Noci, G., Welford, R., & Young, W. (1997). A stakeholder's view of environmental reporting. *Long Range Planning, 30*(5), 699–709.
Backhaus-Maul, H. (2008). USA. In A. Habisch, R. Schmidpeter, & M. Neureiter (Eds.), *Handbuch corporate citizenship – corporate social responsibility für manager.* Berlin/Heidelberg: Springer Verlag (485–492).
Brammer, S., & Pavelin, S. (2008). Factors influencing the quality of corporate environmental disclosure. *Business Strategy and the Environment, 17,* 120–136.
Chen, S., & Bouvain, P. (2009). Is corporate responsibility converging? A comparison of corporate responsibility reporting in the USA, UK, Australia, and Germany. *Journal of Business Ethics, 87,* 299–317.
Da Silva Monteiro, S. M., & Aibar-Guzmán, B. (2010). Determinants of environmental disclosure in the annual reports of large companies operating in Portugal. *Corporate Social Responsibility and Environmental Management, 17,* 185–204.
Dando, N., & Swift, T. (2003). Transparency and assurance: Minding the credibility gap. *Journal of Business Ethics, 44,* 195–200.
Elkington, J. (1997). *Cannibals with forks: The triple bottom line of 21st century business.* Oxford: Capstone.
Fifka, M. S. (2011a). *Corporate Citizenship in Deutschland und den USA – Gemeinsamkeiten und Unterschiede im gesellschaftlichen Engagement von Unternehmen und das Potential für einen transatlantischen Transfer.* Wiesbaden: Gabler.
Fifka, M. S. (2011b). Corporate responsibility reporting and its determinants in comparative perspective – a review of the empirical literature and a meta-analysis. *Business Strategy and the Environment.* doi:10.1002/bse.729.
Fifka, M. S., & Drabble, M. (2012). Focus and standardization of sustainability reporting – a comparative study of the United Kingdom and Finland. *Business Strategy and the Environment.* doi:10.1002/bse.1730.
Gray, R. (2002). The social accounting project and accounting, organizations and society: Privileging engagement, imaginings, new accountings and pragmatism over critique? *Accounting, Organizations, and Society, 27*(7), 687–708.
Gray, S. J., Radebaugh, L. H., & Roberts, C. B. (1990). International perceptions of cost constraints on voluntary information disclosures: A comparative study of U.K. and U.S. multinationals. *Journal of International Business Studies, 21*(4), 597–622.
Gray, R., Dey, C., Owen, D., Evans, R., & Zadek, S. (1997). Struggling with the praxis of social accounting: stakeholders, accountability, audits and procedures. *Accounting, Auditing & Accountability Journal, 10*(3), 325–364.
Hooghiemstra, R. (2000). Corporate communication and impression management – new perspectives why companies engage in corporate social reporting. *Journal of Business Ethics, 27,* 55–68.
Kolk, A. (2008). Sustainability, accountability and corporate governance: Exploring multinationals' reporting practices. *Business Strategy and the Environment, 17*(1), 1–15.

KPMG. (1993). *International survey of environmental reporting 1993*. n.p.

KPMG (1997). *International survey of environmental reporting 1996*. Stockholm: KPMG.

KPMG/WIMM (1999). *International survey of environmental reporting*. The Hague/Amsterdam: KPMG/WIMM.

KPMG (2002). *International survey of corporate sustainability reporting 2002*. De Meern: KPMG.

KPMG (2005). *International survey of corporate responsibility reporting 2005*. Amsterdam: KPMG.

KPMG (2008). *International survey of corporate responsibility reporting 2008*. Amsterdam: KPMG.

KPMG (2011). *International survey of corporate responsibility reporting 2011*. Amsterdam: KPMG.

Morhardt, J. E. (2010). Corporate social responsibility and sustainability reporting on the internet. *Business Strategy and the Environment, 19*, 436–452.

Owen, D. (2008). Chronicles of wasted time? A personal reflection on the current state of, and future prospects for, social and environmental accounting research. *Accounting, Auditing & Accountability Journal, 21*(2), 240–267.

Perrini, F., & Tencati, A. (2006). Sustainability and stakeholder management: The need for new corporate performance evaluation and reporting systems. *Business Strategy and the Environment, 15*, 296–308.

Reynolds, M. A., & Yuthas, K. (2008). Moral discourse and corporate social responsibility reporting. *Journal of Business Ethics, 87*, 279–288.

Schaltegger, S., & Burritt, R. (2006). Corporate sustainability accounting: A nightmare or a dream coming true? *Business Strategy and the Environment, 15*, 293–295.

Schaltegger, S., & Burritt, R. (2010). Sustainability accounting for companies: Catchphrase or decision support for business leaders? *Journal of World Business, 45*(4), 375–384.

Spence, C. (2009). Social and environmental reporting and the corporate ego. *Business Strategy and the Environment, 18*, 254–265.

Stanny, E., & Ely, K. (2008). Corporate environmental disclosures about the effects of climate change. *Corporate Social Responsibility and Environmental Management, 15*, 338–348.

Ulrich, P. (2008). Corporate citizenship oder: Das politische Moment guter Unternehmensführung in der Bür-gergesellschaft. In H. Backhaus-Maul, C. Biederman, S. Nährlich, & J. Polterauer (Eds.), *Corporate citizenship in Deutschland – Bilanz und Perspektiven* (pp. 94–100). Wiesbaden: VS Verlag.

Welford, R., & Gouldson, A. (1993). *Environmental management and business strategy*. London: Pitman.

Yongvanich, K., & Guthrie, J. (2006). An extended performance reporting framework for social and environmental accounting. *Business Strategy and the Environment, 15*, 309–321.

Part IV
Statements

Plant-for-the-Planet: A Worldwide Children's and Youth Movement

Felix Finkbeiner

1 Climate Justice and Climate Neutrality: The Central Demands of a Worldwide Children's and Youth Movement

We children see a fundamental problem: we will not be able to hold adults liable for the problems they have caused and not resolved, because they will be dead by the time we have to deal with them. If that weren't so, part of the adult population would behave differently. Imagine if children could sue adults for unsustainable behaviour and businesses would have to set up provisions in their annual financial statements for these litigation risks! Studies carried out by Bertelsmann (see Berliner Morgenpost 2012) and Shell (2012) show that three fourth of all children and teenagers in Germany view the climate crisis and global poverty as the two principal challenges of humanity.

Currently, there are legal proceedings instituted by teenagers underway in Washington DC, USA (The iMatter March 2012). These youngsters demand a guarantee from both the US federal government as well as from the states that they will adopt sufficient measures to reverse the climate damage, on the grounds that the atmosphere is common property that belongs to all citizens. On 1 March 2012, President Barack Obama declared before students of Nashua Community College in New Hampshire: "Let's put every single member of Congress on record:

The following Plant-for-the-Planet members and ambassadors for climate justice contributed towards this text (in alphabetical order): Alexandra, Alina, Analea, André David, Anna Be., Anna Br., Anna-Lena, Anna-Maria, Antonia B. Antonia Sch., Carolin Alexa, Cécilia, Clara Helene, Clara Madeleine, Davide Cosimo, Emilia, Felix R., Felix F., Fidan, Franziska, Jeanne, Jonathan, Jule, Julius, Kai, Laura, Lea, Lena Sch., Lena A., Lennart, Lykke Luzia, Maja, Manfred, Marc, Marie-Anna, Mia, Moritz, Niklas, Noah, Noam, Paula, Raphael, Rebecca, Rhoda, Rufat, Silja, Sophie Juleska, Svenja, Tamina, Timon and Timothe.

F. Finkbeiner (✉)
Lindemannstr. 13, Tutzing 82327, Germany
e-mail: felix@plant-for-the-planet.org

C. Weidinger et al. (eds.), *Sustainable Entrepreneurship*, CSR, Sustainability, Ethics & Governance, DOI 10.1007/978-3-642-38753-1_17,
© Springer-Verlag Berlin Heidelberg 2014

You can stand with oil companies, or you can stand with the American people. You can keep subsidising a fossil fuel that's been getting tax payer dollars for a century, or you can place your bets on a clean energy future"(CNN 2012).

The cry for climate justice can be heard from many children and teenagers the world over.

2 The Darkest Day: 12/11

For many adults 9/11 was the darkest day in living memory. This act of terror took the lives of 3,000 people and we're still fighting wars because of it. Every day 30,000 people die of starvation and we ask ourselves: who fights for them? For us children the darkest day was 12/11. For 17 years – longer than we've been born – adults have been negotiating about the climate with the express aim of concluding a follow-up agreement to replace the Kyoto Protocol, which will expire at the end of 2012. On 11 December 2011 they announced the result of their negotiations: there will be a new agreement in 2020. This means that in the years of 2013, 2014, 2015, 2016, 2017, 2018 and 2019 there will be no agreement, and everyone can emit as much greenhouse gas as they like. On 12/11 the adults broke their promise that the average temperature would not rise by more than 2 °C. But this goal is essential for survival, as scientists explain that a rise of 2.3 °C or 2.4 °C will exceed the threshold above which Greenland's ice will begin to melt away. If this 2–3 km thick ice sheet melts, the sea level will rise by up to 7 m. Forty percent of the world's population lives in coastal regions.

Some of us children already get involved at UN conferences because we have understood that we need binding worldwide agreements to solve global problems. At the 2010 Climate Change Conference in Cancún, we were impressed by the small island states who refused to support the 2 °C goal, as their islands would otherwise already have disappeared. They demand a maximum of 1.5 °C. Anote Tong, the president of Kiribati, explained to us children that he has concluded agreements with Australia and New Zealand regarding the immigration of 600 families annually, because he knows that the Kiribati islands will soon be under water. After Cancún, talk was of a 2/1.5 °C goal.

At our Plant-for-the-Planet Academies, children are taught a mnemonic for the relation between the CO_2 ton per head emission and temperature rise. In order to keep the temperature from exceeding 2 °C, every person on the planet may emit no more than 2 t of CO_2 per year; and only 1.5 t of CO_2 if we want it to rise no more than 1.5 °C. Today our emissions per person per year are 5 t. No one can say what a 5 °C rise in average temperature will mean in real terms, but we do know that back when our average temperature was only 5 °C lower than it is now, there was a 2 km thick ice cap above us.

On 7 December 2011, when the Canadian minister for environment addressed the plenary session of the Durban Climate Change Conference, six young Canadians stood up and turned around, showing the backs of their T-shirts printed with the message "Turn your back on Canada". All six of them were escorted out of the auditorium and banned from the conference. Less than a week later, on 13 December 2011, the Canadian environmental minister withdrew from the existing Kyoto Protocol for the following reason: instead of reducing its CO_2 emissions by 6 % compared to 1990, Canada had increased them by 35 % and would therefore have had to pay a fine of 11 billion euros. In order to avoid having to pay, Canada bowed out of the international agreement.

3   Lessons Learned from 12/11

That's how easy it is. The future of the world's children isn't worth 11 billion euros. Much more money has been and is being shelled out on saving individual car companies, banks or countries. If you consider that we, the children, will be the ones that have to pay back these debts one day, then we are even more dumbfounded. For at least 40 years, ever since the Club of Rome warned of the limits to growth, nobody has been able to claim ignorance when asked "what have you done?" But why is so little being done? Is it because of a differing perception of the future? Or can a simple experiment with a monkey explain this highly complex situation? If you let a monkey choose between a banana now and six bananas later, it will always take the banana now. If a lot of adults think like monkeys, then we children have a big problem.

4   Sustainability

For us, sustainability isn't just a hollow phrase for financial statements and political speeches. Sustainability is the only concept for our survival. Companies don't need special departments for sustainability, but should rather make sustainability their corporate goal. And quickly, or else we children have no future. Adults should learn from the foresters who coined the term 300 years ago. Everything they reap is thanks to their ancestors' efforts, and all the work they do throughout their lives is done for subsequent generations. Some companies are proud of their profits. But is it an accomplishment to make profits at the expense of us children – like chopping down trees without reforestation? Chief Shaw, the chief of a Native American tribe, told us about their Council of Elders. This council examines every major decision as to whether it will still bring benefits seven generations later.

If we had such a 'sustainability council', then there would be neither nuclear power, nor the burning of fossil fuels, we wouldn't have so many financial instruments that no one understands anyway, and we wouldn't have people who speculate with food while others are starving. No one has yet managed to get us children to understand why we need speculators.

Over 2 years we carried out a number of consultations (UNEP/UNEF 2008-2011) with several thousand children and teenagers from more than 100 countries and we summed up the result in four words: Stop Talking. Start Planting. We also formulated a three-point plan to save our future (Rio + 20 – United Nations Conference on Sustainable Development 2012):

(a) Leave the fossil fuels in the ground – climate neutrality by 2050

Today we are taking as much carbon – in form of crude oil, natural gas and coal – from the earth in 1 day as the sun stored there in a million days. This CO_2 caused by our energy production is one of the primary reasons for global warming.

We children call on all of the world's leaders, politicians, especially national governments, provincial governments, mayors, corporate executives and all people with influence in society to do everything in their power to bring about 100 % climate neutrality immediately, at the latest by 2050 worldwide.

In order to send out a small signal, we put our own product on the market in January 2012 – a product like we wish all products of the world were: namely, both fair and climate neutral. We started with our favourite product and called it the 'Change Chocolate' or the 'Good Chocolate'. The cocoa farmers get enough money to be able to plant precious wood species among their cocoa trees, thereby increasing their income from 4,000 US dollars to 20,000 US dollars. The farmers' children are able to go to school and do not have to harvest our cocoa beans. In 1 year we sold more than a million chocolate bars in Austria and Germany alone (The Good Chocolate 2012).

(b) Fight poverty with climate justice

In order to limit further global warming to the pledged 1.5–2 °C, we may only emit another 600 billion tons of CO_2 until 2050 (WBGU 2012). If we emit more than that, the temperature will rise above the 2 °C mark. If we divide these 600 billion tons by 40 years, then we are left with 15 billion tons of CO_2 per year for all of us together. The question is: how do we share out these 15 billion tons of CO_2 among the world's entire population? 60 % for the USA and Europe, like it is today? As far as we children are concerned, there is only one solution: everyone gets the same share, namely 1.5 t of CO_2 per person per year, assuming a world population of 9–10 billion by 2050.

And what happens to those who use more or want to use more? It's simple: whoever wants more must pay. If a European wants to continue gushing out 10 t of CO_2, he can, but he has to buy the right to do so from other people, e.g. in Africa, who only emit about 0.5 t. In this way, the principle of climate neutrality also makes poverty a thing of the past. This money can be used by the Africans to pay for clothing, education, medical care and technology. They also don't need to copy our silly behaviour by using coal, crude oil and other fossil fuels for energy, but can rather produce their energy with the help of the sun and other renewable energy sources.

(c) Let's plant 1,000 billion trees by 2020

The best news for humankind: there is a 'machine' that can break down CO_2, turn it into oxygen, store the carbon and, on top of that, process it into delicious sugar. A single 'machine' of this kind is called 'tree' and a whole factory 'forest'.

We children appeal to each and every person to plant an average of 150 trees by 2020 as a first step. If everyone cooperates, that makes 1,000 billion new trees in total. Planting and taking care of trees is child's play. In the past 6 years, adults and children together have already planted 12.6 billion trees. In the next 8 years, we need a lot more citizens, governments and companies to plant the remaining 987.4 billion trees with us. There are enough accessible regions around the world where it is possible to plant trees without interfering with agricultural or residential areas, and without having to plant in arid areas.

These new trees will bind 10 billion tons of CO_2 each year, i.e. about a quarter of all human-induced CO_2 emissions. In this way, we will firstly buy ourselves time to make the transition to a sustainable, fully CO_2-free lifestyle, and moreover, in a few years we can fell these 1,000 billion trees, bind the carbon (C) in furniture, houses, bridges, etc. for many years to come or process them into organic charcoal, which would in turn enrich our soil with carbon. Of course we would reforest these 1,000 billion trees and repeat this process again and again. Like a sponge, we actively absorb part of the CO_2 from the air and store it intelligently and sustainably.

5 What Would We Children Like to See in the Coming Years?

We children from Plant-for-the-Planet are linked with thousands of other children's and youth organisations around the world. We meet up with them at physical conferences, but as digital natives we mainly exchange views in cyber space and network in that way. Every day sees an increase in the number of young people throughout the world who understand that the complacency and ignorance of adults is threatening to destroy their future. We know, of course, that it will cost a lot of money to transform an economic and financial system into a sustainable global economic system. But we believe that this money is an excellent investment. Severn Suzuki, a 12-year-old girl from Canada, spoke about the environment and development at the UN conference in Rio de Janeiro in 1992. Severn Suzuki (2012) we children considered repeating her exact words during Rio + 20 in the summer of 2012 to show that every one of her words is still applicable today. Unfortunately, humankind neglected to make use of these 20 precious years to shift towards sustainability.

But there's also a positive side: all of the videos of Severn's speech on the internet have had a total of about 20 million views in 20 years. Today, good videos get 20 million views in just 2 days.

Let us change the future together. Planting trees is not only important to bind CO_2 and slow down and stop global warming. It is also a first important, emotional and very symbolic step: everyone can plant a tree – old or young, rich or poor, sick or healthy, black or white – and almost everywhere on the planet at that. If we, as a global family, tackle the global challenges in a concerted effort, then we children and teenagers can again see a future for ourselves. Together and in solidarity we can overcome all challenges.

Two days before we presented our three-point plan before the UN general assembly, on 31 January 2011, we held a talk in front of 400 pupils at the United Nation International School in New York. At the end, a 10-year-old boy named Theo got up and said: "Felix, we can do it! The Egyptians are also doing it!" That was the seventh day of the revolution.

One year later 16 children of Plant-for-the-Planet met with Waleed Rashed, one of the revolutionary leaders in Egypt. Planting trees is how we express our fight for our future. We children know that a mosquito can't hurt a rhino, but we also know that a thousand mosquitoes can get a rhino to change direction.

Literature

Berliner Morgenpost – Berlin. Current news Berliner Morgenpost – Berlin. N.p., n.d. Web. http://www.morgenpost.de/ printarchiv/familie/article1150362/Felix-und- die-Sorge-um-die-Zukunft.html. Accessed 9 Apr 2012.
CNN. (2012). Obama calls for congress to vote on oil subsidies – CNN.com. CNN.com – Breaking News, U.S., world, weather, entertainment & video news. N.p., n.d. Web. http://www.cnn.com/2012/03/01/politics/obama-energy/index.html. Accessed 9 Apr 2012.
Rio + 20 – United Nations Conference on Sustainable Development. (2012). N.p., n.d. Web. http://www.uncsd2012.org/rio20/content/documents/63Plant-for_the_planet_sub-mission_rio20_20111028.pdf. Accessed 7 Apr 2012.
Severn Suzuki. (2012). *Link to Severn Suzuki's speech.* www.youtube.com/watch?v=uZsDliXzyAY. Accessed 7 Apr 2012.
Shell. (2012). *Climate change/Germany. Shell in Germany/Germany.* N.p., n.d. Web. http://www.shell.de/home/content/deu/aboutshell/our_commitment/shell_youth_study/2010/climate_change/. Accessed 9 Apr 2012.
The Good Chocolate. (2012). *Plant-for-the-Planet/Stop talking.* Start planting. N.p., n.d. Web. http://www.plant-for-the-planet.org/de/ node/414. Accessed 7 Apr 2012.
The iMatter March. (2012). *Kids vs global warming.* N.p., n.d. Web. http://www.imat-termarch.org/#!lawsuit. Accessed 9 April 2012.
UNEP & UNEF. (2008–2011). Four children's and youth consultations took place in Stavanger, Norway, in 2008, in Daejeon, South Korea, in 2009, and in Bad Blumau, Austria, and Nagoya, Japan, in 2010, mostly on behalf of UNEP. These consultations with several thousand children and young people from more than 105 countries were condensed into a three-point plan in an online consultation on behalf of the UN Forest Forum (UNFF) between November 2010 and January 2011 and presented at the 9th meeting of UNFF in New York on 2 February 2011.
WBGU. (2012). *SG 2009 Budgetansatz.* WBGU: Home. N.p., n.d. Web. http://www.wbgu.de/sondergutachten/sg-2009-budgetansatz/. Accessed 9 Apr 2012.

Doing Sustainable Business Through a Strong Set of Values

Walter Rothensteiner

1 Introduction

Thanks to their role as loan providers and savings managers, banks are in a position to contribute towards a sustainable and viable development. The independent Raiffeisen banks and their offices throughout Austria still adhere to the values and principles of their founding father Friedrich Wilhelm Raiffeisen. Without the consistent application of ethical values in their day-to-day business, the development of the Raiffeisen organisations would not have taken the form it did.

One of these social reformers was Friedrich Wilhelm Raiffeisen, the founder of Raiffeisen. The second half of the nineteenth century was characterised by unbridled economic liberalism. In his capacity as the mayor of several towns in the Westerwald in Germany, Friedrich Wilhelm Raiffeisen was directly confronted with people's hardships and tried to find long-term solutions. In alignment with his motto, 'What one cannot do alone, many can do together', he took significant steps towards stabilising the economy: the joint purchase of supplies such as seeds, the storage and sale of agricultural products. Austria's first Raiffeisenkasse opened in Mühldorf near Spitz an der Donau in Lower Austria in December 1886. Farmers, craftsmen, workers and tradesmen belonged to this first Austrian Raiffeisen cooperative.

The 513 independent Raiffeisen banks and their 1,682 offices throughout Austria still adhere to the values and principles of their founding father. If the term 'Sustainable Entrepreneur' had existed more than 125 years ago, Friedrich Wilhelm Raiffeisen would have been a textbook example. Since the very beginning, societal solidarity, self-help and sustainability have been the guiding principles for doing business at Raiffeisen.

W. Rothensteiner (✉)
Raiffeisen Zentralbank Österreich AG, Am Stadtpark 9, 1030 Vienna, Austria
e-mail: sustainabilitymanagement@rzb.at

C. Weidinger et al. (eds.), *Sustainable Entrepreneurship*, CSR, Sustainability, Ethics & Governance, DOI 10.1007/978-3-642-38753-1_18,
© Springer-Verlag Berlin Heidelberg 2014

2 The Challenges of Our Time

The problems of our times are no secret: the population growth, the shrinking middle class, the rapid urbanisation, the constant demand for economic growth, and the resulting increasing demand for food, water, land, energy and other resources. In the past decade, the per capita income in the emerging markets grew by 80 %, which in turn caused a rise in consumption. By 2030 another three million middle-class consumers are expected to push the demand even higher. If the projections by McKinsey (2011) are anything to go by, water consumption will increase by 60 % and energy consumption even by 80 %. This is why the development and implementation of solutions is an extremely pressing matter.

3 The Role of Banks

Thanks to their role as loan providers and savings managers, banks are in a position to contribute towards a sustainable and viable development. As an industry that has less to do with using external resources or raw materials, but where service is the focal point, it is all the more important to take sustainable aspects into consideration when performing these services. It is generally becoming more and more important for all kinds of businesses to assume social responsibility. Especially in areas where the government does not assume responsibility or only to an unsatisfactory degree, companies are called upon to create their own initiatives. This is why we also need entrepreneurs like Friedrich Wilhelm Raiffeisen, who prove that a sustainably managed financial circuit can create value for everyone involved and also give meaning. The central principle of Sustainable Entrepreneurship and viable business is therefore to follow up 'sustainable thinking' with appropriate action.

4 Forms of Evaluation

Defining ecological and societal standards and taking them into account is a prerequisite for companies to manage their sustainable development. One of the biggest challenges in this respect is one of content. Because there is neither a uniform, mandatory understanding of sustainability for companies, nor globally accepted performance indicators and assessment criteria for corporate sustainability performance, the selection of appropriate sustainability criteria is often problematic. Banks and financial services providers usually face the challenge of having to develop their own sustainable economic and performance-related criteria in order to define their commitment and make it measurable.

Most banks develop their catalogue of criteria together with an ethics commission and/or a sustainability rating agency. These provide content-related and scientific input regarding methodological and criteria-specific topics. The evaluation criteria are often based on internationally applicable conventions, protocols, guidelines and standards such as the UN Global Compact or the guidelines of the Global Reporting Initiative, with a focus on the Financial Service Sector Supplement and the ISO 26000 CSR guidelines.

The next milestones that the banking industry is aiming for are the creation of generally binding minimum standards for the sustainable further development of our core business and benchmarks in the banking sector.

5 Raiffeisen's Sustainable Path

Raiffeisen's success can be ascribed to its regional and societal integration, fairness, respect and long-term business relationships. The business model of Raiffeisen states that the focus is always on the individual. Without the consistent application of ethical values in their day-to-day business, the development of the Raiffeisen organisations would not have taken the form it did. The vision of sustainability, as developed by Raiffeisen together with its stakeholders, shows clearly which way the road will lead. In the medium term, Raiffeisen is to become one of the leading groups of companies in terms of sustainability and corporate responsibility. In order to reach this goal and make it visible, Raiffeisen Zentralbank Österreich AG (RZB) set up a sustainability management department at the beginning of 2012. This shows how sustainability management – i.e. orienting business activities towards their long-term economic, ecological and social compatibility – is being implemented and put into practice.

The sustainability strategy of the RZB Group is put into action in three areas: as a responsible banker, a fair partner and a committed citizen. This strategy is embedded in the values and principles of Raiffeisen.

We at Raiffeisen can achieve the greatest effect with a sustainable approach in our core business. This is why the term 'responsible banker' has particular significance. As a fair partner, we cultivate an active, transparent and open dialogue with all our stakeholders. As a committed citizen, we assume responsibility for society and the environment.

One recent example of sustainability in our core business was the Raiffeisen Klimaschutz-Initiative's development and introduction of the new Raiffeisen BioCard^{TM} – a world first – in an exclusive limited edition in 2012. This PRE-LOAD card is made of compostable biological polymer based on corn starch. As the packaging is made of cardboard, both of the main components are biodegradable.

Showing one's commitment to the principles of sustainability has undoubtedly become an economic factor. Violations of ecological and social standards can quickly lead to a damaged reputation or loss of image – at least in those areas where customer loyalty plays a large role. The role of banks as an integral part of a sustainable development cannot be denied: on the one hand, because their core activities and their possibilities for action allow them to finance sustainable solutions and business models; on the other hand, because their own sustainable business behaviour makes them co-creators of sustainable systems. Raiffeisen is aware of this responsibility and bases its conduct on the motto 'We create sustainable value'.

Literature

Online Document

Dobbs, R., Oppenheim, J., Thompson, F., Brinkmann, M., & Zornes, M. (2011). McKinsey Global Institute, McKinsey sustainability, & resource productivity practice resource revolution: Meeting the world's energy, material, food and water needs. http://www.mckinsey.com/features/resource_revolution

Sustainable Entrepreneurship: Europe Should Market Its Expertise Better

Markus J. Beyrer

The European economy, as well as the global economy as a whole, faces stark challenges. The questions we are trying to answer since the beginning of the economic crisis are difficult ones. But along with the hardships states, companies and citizens suffer in bad times there come new opportunities.

Sustainable entrepreneurship is one of the greatest opportunities for the European economy if it wants to remain in the race of global competition. It contributes to the efforts Europe makes in order to grow out of the crisis; to return to the path of sustainable growth, to create new jobs. By increasing social as well as business value sustainable entrepreneurship cuts both ways: it contributes to the development of business solutions and answers the most urgent social and ecological challenges.

European companies are already world leaders in sustainable technology and production, which has positive economic, social and environmental impacts for society as a whole. Over 43 % of developed countries' government R&D into energy and environment takes place in the EU. This leading position in sustainable technology and production is also a core element of Europe's competitiveness in the global market.

European companies apply sustainable and cutting edge technologies in advanced recovery for oil and gas, deep sea drilling and extraction, alternative fuels development, energy efficiency measures, environmentally sustainable mining techniques or recycling of minerals. Our manufacturing industries use energy-efficient and more environmentally friendly transport or power equipment; have products and processes for more efficient urban development. Europe's chemicals, metals and construction materials industries are at the forefront of innovation for new industrial applications to create cleaner and more efficient products. The European food and drink sector takes social and environmental considerations really seriously. European service providers run information

M.J. Beyrer (✉)
BUSINESSEUROPE, 168 avenue de Cortenbergh, Brussels 1000, Belgium
e-mail: m.riegelnegg@businesseurope.eu

C. Weidinger et al. (eds.), *Sustainable Entrepreneurship*, CSR, Sustainability,
Ethics & Governance, DOI 10.1007/978-3-642-38753-1_19,
© Springer-Verlag Berlin Heidelberg 2014

systems which improve energy efficiency, have new financing models for sustainable investments and put cleaner energy services and more sustainable transport services in place.

The environmental performance of companies is inextricably linked to their competitiveness. Business benefits can be drawn from rationalising use of resources, reducing production costs and increasing energy efficiency. Environmental elements are now an integral part of corporate social responsibility, as companies understand the need but also the benefits of conducting business in an environmentally sustainable way.

All these – and numerous other examples – illustrate that European entrepreneurs, large, medium-sized and small companies deliver really successfully when it comes to the overall social and environmental element of their corporate strategies. European employers and entrepreneurs were at the forefront of engaging in corporate social responsibility and sustainability long before it became part of EU policy. Many companies also work with stakeholders to address major challenges, for example energy efficiency, supply-chain management or human rights. Most of them no longer need a wake-up call any longer and are ready to look at the broader dimensions of sustainable entrepreneurship.

Europe should turn this knowledge and experience to its advantage as there are huge opportunities in marketing it through closer cooperation with the EU's trade partners. Europe should incorporate the Sustainable Entrepreneurship approach into its international cooperation with non-EU countries. This could be the basis for new partnerships with emerging markets that have higher economic growth rates than Europe but face significant sustainability challenges over the medium to long term.

European companies – represented by BUSINESSEUROPE through its 41 member federations from 35 countries – are among the world's leading foreign direct investors. These investments are seen as part of a long-term commitment to the workforce and the economic development of the partner country. In addition, European businesses are leaders in voluntary corporate social responsibility which more often than not goes beyond the legal requirements of a partner country to improve social and economic conditions as well as the protection of the environment there.

The Sustainable Entrepreneurship approach brings benefits for European companies trading or manufacturing abroad, and makes good business sense as entrepreneurs have a particularly acute need to retain staff with the necessary skills for the long-term success of the company. The vast majority of European companies are already well aware of the importance of having good relations with their employees and the society to remain globally competitive. Being on good terms with their customer base is equally important both for larger and smaller companies, as well as keeping a good profile with the local community.

BUSINESSEUROPE has been working successfully on different issues related to international trade negotiations and environmental sustainability for many years. The objectives of EU trade policy are to create partnerships for growth and jobs and strengthen competitiveness, bringing benefits not only for the EU but also the developing economies in which European companies conduct trade activities.

European companies are convinced that Sustainable Entrepreneurship and corporate social responsibility should be an element of the dialogue with partner countries and regions with which the EU is conducting trade negotiations. This is already the case in some instances; the Free Trade Agreements with Korea, Colombia and Peru for example include a sustainability chapter which requires both sides to cooperate on jointly-agreed social and environmental standards. By generating new investments and supporting the sustainable business strategies of European companies, Free Trade Agreements encourage EU companies to invest in developing markets. Without these investments, developing countries would remain poor and miss out on the tremendous opportunities that economic cooperation with the EU can provide in terms of job creation and environmental sustainability.

European enterprises provide part of the solution by safeguarding human rights, making a particularly positive contribution by increasing prosperity, social standards and improving education in countries where governance is weak. Many of them have already committed to taking action by adhering to international initiatives. Of course, companies have neither the political or societal mandate, nor the capacity and resources to substitute the actions of governments where human rights legislation is not adequately implemented or enforced. They do, however clearly have a responsibility to respect human rights in their business activities and they already do so.

European enterprises also take measures on a voluntary basis, for example developing codes of conduct for procurement of goods or combating child labour. There are huge challenges for businesses in this area as large companies often have very long supply chains. However, some of them already place obligations voluntarily on subcontractors and suppliers, including specific requirements in their contracts and asking them to take similar actions with their own suppliers.

In today's global economy, production is increasingly organised along global supply chains. They have become an important factor in ensuring companies' competitiveness on domestic as well as global markets. As a result, open trade helps embed local companies in global production chains, makes them more competitive and creates more jobs. Trade and investment flows are complementary, create jobs and promote the transfer of technology.

People, of course, may be wary about the impact of open trade on their job security and income, and the environmental impact of the way we do business, for instance in terms of resource use and climate change. Europe's entrepreneurs must show the world that Sustainable Entrepreneurship can provide the right answers to these concerns.

Sustainability and SMEs: The Next Steps

Almgren Gunilla

1 Introduction

Being, becoming and remaining sustainable is one of the key challenges facing the small and medium-sized enterprises that my association represents. Sustainable Entrepreneurship is in reality more than just an opportunity – it is a pre-requisite for a successful modern business.

As the President of a Europe-wide intermediary organisation and as a small entrepreneur myself, I have often wondered what can and should be done to further promote sustainable entrepreneurship among our companies, which represent a wide pool of potential in this respect. Not all this potential is untapped, as many SMEs are at the forefront of sustainability. However, much remains to be done – by our companies themselves, by policymakers at all levels and by society at large. Before delving into what should be done, let us see where we are.

2 Sustainable Entrepreneurship in SMEs: What It Is, Where We Are

2.1 Sustainable Development and Sustainable Entrepreneurship

The concept of sustainable development is rooted in the 1970s, when the oil crisis threw a spanner in the idea of perennial economic growth. The famous report on "The limits to growth" published in 1972 and commissioned by the Club of Rome is

A. Gunilla (✉)

The European Association of Craft, Small and Medium-Sized Enterprises, Rue Jacques de Lalaing 4, Brussels 1040, Belgium

e-mail: g.huemer@ueapme.com

C. Weidinger et al. (eds.), *Sustainable Entrepreneurship*, CSR, Sustainability, Ethics & Governance, DOI 10.1007/978-3-642-38753-1_20,

© Springer-Verlag Berlin Heidelberg 2014

among the first publications to mention the word "sustainable" in relation to our economic activities. The authors were looking for an economic model that is "sustainable without sudden and uncontrolled collapse". The language clearly reflects the epoch, but also represents the embryonic form of the concepts of sustainable development and sustainable entrepreneurship.

In 1987, the United Nations World Commission on Environment and Development published "Our common future", a widely cited publication also known as the "Brundtland Report", which gave the following definition of sustainable development: "Sustainable development is development that meets the needs of the present without compromising the ability of future generations to meet their own needs." The Brundtland Report introduced an important element in the discussions on sustainable development: the time factor.

In its first phase and unfortunately in many cases even now, the supporters of sustainable development saw enterprises as an enemy rather than an ally. Greedy private businesses were depleting the world resources and giving nothing in return, according to many. As time passed, however, it became clear that private companies had a key role to play to increase sustainability. Initiatives such as the World Business Council for Sustainable Development in the 1990s and the United Nations Global Compact in 2000 brought this perspective into the mainstream.

The concept of sustainable entrepreneurship is therefore rooted in sustainable development. However, it would be too simplistic to define sustainable entrepreneurship as just private companies caring about sustainable development. In fact, sustainable entrepreneurship is acting in a way that makes companies economically sustainable, socially sustainable and environmentally sustainable at the same time.

Therefore, my own definition of Sustainable Entrepreneurship is slightly more complex. Sustainable entrepreneurship to me means linking the entrepreneurial spirit with values and with responsibility towards the present and future generations. Personally, I am in a fortunate position, since nowhere other than in small and medium-sized enterprises is this link stronger and more visible. In fact, closer ties with the local community and long-lasting interests in its development make acting in a socially responsible and sustainable manner one of the key features of European SMEs' business model.

2.2 Sustainable Entrepreneurship in SMEs: The Status Quo

If we take my own definition as a starting point, sustainable entrepreneurship means matching the wealth and economic prosperity created by entrepreneurs with social cohesion, environmental protection and long term sustainability concerns. However, these "pillars" must be balanced very carefully – for instance, if environmental regulations are too strict and not adapted to the business reality, the entrepreneurial spirit is stifled and sustainability cannot ultimately be achieved.

Sustainable entrepreneurship is, I believe, already firmly rooted in the actions of companies in general and SMEs in particular. Looking at the last 10 years, it is clear to me that more and more small enterprises, which make up the vast majority of businesses, are taking up sustainable entrepreneurship not as a passing fad but as a long-term commitment.

In particular, all sectors linked to the so-called "green economy" are certainly leading the way towards a more sustainable entrepreneurship. SMEs are at the forefront of this venture as users, installers, creators of new technologies, advisers on energy efficiency and micro-generators of renewable energy. However, it would be wrong to assume that SMEs in other sectors are trailing behind. For instance, at a conference we organised some years ago with the then Belgian Presidency of the European Union and our Belgian member organisations, we heard of a German baker who thought about using his own stale bread as combustible to burn in his ovens. He then moved on and converted all of his trucks to hybrid or biofuels. Step by step, he halved his company's energy consumption and reduced its carbon footprint by more than 90 %, while the generated savings quickly repaid the investments he made.

In addition, sustainable entrepreneurship has become even more relevant in the wake of the financial and economic crisis that we are living. In a way, the crisis has acted as a catalyst and increased the attention not only towards sustainable entrepreneurship, but also to sustainability as a whole. It has demonstrated that we need not only sustainable companies, but also fiscal and financial sustainability from our States and our banks. Together with the pre-existing challenge of sustainable development and with the ever-present economic quest to make the most out of scarce resources, this has led to an increased relevance of the sustainability concept. For instance, sustainable growth is one of the three pillars of the Europe 2020 Strategy, the EU's policy blueprint for the coming decade.

3 The Challenges Ahead

Despite the progress made, there is definitely room for improvement on sustainable entrepreneurship, both for companies and for society at large. Unfortunately, in some people's mind thinking "sustainable" and "entrepreneurship" are still irreconcilable antonyms. To redress this, action is needed at three levels: at company level, at policy level and at social level.

At company level, a serious business case must be made in favour of sustainable entrepreneurship. This requires first of all awareness raising activities towards entrepreneurs, as well as the exchange of good practices among companies. It also requires the full involvement of the employees. It is demonstrated that enterprises where the staff understand, share and respect the company's goals are the ones that thrive the most and that succeed in reaching the goals they have set for themselves. Again, small and medium-sized companies have a comparative advantage in this respect, since their staff does not work "for" the entrepreneur, but "with" the entrepreneur.

The policy level is also important. Although entrepreneurs have clearly a firsthand responsibility, they do not act in a vacuum and they cannot be expected to change the world on their own. Therefore, the first fundamental task for policymakers in this respect is to ensure that the right regulatory environment is created. Although there is of course no one-size-fits-all solution for all countries and all companies, some of the defining elements of a favourable regulatory environment are the lack of excessive regulatory burdens, the availability of affordable technical assistance, as well as easier and affordable access to finance for sustainable investments. The second but not less important role played by policymakers is to ensure that public undertakings and public finances are themselves sustainable.

Last but not least, society at large is also a decisive factor. Nowadays, we are all linked to private companies in one way or another. There are entrepreneurs, business organisations, service providers, regulators and last but not least there are customers and consumers. Each individual can therefore encourage Sustainable Entrepreneurship from multiple angles, from creating the right framework conditions, as I mentioned above, all the way down to "voting with our wallets" as consumers.

The time factor is an additional challenge. Sustainable Entrepreneurship means looking not only at the short, but also at the medium and long term. Unfortunately, most of us think mainly short term. People want it all, and people want it now. That is why individualism, selfishness and nationalism thrive especially in times of crisis, when people seem to refuse to see that we are all interconnected, across countries and across generations. The bad news is that this inconsistency will always be present in our societies to a certain degree. After all, it is a by-product of the freedom of individual expression that we cherish and defend. The good news is that the more we do to promote sustainability, the more people will understand its benefits and the less inclined they will be to resort to the opposite end of the spectrum.

Literature

Meadows, D. H., Club of Rome, et al. (1972). *The limits to growth: A report for the Club of Rome's project on the predicament of mankind.* New York: Universe Books.
World Commission on Environment and Development. (1987). *Our common future.* Oxford: Oxford University Press.

CSR Europe: Sustainability and Business

Stefan Crets

Forward-looking companies no longer see social and environmental challenges as only obstacles but as opportunities for innovation and growth. In the fast-developing field of corporate social responsibility, the focus is shifting away from risk management towards a more visionary and entrepreneurial approach that seeks to create shared value and identify opportunities for innovation that can benefit business, society and the environment alike. However, for many companies, the challenge still remains as to how they can stimulate entrepreneurship and innovation both inside their own companies and beyond. This publication is intended to provide some guidance on this challenge by bringing together the latest best practice, political frameworks and theoretical approaches towards sustainable entrepreneurship.

With today's companies facing growing expectations of their social and environmental behaviour and at the same time, the most dramatic economic crisis the world has seen for over 85 years, there is both a need and opportunity for business to harness the entrepreneurial talent that exists within both them and society.

Numerous companies are tackling this challenge directly through innovative and inclusive programmes designed to harness the creativity in society while nurturing future generations of entrepreneurs. Within our network we see examples of companies using their existing infrastructure to leverage new business concepts and ideas. For example, **Coca Cola Enterprises** 'Passport to Employment' initiative helps to prepare *around 2,700 young people* into the world of work, a further *29,000 people* have gained access to paid employment through **L'Oréal's** Solidarity Sourcing Programme and **Telefónica's** Think Big Youth Programme has already seen *3,500 projects* launched by young social entrepreneurs to the benefit of more than *65,000 young people*. These projects have encouraged innovative thinking, entrepreneurship and new business across Europe.

S. Crets (✉)
CSR Europe, Rue Victor Oudart 7, Brussels 1030, Belgium
e-mail: sc@csreurope.org

C. Weidinger et al. (eds.), *Sustainable Entrepreneurship*, CSR, Sustainability, Ethics & Governance, DOI 10.1007/978-3-642-38753-1_21, © Springer-Verlag Berlin Heidelberg 2014

Although these individual efforts are impressive, it is important for businesses and societal actors alike to recognise that by working collaboratively they will create the necessary change to transform and scale up the way businesses approach entrepreneurship. For example, The Swedish Jobs and Society Foundation, is setting an example by working together with corporate companies to help *more than 10,000 people every year to start-up successful and viable enterprises* in Sweden.

While is it is important reach to out and engage with the community on entrepreneurship projects, internal employees and innovators should not be overlooked. Many companies are now incorporating corporate social entrepreneurship at the heart of their innovation and improvement strategies. For example, recognising that the manufacturing process is imperfect, **Toyota** empowers its employees to identify deviations and resolve them. In doing so, they are creating the right environment for employees to come up with new ideas that challenge the conventional attitudes and processes of their company whilst at the same time creating a lean manufacturing system.

Other companies, such as **3M**, have embedded intrapreneurship into their business strategy and the company encourages employees to explore the development of innovative products and services through formal corporate intrapreneurship programs. A classic example of a successful Intrapreneurial creation of a new product is 3M's profitable product line "Post-it Notes" (TM). This example highlights how valuing and developing a company's employees' diverse talents, initiative and leadership will eventually lead a company to higher levels of innovation, productivity, financial reward and value for society at large.

From a policy perspective, the European Commission's Entrepreneurship Action Plan is a blueprint for decisive action to unlock Europe's entrepreneurial potential, to remove existing obstacles and revolutionise the culture of entrepreneurship in Europe. The plan of action puts forward a series of proposals that stress the importance of education and training initiatives and aims to promote the growth and development of new-generation entrepreneurs. The European Commission will now work closely with member states, business organisations and stakeholders to see that the action plan is implemented.

For innovation and entrepreneurship to reach the next level it is necessary for all players to embark on this journey together. It is only through the development of new partnerships that span multiple-business sectors, governments and the public that we can collectively develop solutions for the shared challenges that we face. I hope that many of the outstanding examples outlined in this book will serve as a wake-up call and catalyst for organisations to think about what they can do to harness the strong pool of entrepreneurial talent that exists.

Future-Oriented Actions

Jakob von Uexkull

I founded the World Future Council in Hamburg in May 2007 to give a voice to the interests of future generations. Environmental challenges such as enacting effective laws to support the accelerated use of renewable energies have remained a core concern in our work. We also focus on social justice issues: How do we share our resources, our planetary wealth and our responsibilities? The financial and economic crisis has caused slowly festering problems to escalate. We need financial markets that facilitate real wealth creation. The World Future Council has shown how to create new financial resources in the public interest without causing inflation.

Our current economic system is to a large extent creating illusory wealth at the expense of the environment and of future generations. A large portion of the profits that companies make is derived by ignoring environmental and social costs. We need policies which prevent this. In the Top Runner programme in Japan, the most energy-efficient appliance becomes the binding minimum standard for all manufacturers. Statutory minimum and maximum wages could also create the foundation for a social economy.

1 EU Must Make Better Use of Opportunities

The EU would be perfectly suited to assuming a global active pioneering role and defining new policy frameworks. But at the moment we are regressing into nationalistic thinking in many areas because irresponsible media are spreading a lot of nonsense about the euro crisis and because most people don't understand money creation and finance.

We need rules and laws that create future-oriented incentives and encourage the development of active citizenship, entrepreneurship and innovation. Neoliberal

J. von Uexkull (✉)
World Future Council: Head Office, Mexikoring 29, Hamburg 22297, Germany
e-mail: lea.strack@worldfuturecouncil.org

C. Weidinger et al. (eds.), *Sustainable Entrepreneurship*, CSR, Sustainability,
Ethics & Governance, DOI 10.1007/978-3-642-38753-1_22,
© Springer-Verlag Berlin Heidelberg 2014

ideologues still believe that the market can solve problems better the less government involvement there is. I always ask them why they are not investing in Somalia where there are practically no laws or government power?

2 Fair Entrepreneurship Demanded

For me Sustainable Entrepreneurship means fair and diverse business practices that promote human welfare while respecting natural limits. What is needed now is a public service campaign to educate about the way our money and financial system works, because we need it to finance a new economy which can rise to the global challenges posed by growing inequalities, climate change and the depletion of natural resources.

In this context I welcome the initiative of the **sea**. The World Future Council also presents an award. Our Future Policy Award honours policies which improve the living conditions for current and future generations. We work to spread these "best policies" and thereby support the creation of more equitable, sustainable and peaceful societies. The Future Policy Award is the first prize that honours policies rather than people on an international level. Since 2009 we've been awarding laws and regulations enhancing food security and protecting biodiversity, forests and oceans.

3 Ideas with an Impact

There are many ideas and individuals working to secure our future. For example, Professor Michael Braungart in Hamburg invented the concept 'cradle to cradle', which proves that business can radically improve ways of production and still make a profit. But when sustainable practices cost a lot more than non-sustainable ones, which are still massively subsidized at the cost of future generations, then even the most conscientious entrepreneur will not be able to compete and survive. This is why we need governments to step in and level the playing field for all participants in the market, so that the costs of unsustainable business practices become unaffordable.

4 Fair Competition Vital

Sustainable entrepreneurship requires that the externalisation of environmental costs is considered by law as unfair competition. Such laws should stipulate that companies that do not pay the full societal and environmental costs of doing business are guilty of gaining an unfair advantage over their sustainably operating competitors. An advantage in price or quality gained by doing damage to common goods is no less unfair than deceptive practices such as misleading advertisements or taking advantage of someone's inexperience. If externalisation is treated as unfair competition, then corporations that externalise can be sued for misrepresenting their competitive advantages (lower prices, better quality etc.) as better market performance. As a corollary, agreements between companies to internalise costs that had previously been passed on should be exempt from the prohibition of restrictive practices and cartels.

5 Everyone Must Join in

Building a future-oriented and sustainable world requires all of us to become more strongly engaged in public debates and political processes. I do not mean us all becoming life-long career politicians. We need more persons who, after their time in office is over, enter other spheres of life to show how politics works, and who can restore trust in political solutions, because there is no quicker or more effective way to bring about change than through binding rules and laws.

One project that has left a lasting impression on me is the No Problem Orchestra with its NO PROBLEM MUSIC THERAPY concept for people with severe disabilities, another area the World Future Council works in. We started this project with the Essl Foundation in January 2012 with the international Zero Project Conference on best practices and policies for persons with disabilities and are assisting the implementation of the UN Convention on the Rights of Persons with Disabilities in countries throughout the world. We have a new thematic focus each year, highlighting exemplary laws and examples of good practice that protect and strengthen the rights of people with disabilities.

We Are Living Beyond Our Means

Claudia Kemfert

The current economy is not sustainable as such. We are living beyond our means. If we continue to waste as many scarce resources as before, we would need three more reserve planets to exploit. This means we need to handle resources much more efficiently, as well as replace fossil fuel resources. In my view, sustainable entrepreneurship distinguishes companies, projects or persons that not only address the topic of sustainability and social responsibility, but also actively implement it and act as role models. We need many companies and people who are already working on the world of tomorrow today. Sustainability plays a central role in the energy sector because, in addition to modern infrastructures and power stations, we also need facilities for building and storing renewable energy and also materials for effective improvements in energy efficiency.

1 Exemplary Sectors

The energy industry, but also the infrastructure, sustainable mobility or urban planning and efficiency sectors are fundamental in the development of sustainable technologies. But the chemical industry also constitutes a key player; it researches technologies and substances for the replacement of fossil energy, cutting-edge storage technologies and materials for improving efficiency in the automotive industry or in the field of building energy, e.g. with insulating materials. Thanks to my capacity as a judge in a number of juries, I see many great projects featuring

C. Kemfert (✉)
Hertie School of Governance, Berlin, Germany

Department Energy, Transportation, Environment, German Institute for Economic Research, Mohrenstraße 58, Berlin 10117, Germany
e-mail: sekretariat-evu@diw.de

C. Weidinger et al. (eds.), *Sustainable Entrepreneurship*, CSR, Sustainability, Ethics & Governance, DOI 10.1007/978-3-642-38753-1_23,
© Springer-Verlag Berlin Heidelberg 2014

Sustainable Entrepreneurship. Time and again, I am fascinated by people, companies or projects that combine economic, ecological and social sustainability in an ideal way. These can be regional and local projects within the scope of the energy turnaround in Germany, or successful projects in the food, health, or water supply sector in less developed regions. I find it important that these solutions are innovative and above all sustainable, and can be a model for a new development.

2 Long-Term Project: Energy Turnaround

In this context, the energy turnaround is primarily a long-term project. All remaining nuclear power plants will be shut down by 2022, being partially replaced by renewable energy. We need new power plants, since many old coal-fired power plants will also be shut down during this period. Gas-fired plants are better suited for combining with renewable energy, as they are flexible and can easily be switched on and powered down. Since the addition of renewable energy reduces the price of electricity on the stock exchange, such power plants are less economical. We therefore require a smart market design that not only offers sufficient financial incentives for energy supply, but also factors in demand. For example, energy-intensive industries can adapt their demand to the market conditions if they have financially attractive conditions to do so. Reducing the consumption of energy in buildings is also important. As is network expansion from north to south, in other European countries, as well as decentralised, intelligent networks. With regard to all of these issues, we are only just starting out – after all, there are still four decades to go.

3 Climate Protection as an Economic Driver

The growth potential of the 'green industry' is increasingly being recognised by German companies. Germany continues to be on a good path and leads many 'green' market areas. Other countries have certainly also recognised the economic potential. China, for example, is investing heavily in renewable energy and environmental technologies. But the United States is also investing in innovative technologies. Germany continues to lead the market, particularly in classic environmental protection technologies such as, for example, water treatment, recycling, raw material recycling and renewable energy. But China makes it clear that other countries have recognised the economic opportunities of the important future markets. The competition is not harmful, however, but inspiring.

4 SEA as an Indicator of Sustainable Development

The Sustainable Entrepreneurship Award (SEA) initiative is superb, in my view. The SEA honours companies or individuals that implement the themes of sustainability and social responsibility in an unparalleled way. Successfully implementing and effecting new projects, ideas or social responsibility often involves fighting past and conventional models. Not everyone wants to or can cope successfully with this; often resistance prevents major changes towards a sustainable future. I find it important and appropriate to distinguish leading figures, companies and projects that are already successful or are planning a change that should be supported. This could be an innovative idea or an established sustainability project. Raising awareness for such significant ideas and projects and honouring individuals plays a crucial role in this matter. This is what makes an award for these achievements so important, and the SEA so very meaningful.

Responsible Entrepreneurship

Katherina Reiche

For me, Sustainable Entrepreneurship is responsible entrepreneurship that takes into account the environmental, social and economic consequences in core business along the entire value chain in entrepreneurial reflection and decision-making. Sustainable Entrepreneurship means accepting responsibility for one's own actions.

Although the term sustainable entrepreneurship primarily designates companies as key players, politics and society also bear responsibility in this area: one of the duties of the state lies in formulating long-term goals and establishing an environment conducive to tackling global, national and local challenges such as climate change, scarcity of resources and global loss of biodiversity. On the basis of these objectives, the state must provide a framework for sustainable entrepreneurship, create opportunities and provide impetus in the desired direction so that the companies are able to exercise their responsibility themselves as far as possible and the market develops its innovate energy. Naturally, investors should also assume their responsibility – we already have sustainability rankings, the new sustainability code presented by the Council for Sustainable Development and the international standard ISO 26000. Consumers are called to exercise responsibility in their demand for products and services. Orientation is not always easy. Therefore, labelling and certification are useful if they help consumers to choose. From an environmental point of view, I would like to mention the Blue Angel, which we promote.

The European Commission took up the CSR issue in its October 2011 communication. However, the governments still require much convincing. In Germany, we encourage companies to voluntarily assume social responsibility in their core business. With its action plan 'CSR in Germany', the German federal government wishes to more strongly anchor CSR in business and public administration and interest more small and medium-sized companies in CSR. The federal government has therefore launched a funding programme for SMEs titled 'Corporate Social

K. Reiche (✉)
German Federal Ministry for the Environment, Nature Conservation and Nuclear Safety,
Stresemannstraße 128-130, Berlin 10117, Germany
e-mail: Helge.Heegewaldt@bmu.bund.de

C. Weidinger et al. (eds.), *Sustainable Entrepreneurship*, CSR, Sustainability, Ethics & Governance, DOI 10.1007/978-3-642-38753-1_24,
© Springer-Verlag Berlin Heidelberg 2014

Responsibility in SMEs', which offers SMEs the best possible practical assistance on CSR issues. In addition, strategic partnerships and networks are to be built. With regard to international cooperation, the German federal government will support the strengthening of the dialogue on the CSR framework in the relevant international forums, such as the UN, the G8 and G20 and the EU. In this context, the federal government will also promote the further development of the OECD guidelines for multinational enterprises.

I welcome the Sustainable Entrepreneurship Award, as this award makes exemplary approaches in sustainable entrepreneurship visible to a broader public.

Part V
Looking Ahead

We Have to Embed Egoism

Ernst Ulrich von Weizsäcker

Interview with Ernst Ulrich von Weizsäcker

Co-President of the Club of Rome and honorary member of the World Future Council

As an expert for biology, the environment and ecological efficiency, you have contributed significantly towards creating greater awareness of the issue of sustainability. In your book *Factor Five: Transforming the Global Economy through 80 % Improvements in Resource Productivity*, you explain how global resource productivity can be improved by up to 80 % and politically implemented. Within which time frame could you see that happening?

In 60 years, if one lets everything continue as before; in 30 years, if one changes the framework conditions in a sensible way. And, incidentally, the developing countries will probably be quicker than us, because we have already harvested many of the 'low-hanging fruits'.

You are calling on industry to follow new paths and promote, among other things, the principle of remanufacturing, which is only in its beginnings in Europe. Have you already been able to find supporters for this in large European or German industrial firms?

To be honest, I am not a preacher looking for supporters. But remanufacturing is entering the global markets from Asia, and there's probably no stopping its advance in the automotive industry. When some metals become really scarce and expensive, people will realise that the route via the scrap trade, melting and recycling is less elegant and less cost effective than remanufacturing.

How far has Europe travelled on the path towards a sustainable future, in your opinion? And what role can, should or must European – but also national – politics play?

Europe – especially Germany, Scandinavia and the Benelux – is better than most other regions. Japan and South Korea are on a par with us; China is rapidly moving

E.U. von Weizsäcker (✉)

Head Office Hamburg, World Future Council, Mexikoring 29, Hamburg 22297, Germany

e-mail: ernst@weizsaecker.de

C. Weidinger et al. (eds.), *Sustainable Entrepreneurship*, CSR, Sustainability, Ethics & Governance, DOI 10.1007/978-3-642-38753-1_25,

© Springer-Verlag Berlin Heidelberg 2014

in the same direction, but is still consuming vast amounts of energy and minerals for the development of its infrastructure at present. Europe can successfully make it its trademark to think ahead and act in an environmentally friendly manner. Chinese people have told me that they abide by REACH, the EU's chemicals directive, because they believe that soon only REACH-certified products will be approved for sale.

One of your primary concerns is a permanent green tax reform. What specific form should this take on? And how can these changes be prevented from causing the income gap between the rich and the poor to widen even more?

I am indeed in favour of an active, gentle increase in the price of energy and primary resources in proportion with the documented improvements in efficiency, so that the monthly costs for energy and minerals remain the same on average. This is analogous to the industrial revolution, which saw the hourly gross wages and the labour productivity mutually stimulating each other, ending up with at least a twentyfold increase. What a fantastic wealth generator! To prevent fractures in society, I suggest – incidentally, this is copied from South Africa – concessionary rates for the poor. Furthermore, industry should come to enjoy neutrality regarding revenue: the money collected there should flow back into that particular sector – on a per job basis. This creates a twofold incentive to become more efficient and to maintain/generate jobs. So nobody needs to emigrate.

You wrote that early human civilisations in which egoism was dominant simply died out, and that in the surviving civilisations egoism was always embedded in social obligations, ownership for example. But are people not inherently egoistical? How can we succeed in finding a replacement for egoism as the decisive driving force in trade?

I cannot and do not want to abolish egoism. I want to embed it. Adam Smith's notion that the egoism of the individual creates prosperity for the country was firmly embedded, to him (he was a moral philosopher), in the laws and social conventions. The model turned nasty with the advent of globalisation: now the market that rewards egoism is global, while the law and sense of decency remain national. Europe must fight for global rules!

What does Sustainable Entrepreneurship mean to you personally?

Promoting ecological sustainability in one's own business, in addition to economic and social sustainability, which are integrated out of self-interest. And cooperating with the government and with society, when the general conditions are to be changed so that sustainability becomes more and more viable.

Which sectors do you think are doing the most in terms of Sustainable Entrepreneurship, and which have a lot of catching up to do? Which is the most memorable SE project you've ever heard of – and why?

In general, the closer to the customer, the more exemplary it is. In the case of foodstuffs and personal hygiene, the top brands cannot afford to have a bad image. The biggest need for catching up is in the financial markets with their ruthless battle for return on capital. This battle makes it almost impossible for entrepreneurs of other sectors to make long-term decisions. The crazy thing is that the financial markets under the leadership of the Anglo-Saxons repeatedly manage to be excluded from regulations that the chemicals or toy industries are subjected to as a matter of course!

How do you rate the Sustainable Entrepreneurship Award (SEA) initiative?

Positively. It should also have no qualms about comparing sectors with each other and not only choose the 'best in class' within a particular sector. Otherwise the villains in the financial market will continue to get away with their misdeeds.

Business Success Through Sustainability

Christina Weidinger

1 Introduction

Societal challenges have grown continuously over the last few decades. Today we already count more than seven billion Earth-dwellers (Handelsblatt 2011), and in many parts of the world we are facing dynamic leaps in development. We are living in times of change. The still difficult economic environment in Europe is now showing to many what has already been known for a long time: we need new management approaches and economic innovations to stay fit for the future in this dynamic environment.

'Confidence in the economy' and 'entrepreneurial creativity' have become particularly scarce resources (Edelman 2012). In the past decade, we could still draw on almost unlimited trust in the economy. Now we must generate greater social added value through increased transparency as well as through innovation. This is the only way to make it clear to all that businesses are part of the solution, not part of the problem. This is an important approach to clarifying the concept of Sustainable Entrepreneurship: it means taking greater advantage of the positive effects and creative potential of entrepreneurship than ever, in order to achieve a sustainable development of our society (Fig. 1).

Because the current crisis shows: we need more ecological, social and, above all, greater economic sustainability in our actions. This is where entrepreneurs can make an important contribution by aligning their business models so that their actions create both a business and a social added value (Porter and Kramer 2011). Only in this way will Europe and our businesses succeed in overcoming the crisis and restoring the innovative function to our continent that it held for centuries. We must become innovation leaders once again, instead of remaining innovation followers or even

C. Weidinger (✉)
(SEA) Sustainable Entrepreneurship Award, Karlsplatz 1/17, Vienna 1010, Austria
e-mail: christina.weidinger@diabla.at

C. Weidinger et al. (eds.), *Sustainable Entrepreneurship*, CSR, Sustainability, Ethics & Governance, DOI 10.1007/978-3-642-38753-1_26, © Springer-Verlag Berlin Heidelberg 2014

Fig. 1 Sustainable
Entrepreneurship and shared
value

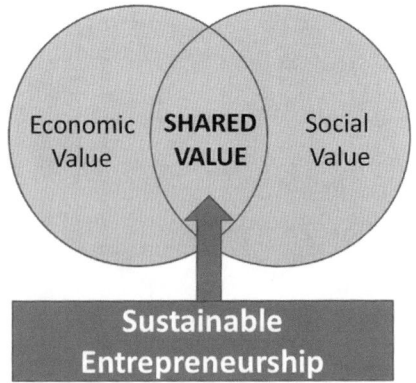

Fig. 2 From laggards to
leaders

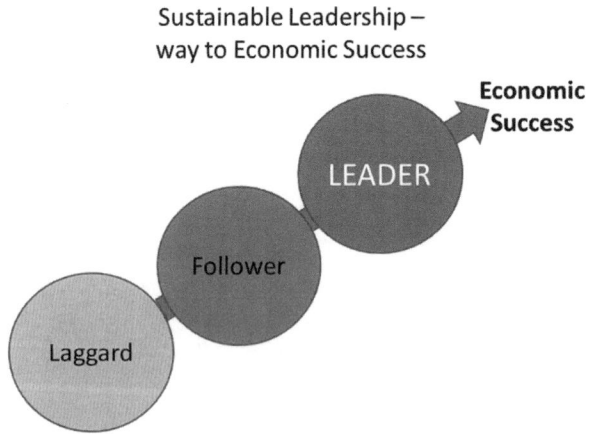

laggards. This is the only way to economic success in an increasingly unstable and dramatically changing global environment! (Fig. 2)

The assertions of the American sociologist and economist Jeremy Rifkin may be controversial, but they have contributed greatly to a rethinking and reorientation process in industry and society (Rifkin 2011). Rifkin uses the term 'access society', which refers to a radical cultural shift that was triggered by the internet, among other things. This new model of society is based on two pillars (Rifkin 2011): (1) Commercial offers, provided by businesses. The main motivation here is to make money. (2) Free offers, created by individuals or communities in their free time. The main motivation here is giving, sharing, expressing oneself, creativity. Moreover, Rifkin speaks of the third industrial revolution, which we are currently in the midst of. It is characterised by a combination of new communication technologies with new energy systems. This creates a system that is no longer centralised, but rather decentralised, like one big network (Rifkin 2011).

2 Entrepreneurship and Sustainable Growth

2.1 *Small and Medium-Sized Enterprises: Backbone of Economy*

Networks are only as strong as their strongest link and as weak as their weakest. Correspondingly, when this notion is applied to the economy, small and medium-sized enterprises (SMEs) play a key role on the road to a new, sustainable and decentralised economy. Decentralisation offers big opportunities in Europe in particular, owing to its economic structure, which is characterised by small and medium-sized businesses. This means a first step will involve a(nother) renaissance for SMEs, which set the tone in the European economy, albeit often behind the scenes. After all, Europe is not a continent of corporations, but first and foremost a continent of SMEs. In 2005 there were almost 20 million companies active in the EU-27 countries in non-financial industrial sectors. The overwhelming majority of these, namely 99.8 %, were SMEs with no more than 250 employees. Their relative importance, however, was lower in terms of their contribution to providing jobs and wealth, as 67.1 % of the non-financial business economy workforce in the EU-27 was employed in SMEs, while only 57.6 % of the added value in these sectors was generated by SMEs. (Eurostat 2008) (Fig. 3)

Small and medium-sized enterprises are particularly aware of the importance of a stable environment including, among others, good educational, health and social systems. They are frequently firmly rooted in their region, often for decades, and invest in their social environment (Bertelsmann Foundation 2011). This commitment is now more important than ever. Europe would do well to remember the strength of SMEs if it wants to regain its role as an innovation leader. Presently, innovation is taking place on continents other than Europe. China and Korea are no longer copying and plagiarising, but rather developing, researching and innovating (Business People 2013), and Asia is also a frontrunner in terms of adapting business models towards social added value (MIT Sloan Management Review 2012) – in technology, in management, in growth, in location attractiveness. But the question is, which concept will be more sustainable in the long run: Growth at any price – without considering societal development or climate change? Or sustainable growth? The answer is unwavering: without sustainable business there is no future. And sustainable business is more than corporate social responsibility.

2.2 *Entrepreneurship: Engines of Prosperity*

European history shows that in all times of change entrepreneurs were particularly important social innovators: company pension funds, internal training and much more remind us of that (Schmidpeter 2013a, b). Entrepreneurs are often the first to

Structure of the European Economy

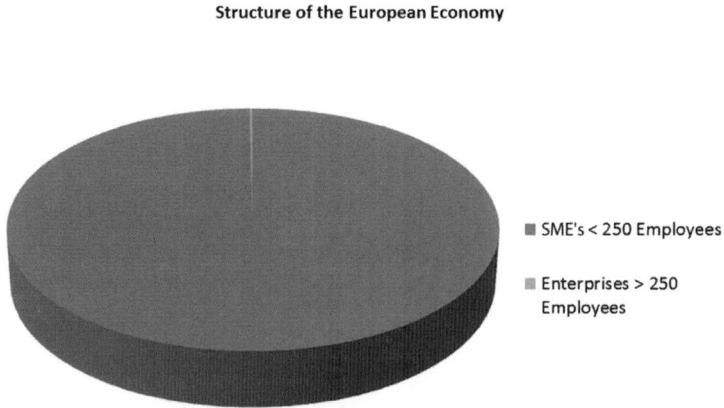

- SME's < 250 Employees
- Enterprises > 250 Employees

Fig. 3 Structure of the European economy

recognise where the problem lies and are willing to invest in their social environment. These entrepreneurial approaches are needed more than ever in order to find new ways to tackle the current challenges – resource scarcity, demographic trends, and much more. We can only survive if companies are part of the solution and not part of the problem; that is, if companies think of themselves as sustainable entrepreneurs.

Entrepreneurs are the engines of our economy and contribute significantly to the prosperity and development of society (cf. articles in Osburg and Schmidpeter 2013). In this function, it is important that they also accept the societal challenges we are facing now and will face in the future. This means pursuing sustainable solutions for environmental problems, social developments such as demographic change, and economic crises. Sustainability is increasingly permeating all areas. In the face of stiffer competition, consumers increasingly require and also expect sustainability as a unique selling point. Business is responding to this market change. This is demonstrated, for example, by the increasing supply for fair trade products, ecological production and social compatibility.

Large profit-oriented companies such as REWE or Siemens, to name but two prominent examples, are reacting (Company report REWE Group 2012, Company report Siemens 2012). It is a matter of combining the pursuit of necessary gains with social responsibility – because no one likes to collaborate with a business that is considered untrustworthy. For the majority of Austrian companies, trust and responsibility are already part of the strategy and core business – and an important competitive factor in international business. Naturally, Sustainable Entrepreneurship does not occur out of pure charity, and should not and must not do so.

Relation between Business and Society

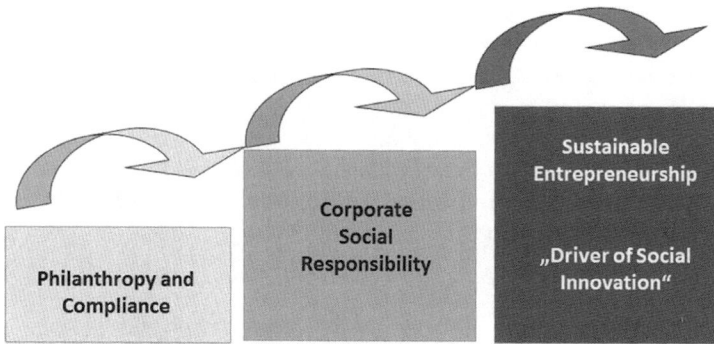

Fig. 4 Development of business-society relation

2.3 Sustainable Entrepreneurship: Next Stages of Business

There is a difference between Sustainable Entrepreneurship (SE) and corporate social responsibility (CSR). While CSR has been mainly implemented by large companies and often is only a reaction to stakeholder demands, Sustainable Entrepreneurship is a chance for all businesses – especially those in the large and significant group of SMEs – to truly set themselves apart from their competitors in the future. Sustainable Entrepreneurship can be considered the next step of business strategy (Fig. 4).

Sustainable Entrepreneurship is becoming the most important pillar on which our economic future rests. It is the foundation on which we are building a new European economy, leading business out of the current struggles. Europe can only emerge from the crisis and overcome future challenges if we commit to Sustainable Entrpreneurship and act accordingly. Those companies that orient their core processes and strategies on it will be the winners. In short: Sustainable Entrepreneurship is the chance for Europe to re-establish itself as a trailblazer (MIT Sloan Management Review 2012). We still have a lot of work ahead before we get there. We have to develop ideas and strategies, as well as new products and services. And nowadays, when we speak of innovation, we have to use the term in a much broader sense than before. Innovation is not only an important concept for products and services, but also in companies' strategies.

Sustainable Entrepreneurship is not in the least a mere environmental issue, but rather consists of many facets. Economy, ecology and social innovations stand side by side on an equal footing. Sustainable Entrepreneurship is more than just an idea – it is, upon close examination, a kind of lifestyle, a very particular way of life.

The basis for this new thinking and for the future of business activities has already begun. We are already immersed in the process of change. We are witnessing the third industrial revolution (Rifkin 2011). This will be the first industrial revolution that does not arise from technical innovations – such as mass production, and later the internet. We are witnessing a revolution of mentality and economic thought. Pure growth alone can no longer create the conditions to deal with future crises, but above all, it does not help to safeguard a business for the next few decades. It is, therefore, time for new business ethics so that sustainability is not only interpreted as a marketing gimmick, but really receives the recognition it deserves. Sustainability also becomes an innovative business model that focuses on economic, social and environmental responsibility, thus ultimately providing businesses with a clear competitive advantage!

3 Sustainable Entrepreneurship and Business Success

Sustainable innovation must be understood as a corporate value. It is not a fig leaf, a job for do-gooders or idealists, but rather an essential strategic decision. At its core is the enterprise's definition and assessment of opportunities arising from this unique selling proposition – in the social, ecological and economic dimension. Innovation is the best strategy to constructively solve the supposed contrast between business and social/ecological challenges, in a way from which all sides can benefit (cf. also articles in Osburg and Schmidpeter 2013). This means that everyone can subscribe to this innovation strategy of intelligent entrepreneurs, because resources are optimally used and investments in the future are made. This approach not only has majority appeal in society, but is also economically viable if properly implemented.

Roland Berger Strategy Consultants state: "Whoever believed that pressing cost programmes and ambitious growth programmes would drive the topic of sustainability off the agenda of top management or expose it as a luxury problem has another thing coming: sustainability is no longer simply a fad or marketing trend" (Roland Berger 2010). It will increasingly become an integral part of the philosophy and operational alignment of future-oriented companies. Sustainability, therefore, does not mean sacrificing profitability, but is rather an opportunity for (new) profitability. In conclusion: we need more entrepreneurship (but of the innovative, true kind!) as the key to societal sustainability.

At the end of the day, Sustainable Entrepreneurship must pay off. Therefore, Sustainable Entrepreneurship is an individual process arranged differently for each business, as the stakeholders have different demands. The economy is subject to increasing change. It is, therefore, not necessarily to apply one way to different industries, because the challenges in the chemical and automotive industries are different than in the leisure industry, or in the service sector (Bertelsmann Foundation 2013). At the same time, sustainability offers specific opportunities in certain sectors: Austrian companies possess competitive expertise especially in the forward-looking areas of urban technologies, environmental and resource protection, as well as renewable energy, which is increasingly in demand abroad and contributes to sustainable development.

4 Taking Control of the Sustainability Agenda

4.1 SE Implementation: From Concept to Corporate Management

How can entrepreneurs clarify whether or not they are actually operating on a sustainable basis? It is actually quite simple: by reflecting on whether or not one's everyday actions and decisions sustainably achieve a social added value in the long term. The German economist and sustainability expert Andreas Suchanek expresses it thus: "Invest in conditions of social cooperation for the mutual benefit!" (Suchanek 2007) How can this philosophy of Sustainable Entrepreneurship be integrated in both the strategy and the management of a business enterprise?

For the business consulting firm Bain & Company (2012), the term 'sustainability' "encompasses all aspects of ethical business practices, addressing social, environmental, regulatory and human-welfare issues responsibly and profitably. Suppliers, employees, customers, shareholders, governments and communities all have specific agendas that need to be understood and managed." The logical conclusion: companies are called upon not only to assume responsibility, but also to take control of their sustainability agenda. According to Bain, sustainability is rapidly becoming a prerequisite for profitable growth, as shown in a recent study (Bain 2012): "[...] at least two-thirds of 25,000 consumers in the US, Canada and western Europe form impressions based partly on a business's ethics, environmental impact and social responsibility. Yet, ironically, many consumers are unwilling to pay more for sustainable products" (cf. Bain 2012) (Fig. 5).

However, it is clear that every strategy first of all needs a basic idea. So the first step for business is to prioritise their sustainability issues. These need to be looked at from a variety of perspectives, as sustainability is a process that involves business in their entirety. Every change that arises from this process is, at the same time, a part of the future strategy. Bain cites the Wal-Mart example: Wal-Mart's package-reduction initiative will result in immense savings in the coming years. 213,000 fewer lorries on the road means savings in the range of USD 3.4 billion (Bain 2012). Many consumer-based companies are incorporating sustainability practices throughout their business practices, according to Bain: "In the process, they are transforming their products' design and assortment, their supply chains, their operational footprints and their messaging to consumers, investors and employees" (Bain 2012). A business that wishes to implement an effective sustainability strategy must, however, also understand the risks and opportunities associated

Sustainability Management

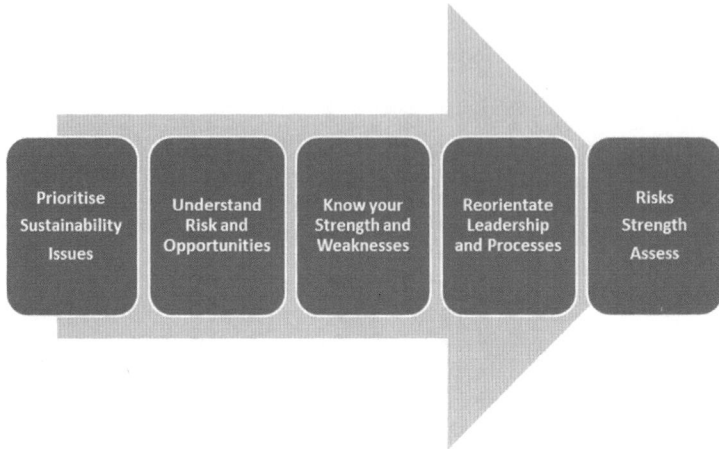

Fig. 5 Implementation of sustainability

with the area of sustainability, emphasise the auditors at PricewaterhouseCoopers (PwC 2012). This involves not only recognising the risk and potential that could result from a believable and effective sustainability strategy, but also orienting the company's management, leadership and performance assessment on it.

Sustainable Entrepreneurship stands for the expansion of corporate responsibility by incorporating social and environmental criteria in an integrated corporate strategy. This is equally important in environmental policy and consumer behaviour. However, we can also open up new international markets with innovative solutions. Europe is a continent marked by export. Our goods and services are in demand all over the world. It is immensely important to preserve this basis for our prosperity: foreign trade is, for example, not only the motor of the German economy, but also of the Austrian economy – half of every ten euros of Austrian GDP is generated abroad (WKO 2012). Sustainability in the areas of economy, ecology and social affairs is a core element, as sustainability is an essential prerequisite for the success of the Austrian export industry.

4.2   Economy as a Cycle: Opportunities, Risks and Profitability

In their much-lauded essay (Harvard Business Manager 1993) published 20 years ago, Meffert and Kirchgeorg pointed out the three most important principles of sustainable development: the responsibility principle, the cycle principle and the cooperation principle (Meffert and Kirchgeorg 1993).

The cycle principle corresponds to the idea of a circular economy that ensures a sustainable development. Only when the complex relationship between the economy, society and ecology work as a circular system can nature and social systems be preserved in the long term. Prominent advocates of the cycle principle William McDonough and Michael Braungart demand we 'remake the way we make things' (McDonough and Braungart 2002) and formulate the cradle to cradle principle to this effect. Aligning our economy with the cycle principle requires both considerable technical progress (breakthough technologies) as well as social innovation.

The responsibility principle has two dimensions: on the one hand, reducing the wealth disparity between industrialised and developing countries, while on the other hand taking into account the needs of future generations as much as the present generation. These two dimensions of responsibility, the intragenerative and intergenerative, can only be fulfilled if all systems are developed sustainably. This calls, above all, for sustainable leadership and innovative solution approaches that generate added value for present and future generations.

The cooperation principle clearly shows that many challenges can only be solved if various actors work together. In order to generate social innovation, they need to coordinate their behaviour or allocate joint resources. Intelligent entrepreneurship and the market opportunities associated with it are an important driver, as is the cooperation between politicians, businesses and civil society in order to jointly create a societal framework that is fit for the future. Of particular import are incentive systems that promote Sustainable Entrepreneurship, the international amplification of innovative approaches and the promotion of breakthrough technologies. Not only will the role of corporations change in this process, but also the role of politics and civil society. This means the question is not whether there should be more market or more social cooperation. Instead, sustainability calls for both more market coordination and more social cooperation.

The concept of sustainable development and its three principles is no less relevant today than it was then. Businesses have to accept their responsibility for future generations and act accordingly. This includes sensible resource planning and utilisation, as well as the avoidance of unacceptable or even irreversible negative impact on the environment. After all, in a cycle everything comes round again. Things have an effect on one another. This hypothesis ultimately leads to the cooperation principle, which aims to coordinate economic processes more strongly towards an ecological orientation (cf. Meffert and Kirchgeorg 1993). This results in cycles above and beyond the business itself, through which the entire product lifecycle can be controlled. All three principles have an interrelation with one another. Businesses that take Sustainable Entrepreneurship seriously must tackle all three (Fig. 6).

Two recent studies carried out in cooperation with WirtschaftsWoche magazine show which companies are perceived by German consumers as sustainable in terms of their social and ecological engagement. The results showed that companies with a responsible corporate orientation have a higher turnover. A consumer survey asked the question, "If you consider the brands that you consume in Germany, how green and social do you think they are?" (WirtschaftsWoche 2012), and found the

Fig. 6 Principles of
sustainability

Principles of Sustainability Management

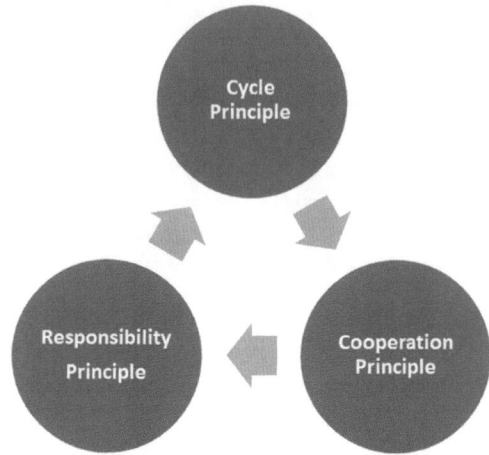

Fig. 6 Principles of sustainability

following: Baby food producer Claus Hipp took the top spot, the baby food brand Alete came in second, and the car manufacturer BMW took third place. The study also made clear that consumers are becoming more and more aware of sustainability issues. They are also becoming more critical and want to be informed of whether companies burden the environment excessively, whether they exploit their employees and how respectful they are of resources. The study also demonstrated that about 15 % of a brand's image is influenced by topics such as environment protection, fair conduct with employees and resource conservation.

This again stresses the fact that sustainability is just as important in the market as quality, attractiveness of the product and economic success of the brand. And what's more: companies with a green image can generate more turnover (WirtschaftsWoche 2012). A responsible corporate orientation adds about 5 % on average to the turnover. A particularly successful example cited by the authors is the frozen foods producer Frosta: 15.6 % of Frosta's turnover can be attributed to a clear sustainability strategy. Sustainable Entrepreneurship has to be more than a management fad. It is a corporate responsibility and philosophy whose systematic implementation increases profitability and has a positive cost-benefit ratio.

4.3 Sustainable Entrepreneurship: Involving Business and Society

Similar to communicating vessels, Sustainable Entrepreneurship is a process consisting of systems that influence one other. Businesses and the public actually possess the strongest pressure and considerable power. Politics and economy respond to them. Some people wish to be re-elected, others care about their sales.

Politics and economics are both part of society. Thus, society is being challenged, in any case. As part of society, civil society is a third force in the matter. In the long term, we need adequate global governance. Its enforcement is ultimately a political challenge. The conditions are, however, not favourable for overcoming this challenge. Therefore, we must currently try to move forward within the triangle of politics, industry and civil society. A critical civil society and critical consumers can move companies in a sustainable direction, particularly companies with brands that are experiencing considerable pressure regarding their reputation. If enough of these companies take action together with civil society in a sustainable direction, conditions may arise under which politics can accomplish what is expected but is currently not being implemented under the present economic conditions due to international competition.

Sustainable economic activity is thus a result of the interaction between society and the economy. Businesses respond to the social requirements expressed by public demand by making corresponding offers. Conversely, some companies deliberately set trends in this area in order to stand out from the competition, as demonstrated by the German retail group REWE in Austria (including Billa and Merkur) with the Ja, natürlich! product range. Policies should support these developments by providing the appropriate conditions. Sustainable development has an economic, environmental and social dimension, and is one of the general objectives of the European Union (EU Commission 2012). Responsible corporate conduct is critical to building trust in the market economy, to opening up trade and to globalisation.

Politics and civil society are changing their roles too. Both realise that no one can solve the current challenges by themselves. Rather, the new aim is to pool all social forces to work together for a sustainable society. Policymakers can adopt a moderating role and shape the respective conditions in dialogue with business and non-governmental organisations so as to create increased incentives for sustainable development. Also, all decision makers in politics, the economy and society can help within their environment to implement the idea of sustainability via concrete measures (Fig. 7).

The EU plays a significant role in this matter: it should strive – and does so, in principle – to create an environment in which the economy has the necessary framework at its disposal in order to view Sustainable Entrepreneurship as an attractive and competitive strategy. With the Competitiveness and Innovation Framework Programme (CIP) of the European Union, which is aimed specifically at SMEs, the EU is promoting innovative activities through better access to financing and through support measures. During the funding period of 2007–2013, the EU is making available a total of EUR 3.6 billion for CIP, which is also benefiting Sustainable Entrepreneurship. Furthermore, the development of SMEs is being facilitated within the scope of the Seventh Framework Programme. Particularly in the field of research, EUR 1.3 billion are being provided in order to strengthen the competitiveness, innovation and sustainability of small and medium-sized European enterprises (Interview Köstinger 2011). However, being or becoming active in this area is also up to the businesses themselves.

Changing Business, Politics and Civil Society

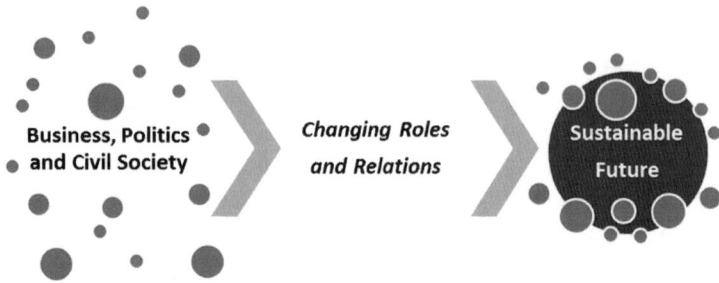

Fig. 7 Changing business, politics and civil society

5 Looking Ahead

After examining current thought on entrepreneurship, innovation and sustainability, one leading question remains: are there any recipes for increasing economic and social value at the same time by integrating sustainability into the core process of a business? This question can, unfortunately, only be answered by each entrepreneur for him- or herself. One thing is certain: Innovations will be core to achieving this alignment between business success and sustainability. In order to bring these innovations to life we need business leaders who invest time and resources in developing new management processes, products and services that contribute to a sustainable development of our society. Therefore, the old shareholder value paradigm must certainly be replaced by shared value thinking. This new ethical perspective can be linked to the tradition of enlightenment and can define a new role of business in society. This thinking will reshape the existing capitalism and sees business as the main driver of social and sustainable development. With this vision in mind, sustainable entrepreneurs take social responsibility one step further. They remodel products, process and whole markets as well as the societies of the future.

By aligning their entire entrepreneurial value creation with the principle of sustainability, business becomes the main driver of sustainable development. Sustainable companies use their entire resources to generate operational and social added value simultaneously. They are successful, but rather for the benefit than at the expense of society. Sustainable Entrepreneurship means constantly reflecting on our own actions and viewing societal problems as an opportunity for entrepreneurship. This view is necessary to develop the breakthrough technologies so desperately needed in the fields of nutrition, energy, health, mobility, etc. The perspective is not 2015 but 2040! A realistic view is needed and it is no good demanding too much in a short time. Entrepreneurship is an open search process that comes up with unique solutions in the long run – solutions nobody can imagine today (Fig. 8).

SEA – Network of Change

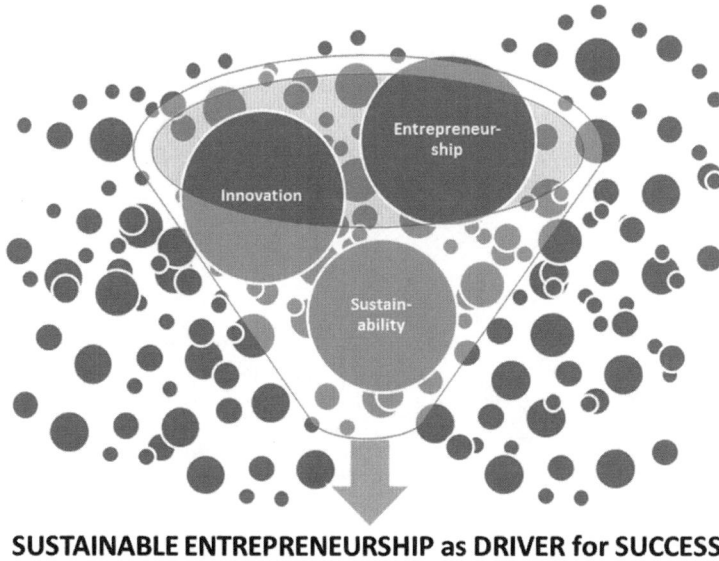

SUSTAINABLE ENTREPRENEURSHIP as DRIVER for SUCCESS

Fig. 8 Sustainable Entrepreneurship Award – Sustainability is in our hands

We need platforms and networks of mutual exchange as well as publicity for the best practices that already exist. The Sustainable Entrepreneurship Award (SEA) incorporates more than previous CSR, environmental or climate awards. This platform honours outstanding projects in the field of Sustainable Entrepreneurship. The initiative helps to raise awareness for flagship projects, thus establishing an incentive to imitate them and to increase societal awareness and recognition for these activities. In its two years, many prominent advocates have supported the idea of the SEA, which speaks for the quality and the necessity of the award. The goal of the SEA is for everyone to become a sustainable entrepreneur and an ambassador for sustainability, in order to generate added value for themselves and their environment by their actions. And to recognise both one's own capabilities but also limits through self-reflection. If you cannot solve problems by yourself, it is time to seek new solutions in cooperation with others.

It is our vision to build a European network of social innovators, Sustainable Entrepreneurship and visionary decision makers. A platform that defines the issue of sustainability from the perspective of innovation, entrepreneurial opportunities and the future viability of our society. Based on concrete examples and solution-oriented cooperation, the opportunities of Sustainable Entrepreneurship are systematically to be made accessible to all European companies. In an open discussion across all ideologies, committed to the competitiveness of our economy and to future generations.

As stated in the introduction, Sustainable Entrepreneurship is still an emerging concept with a highly dynamic development. It may be too early to provide a final definition, but the debate is underway. Many leaders from business, politics and civil society are already discussing the future of entrepreneurship. And one thing seems pretty clear: we can only solve the most urgent problems together with business and not at odds with it. We need innovative business models in order to foster social innovation and make our planet a decent place for our children and grandchildren. We can change the future – Sustainability is in our hands.

Literature

Austrian Chamber of Commerce/AWO (2012). Österreichs Außenhandel 1.-3. Quartal 2012, AUSSENWIRTSCHAFT-Marktanalysen (Austria's Foreign Trade Q1–Q3 2012, AUSSENWIRTSCHAFT Market Analyses), published on 10 December 2012.
Bain & Company Deutschland. (2012). Nachhaltigkeit http://www.bain.de/managementkom petenzen/strategie/nachhaltigkeit.aspx
Bertelsmann Foundation (2011). Verantwortungspartner (Partners in responsibility). Gütersloh.
Bertelsmann Foundation (2013). CSR WeltWeit – Ein Branchenvergleich (CSR worldwide – a cross-sectoral comparison). Gütersloh.
Biesalski, A., & Matthes, S. (2012). Die Gut-Geschäftler. Wie sich Nachhaltigkeit auf den Umsatz auswirkt ("The idealistic business people. How sustainability affects turnover"), WirtschaftsWoche of 4.6.2012. htttp://www.nachhaltige-konzepte.de/aktuelles/102-mehr-umsatz-durch-nachhaltigkeit.html
Bain & Company Germany (2012). http://www.bain.de/managementkompetenzen/strategie/nachhaltigkeit.aspx
Company report REWE Group (2012). http://www.rewe-group-geschaeftsbericht.de/2012/down-load/11817_REWE_GB2012_gesamt.pdf
Company report Siemens (2012). Trust unites us. http://www.siemens.com/annual/12/en/down-load/pdf/Siemens_AR_2012.pdf
Edelman (Ed.) (2012). Edelman trust barometer 2012. http://trust.edelman.com/
EU Commission (2012). Social innovation. http://ec.europa.eu/enterprise/policies/Innovation/policy/social-Innovation/index_en.htm. Accessed 9 Dec 2012.
European Commisison (2008). Enterprises by size class – Overview of SMEs in the EU. Brussels.
Eurostat regional yearbook 2008. (2008). Luxembourg: Office for Official Publications of the European Communities. http://www.groupedebruges.eu/pdf/Eurostat_regional_statistics_2008.pdf
Handelsblatt (2011). Und die Erde könnte auch noch mehr (And the earth could still do more). 31 October 2011.
Hornacek, H. (2011). Interview with Elisabeth Köstinger, MdEP, interviewed on 29 November 2011, published at www.se-award.org
Hauff, V. (Ed.) (1987). Unsere gemeinsame Zukunft. Der Brundtland-Bericht der Weltkommission für Umwelt und Entwicklung (Our common future. The Brundtland report from the world commission on environment and development). Greven: Eggenkamp Verlag.
Hornacek, H. (2011). Interview with Christiane Brunner, member of the Austrian National Council: Consumers can make a big difference; interviewed on 9 December 2011, published at www.se-award.org
Hornacek, H. (2013). Expertentalk: Mut zu Reformen und neuen Wegen (Expert talk: The courage to implement reforms and go down new paths). published in Business People 1/2013.

Kriener, K., Berg, C., & Grimm, J. (2011, June). Nachhaltigkeit im Handel: Herausforderungen – Strategien – Umsetzung ("Sustainability in commerce: Challenges – strategies – implementation"), SAP and the chair for logistics management at the University of St Gallen.

McDonough, W., & Braungart, M. (2002). *Cradle to cradle*. North Point Press: New York

Meffert, H., & Kirchgeorg, M. (1993). Leitbild des sustainable development ("Guideline for sustainable development"). *Harvard Business Manager, 2*, 34–45.

MIT Sloan Management Review (2012). *Social business: What are companies really doing?*.

Osburg, T., & Schmidpeter, R. (Eds.), (2013). *Social innovation*. Springer: Berlin.

Porter, M. E., & Kramer, M. R. (2011). Creating shared value. *Harvard Business Review*. January–February 2011, pp. 3–17

PwC (2012). *Nachhaltiger Erfolg durch sustainable business solutions*. www.pwc.com/at/de/nachhaltigkeit/index.jhtml

REWE Group (2010). *Sustainability report 2009/2010*. http://www.Rewe-group.at/download/PDF/Nachhaltigkeit/Lagebericht_zur_Nachhaltigkeit_2010.pdf

Rifkin, J. (2011). *The third industrial revolution: How lateral power is transforming energy, the economy, and the world*. New York: Palgrave Macmillan.

Roland Berger Strategy Consultants (2010). Sustainability – A profitable business model. *Customer magazine think: Act BUSINESS*.

Schmidpeter (2013a). The Evolution of CSR from Compliance to Sustainable Entrepreneurship. In: C. Weidinger, F. Fischler, & R. Schmidpeter (Eds.) *Sustainable Entrepreneurship*. Springer: Berlin Heidelberg.

Schmidpeter (2013b). Corporate responsibility. In: Economic Austria (Eds.), Entrepreneurship and Values (Wirtschaftspolitiscche leaves). 01/2013, 60 Vintage.

Siemens (2011). *Sustainability report 2011*. http://www.siemens.com/sustainability/pool/de/nachhaltigkeitsreporting/siemens-nb2011.pdf; Siemens AG.

Suchanek, A. (2007). *"Ökonomische Ethik" (economic ethics)* (2nd ed.). Tübingen: Mohr Siebeck Verlag.

Weidinger, C. (2012). Why we need sustainable entrepreneurship, lecture and discussion at the Cigar Club, Vienna, 12 March 2012.

Wirtschaftskammer Österreich/AWO. (2012). Österreichs Außenhandel 1.-3.Quartal 2012, AUSSENWIRTSCHAFT-Marktanalysen, published 10 Dec 2012.

About the Authors

Gunilla Almgren
Gunilla Almgren was appointed as president of UEAPME (European Association of Craft, Small and Medium-Sized Enterprises) on 28 November 2011, after having been vice-president of UEAPME and chairperson of UEAPME's Committee for Sustainable Development. She is an entrepreneur in the sanitary fittings industry, with business contacts all over Europe. She served as first vice-president of the board of Företagarna, the Swedish Federation of Private Enterprises. She also sits on the board of several Swedish business-related organisations as well as private companies. On top of her professional engagements, Ms. Almgren has also actively participated in several committees and working groups set up by the European Commission in the past years on issues like better regulation. Ms. Almgren holds a diploma in business administration from IHM Business School and speaks five official EU languages (English, French, German, Italian and Spanish) in addition to her mother tongue.

José Manuel Barroso
In June 2009 the European Council unanimously nominated him for a second term as President of the European Commission, and he was re-elected to the post by an absolute majority in the European Parliament in September 2009. His political career began in 1980 when he joined the Social Democratic Party (PSD). He was named President of the party in 1999 and re-elected three times. During the same period, he served as Vice President of the European People's Party. As State Secretary for Foreign Affairs and Cooperation he played a key role as mediator in the signing of the peace accords for Angola in Bicesse in 1991, and as Minister for Foreign Affairs he was a driving force in the self-determination process in East Timor between 1992 and 1995. Under his leadership, the PSD won the general election in 2002 and he was appointed Prime Minister of Portugal in April of that year. He remained in office until July 2004 when he was nominated by the European Council and elected by the European Parliament to the post of President of the European Commission. He was named Global Leader for Tomorrow by the World Economic Forum in 1993 and declared "European of the year 2006" by the

newspaper European Voice. He received the Gold Medal of the town of Lamego, Portugal in 2007 and the Honorary Keys to the City of Lisbon in May 2008. He is the author of numerous publications on political science, international relations and the European Union, including, "Le système politique portugais face à l'intégration européenne" (Lisbon and Lausanne 1983), "Uma Certa Ideia de Europa" (1999), "Mudar de Modelo" (2002) and "Reformar: Dois Anos de Governo" (2004). He was awarded the Grand Cross of the Military Order of Christ in 1996 and the Great Collar of the Order of East Timor in 2010. In 2011, he received the Grand Cross of the Royal and Distinguished Spanish Order of Charles III.

Markus J. Beyrer

Markus J. Beyrer has been Director General of BUSINESSEUROPE since the end of 2012. Prior to this he held the positions of CEO of the Austrian Industry Holding ÖIAG, Director General of the Federation of Austrian Industries (IV) and Director for Economic Affairs of the Austrian Federal Economic Chamber, respectively. Before this Mr. Beyrer inter alia served as Chief Economic Advisor to the Federal Chancellor of the Republic of Austria, Dr. Wolfgang Schüssel. In the 1990s he worked in the context of Austria's accession to the European Union and the integration of Austria into the European Single Market (at the Austrian Permanent Representation to the European Union and the European Commission). In addition to his executive positions, Mr. Beyrer has held a number of non-executive board functions in various Austrian industry companies and still is a member of the non-executive board of the Austrian Central Bank. Mr. Beyrer has done studies in law and commercial sciences in Vienna graduating in law at the University of Vienna. Later he followed postgraduate studies in European Law at the Danube University in Krems (Austria) and the Stanford Executive Program at the Graduate School of Business at Stanford University.

Stefan Crets

Stefan Crets is Executive Director of CSR Europe, the European platform for companies and stakeholders to exchange and cooperate to become European leaders in sustainable competitiveness and societal wellbeing. From 2002 onwards, Stefan worked as the CSR leader at Toyota Europe where he developed and implemented a new strategy which informed Toyota's worldwide approach. In 2008, Stefan was appointed General Manager for Corporate Planning and CSR, Toyota Europe. Prior to his experiences at Toyota, he was Programme Advisor at the King Baudouin Foundation, and started his professional career as a research academic at the University of Antwerp. Stefan's deep knowledge of corporate sustainability challenges brings CSR Europe firsthand experience of developing CSR strategy and practice in a company.

Mara Del Baldo

Visiting professor at the University of Vigo (Spain), Department of Finances and Accounting in 2011, Mara Del Baldo is member of different Italian and international scientific associations (European Council for Small Business – ECSB –, Centre for Social and Environmental Accounting Research – CSEAR – University of St. Andrews, and European Business Ethics Network – EBEN – Italia). She is editorial board member of different journals (Piccola Impresa/Small Business; IJSSS (International Journal of Society System Sciences), JBAR (Journal of Business Administration Research), IJBM (International Journal of Business and Management), and JMAA (Journal of Modern Accounting and Auditing)), as well as reviewer for different international journals. Her main research interests include, among others: Corporate Social Responsibility and small entrepreneurs/SMEs' business ethics; entrepreneurial values as drivers for CSR and sustainability; territorial responsibility; SMEs strategies of qualitative development and networking strategies; ethical, social and environmental accounting and accountability (SEAR). She published in Italian and foreign journals (Sinergie, Studies on the Value of Cultural Heritage, JMG, JMAA, International Journal of Sustainable Society – IJSSoc –, Journal of Applied Behavioral Economics, International Journal of Social Ecology and Sustainable Development – IJSESD –) as well as in national and international conferences proceedings and books.

Manuel Escudero
Dr. Manuel Escudero is Director-General of Deusto Business School, University of Deusto, since February 2010. He is founding father of the Principles for Responsible Management Education (PRME), a United Nations-supported initiative, with more than 400 business schools participating around the world. He has been Head of the Secretariat of (PRME) since its inception in 2007 until June 2010 and now remains as Special Adviser of the initiative. Dr. Escudero worked as the Executive Director of the Research Center for the Global Compact at the Levin Institute in the State University of New York (SUNY). Prior to accepting his position as Senior Staff member and Head of Networks at the United Nation Global Compact in 2006, Dr. Escudero was Professor of Macroeconomics at IE Business School in Madrid, Spain. During his career at the IE Business School, he was Research Dean and Faculty Dean, Associate General Director of the IE Business School and founder and Associate Director of the IE Executive College. Dr. Escudero's public sector experience in Spain includes his role as Director of the Ministerial Group of Thought Leaders on Corporate Social Responsibility (CSR), Secretary of the CSR Experts Forum for Corporate Social Responsibility under the Chairmanship of the Minister of Labor, Secretary General – Spanish Network of the UN Global Compact, Senior Advisor for Policy and Programs of the nominated Candidate for Prime Minister of the Spanish Government. Dr. Escudero served in elected office as Member of the Spanish Parliament for the 2000/2004 term. He holds a B.Sc. from the University of Deusto (Spain), and a M.Sc. and Ph.D. from the London School of Economics and Political Sciences.

Matthias S. Fifka
Since 2011, Matthias Fifka holds an Endowed Chair for International Business Ethics and Sustainability at Cologne Business School. From 2001 to 2010, he was teaching International Business and Politics at the School of Economic and Social Sciences of the University Erlangen-Nuernberg. His research and teaching focuses on Sustainability, Corporate Social Responsibility/Corporate Citizenship, Corporate

Governance, and Business Ethics. Since 2007, he has also been working as a visiting professor at the Monte Ahuja College of Business at Cleveland State University. In 2011, he was Member of the German Parliamentary Commission on the Civic Engagement of the Mittelstand. Matthias Fifka has authored and edited 10 books and volumes, and published more than 30 articles in renowned journals and books (e.g., Business Strategy and the Environment, Journal für Betriebswirtschaft, Business Ethics: A European Review). He serves as member of the editorial board and reviewer for several journals, and frequently contributes articles and interviews to a variety of international media. Matthias Fifka speaks German, English, Spanish, Portuguese, and French.

Felix Finkbeiner

Felix Finkbeiner (14) was inspired by Wangari Maathai when, while in the fourth grade, he developed the idea that children could plant a million trees in every country. Children from 100 countries answered the call and founded the child and youth initiative Plant-for-the-Planet. Children empower other children at academies (1-day-workshops) to spread the word of sustainability. Today's 14,000 Climate Justice Ambassadors want to be a million by 2020 and to increase the number of planted trees to 1,000 billion – which means 150 per person on the planet. From a global 2-year consultation they have developed a three-point-plan to save their future. Condensed into four words, the message is: "Stop talking. Start planting". With this effective worldwide campaign, the three-point-plan and the transparency of the promised and the planted trees, the young world citizens are asking heads of government, company bosses and fellow citizens to support their fight for their future. In December 2011, UNEP handed over the responsibility for the Billion Tree Campaign to the children of Plant-for-the-Planet. Now the children are responsible for the official global tree counter, which has already recorded 12.6 billion planted trees.

Franz Fischler

Dr. Franz Fischler is the chairman of the RISE-Foundation, Brussels, and a consultant for several governments and the OECD. He is currently a visiting lecturer at three universities. He holds eight honorary doctorates, the high ranking Austrian medal for special merits, and various awards. He was a member of the EU Commission responsible for agriculture and rural development. He started his career as a university assistant at the Department for Regional Agricultural Planning at the Institute for Farm management in Vienna. Afterwards he was appointed director of the Chamber of Agriculture and became Member at the ALPI dairy factory and Raiffeisen bank, Tyrol. Before he was elected Member of the Austrian National Parliament (Nationalrat), he was involved in the preparations of the Austrian agriculture and forestry sector to join the EU as Federal Minister of Agriculture and Forestry.

Ursula Fischler-Strasak

In March 2013 Ursula Fischler-Strasak joined the Basel-based pharmaceutical company F. Hoffmann-La Roche. Prior to joining Roche, Ursula was leveraging her skills as well as her entrepreneurial spirit with a fellowship at the National Centre for Engineering Pathways to Innovation (Epicenter) at Stanford University, USA. Further, she was a Corporate Sustainability Specialist responsible for external sustainability communication at the agro-science company, Syngenta. She holds a doctoral degree in Strategic Sustainability Management and a Master for Applied Sciences in International Business Management with Finance as major. Her experience and studies combine sustainability strategy, performance management and communication.

Michael Fürst

Michael has more than 13 years of experience in integrity & compliance and corporate responsibility in an academic as well as in a business environment. Since 2005, he is working in different roles with increasing responsibilities for Novartis, one of the leading healthcare companies globally. Michael had responsibilities for the development and management of a behavioral-based integrity management programme, one of the first of its kind. In addition to this, he is responsible for a variety of different corporate responsibility projects from a strategic and operational perspective. Over the last years his work has focused on social business initiatives that are aligned with the strategic priorities of Novartis and with the needs of underserved patient communities. Michael frequently publishes about integrity management and corporate responsibility and speaks regularly at leading universities and business schools such as INSEAD, University of Cambridge, IMD, CBS, etc. He was awarded with the German Max Weber Price for Business Ethics for his scientific work in 2006.

Jane Goodall

In July 1960, Jane Goodall began her landmark study of chimpanzee behavior in what is now Tanzania. Her work at Gombe Stream would become the foundation of future primatological research and redefine the relationship between humans and animals. In 1977, Dr. Goodall established the Jane Goodall Institute, which continues the Gombe research and is a global leader in the effort to protect chimpanzees and their habitats. The Institute is widely recognized for innovative, community-centered conservation and development programs in Africa, and Jane Goodall's Roots & Shoots, the global environmental and humanitarian youth program. Dr. Goodall founded Roots & Shoots with a group of Tanzanian students in 1991. Today, Roots & Shoots connects hundreds of thousands of youth in more than 120 countries who take action to make the world a better place for people, animals and the environment. Dr. Goodall travels an average 300 days per year, speaking about the threats facing chimpanzees, other environmental crises, and her reasons for hope that humankind will solve the problems it has imposed on the earth. Dr. Goodall's honors include the French Legion of Honor, the Medal of Tanzania, and Japan's prestigious Kyoto Prize. In 2002, Dr. Goodall was appointed to serve as a United Nations Messenger of Peace and in 2003, she was named a Dame of the British Empire.

Bradley K. Googins

Prof. Bradley K. Googins, a retired Professor in Organizational Studies at the Boston College's Carroll School of Management. Previously he was the Executive Director of the Boston College Center for Corporate Citizenship from 1997 to 2009. He is also the founder of the Global Education and Research Network, a group of 12 of the leading CSR institutions across the globe from Latin America, Asia, and Europe. He is currently a visiting research fellow at the Catholic University in Milan Italy. He is also serving as a senior research fellow at Deusto University in Bilbao Spain assisting the development of a new Global Center on Sustainable Business.

Estelle L.A. Herlyn

Dr. Estelle L.A. Herlyn is a freelancer at FAW/n and lecturer at FOM (University of Applied Science for Economy and Management). The primary focus of her projects is sustainability, both in research and business environments. Her research interests include sustainability management as well as supply chain management. She has many years of experience in the management and consulting of miscellaneous companies and is a board member of the NGO Ecosocial Forum Germany. Estelle Herlyn holds a sustainability related Ph.D. from RWTH Aachen University and a degree in economics and mathematics from TU Dortmund University. She is a Project Management Professional (PMP).

Samuel O. Idowu

Samuel O. Idowu is the Editor-in-Chief of the Encyclopaedia of Corporate Social Responsibility and the Dictionary of Corporate Social Responsibility. He is a Senior Lecturer in Accounting at London Metropolitan University, UK. He researches in the fields of Corporate Social Responsibility (CSR), Corporate Governance and Accounting. He is on the Editorial Boards of International Business Management Journal and Management of Environmental Quality: An International Journal and he has led several edited books in CSR.

Aileen Ionescu-Somers

Dr. Aileen Ionescu-Somers directs the CSL Learning Platform at the Global Center for Sustainability Leadership (CSL) at IMD business school. She has led large scale research projects on the business case for sustainability, the stakeholder environment and sustainability partnerships, publishing many resulting articles, prize-winning case studies and books. She is on the sustainability advisory boards of several organizations. Dr. Ionescu-Somers previously worked for 12 years with the international conservation NGO WWF, first in program management for the Latin America/Caribbean and Africa/Madagascar regions and then as head of WWF's international project operations, overseeing project financing of projects and program offices globally. She holds a B.A. (First Class Hons), M.A., H. Dip. Ed. and Ph.D. Commerce from NUI-National University of Ireland. She also holds a M.Sc. in Environmental Management from Imperial College London, UK.

Claudia Kemfert

Prof. Dr. Claudia Kemfert is Professor of Energy Economics and Sustainability at the private University, Hertie School of Governance, in Berlin since 2009 and Head of the department Energy, Transportation, Environment at the German Institute of Economic Research (DIW Berlin) since April 2004. Her research activities concentrate on the evaluation of climate and energy policy strategies. Claudia Kemfert advised EU president José Manuel Barroso in a "High level Group on Energy and Climate". She was awarded in 2006 as top German Scientist from the German research foundation, Helmholtz and Leibniz Association. In 2011 she was awarded with the Urania Medaille as well as B.A.U.M. environmental award for best science. She has recently published two highly recognized books to illustrate the economic impacts of climate change and energy policy.

Thomas Osburg

Dr. Thomas Osburg is *Director Europe Corporate Affairs* at Intel Corp. and Board Member for *CSR Europe* and *EABIS*. He is on the Jury for the Sustainable Entrepreneurship Award and holds a Ph.D. (Dr.rer.pol.) degree in Economics and Business Administration. After his graduation, he held several Management positions in the area of International Management and Marketing, CSR, Education and Research at Texas Instruments, Autodesk and Intel, living in France, the U.S. and Germany. Thomas is frequently lecturing on Management, Marketing and CSR/Social Innovation at leading universities in Europe. He is teaching a M.B.A. Module *Technology & Innovation Management* and a module *CSR and Strategic Management* at the University of Geneva.

Marc R. Pacheco

Marc R. Pacheco is a U.S. State Senator representing the First Plymouth and Bristol district in the Commonwealth of Massachusetts. He is the Chair of the Senate Committee on Global Warming and Climate Change, the Senate Chair of the Joint Committee on Environment, Natural Resources, and Agriculture, Vice-Chair of the Joint Committee on Public Health, a Member of the Joint Committee on Telecommunications, Utilities, and Energy, and a Member of the powerful Senate Committee on Ways and Means. As a senior member of the Massachusetts State Senate, Senator Pacheco has over three decades of leadership at the local, state, national, and international level and has made a lifelong commitment to public service in and helping to establish public-private partnerships.

Franz Josef Radermacher

Prof. Dr. Dr. Franz Josef Radermacher is the Director of FAW/n (Research Institute for Applied Knowledge Processing/n), Ulm, and holds a faculty position for Data Bases/Artificial Intelligence at the University of Ulm. Member of the Club of Rome and of several national and international advisory boards as well as President of the Senat der Wirtschaft e.V., Bonn, President of the Global Economic Network (GEN), Vienna, and Vice President of the Ecosocial Forum Europe, Vienna. 1997 Scientific Award of the German Society for Mathematics, Economics and Operations Research. 2005 Laureate of the Salzburg Award for Future Research, Salzburg, Austria. 2007

Laureate of "Vision Award 2007" of Global Economic Network (for Global Marshall Plan Initiative). 2007 Laureate of Karl-Werner-Kieffer Award (Stiftung Ökologie und Landbau, SÖL). Member of the Rotarian Action Group for Population&Development (RFPD) – German Section – e.V. Member of the German National Committee of the UNESCO for the World Decade "Education for Sustainable Development" (2005–2014). 2012 Laureate of the Umweltpreis "Goldener Baum" der Stiftung für Ökologie und Demokratie e.V.

Katherina Reiche
Katherina Reiche has been Parliamentary State Secretary at the Federal Ministry for the Environment, Nature Conservation and Nuclear Safety since 2009. From 2005 to 2009 she had the deputy chair of the CDU/CSU parliamentary group, responsible for education and research policy and environment, nature conservation and nuclear safety. From 2002 to 2005 she had the chair of the spokeswoman of the working group 'Education and Research' of the CDU/CSU parliamentary group in the German Bundestag. She has been a member of the CDU (Christian Democratic Party) since 1996 and a member of the German Bundestag since 1998. In 1997 she was a visiting researcher at the University of Turku, Finland. Between 1997 and 1998 she worked as a scientific staff member at the University of Potsdam. After graduation at high school in Luckenwalde she studied chemistry at the University of Potsdam (1992–1997) and at Clarkson University, NY, USA (1995/1996).

Martin Riester
Martin Riester, M.B.E., is a researcher at the Division of Production and Logistics Management at Fraunhofer Austria and at the Institute of Management Science, Industrial Engineering of the Vienna University of Technology. He is primarily specialized in logistics and production optimization programs focusing on efficiency enhancement and cost-savings. Within this context he is also experienced in executing sustainable continuous improvement projects. In his work he gained widespread experience from international research- and industry projects within different branches. Predominantly in the automotive and electronic industry.

Robert Rosenfeld
Robert "Bob" Rosenfeld is the Founder and CEO of Idea Connection Systems, Inc. For over 40 years, he has been a leader and practitioner in the human dynamics that make innovation happen inside organizations. He created the first Office of Innovation ever to be successfully implemented in Corporate America in 1978 at Eastman Kodak. In 1981, Bob co-founded the Association for Managers of Innovation (AMI). After working with many diverse people and organizations, in 2001, he directed the development of Mosaic Partnerships – a process for breaking down barriers between cultures. Because of Bob's innovation experience, in May 2006, he was named the Center for Creative Leadership's (CCL) first "Innovator in Residence" and in 2008, he was awarded Innovator in Residence Emeritus status. Bob is the author of, Making the Invisible Visible: The Human Principles for Sustaining Innovation (2006). Bob's second book, co-authored by Gary Wilhelmi and written by Andrew Harrison, is titled, The Invisible Element: A Practical Guide

for the Human Dynamics of Innovation. In 2008, Bob spearheaded an ICS team in the creation of the ISPI™ (Innovation Strengths Preference Indicator®), a tool used to highlight peoples' preferences for innovating as well as how they prefer to innovate with others. Bob's efforts in the human dynamics of innovation continue to impact organizations around the globe.

Walter Rothensteiner
Dr. Walter Rothensteiner is Chairman of the Managing Board (CEO) of Raiffeisen Zentralbank Österreich AG and Chairman of the Austrian Raiffeisen Association. Born in 1953; read Commercial Science at the Vienna University of Economics and Business Administration; senior positions at Raiffeisenlandesbank Niederös-terreich-Wien (most recently in management), member of the Managing Board of Leipnik-Lundenburger Industrie AG and sugar industry group Agrana. Joined Raiffeisen Zentralbank in 1995 as Vice-Chairman of the Managing Board; Chairman of the Managing Board and CEO since June 1995. Chairman of the Austrian Raiffeisen Association since June 2012.

Clemens Sedmak
Clemens Sedmak holds the F.D. Maurice Chair in Social Ethics at King's College London and is Director of the Center for Ethics and Poverty Research at the University of Salzburg, Austria. His main areas of research are: Leadership Ethics, Poverty Alleviation, and Theories of Justice.

Wilfried Sihn
Prof. Dr. Wilfried Sihn has been active in the field of applied research and consulting services for more than 25 years, and has been involved in more than 300 industrial projects. He is the vice-president of the International Society of Agile Manufacturing and the international editor of the journal Agility and Global Competition, as well as a guest editor of the International Journal of Technology Management (IJTM). He has written a number of books and more than 200 publications, in which he discusses scientific and practice-related issues. He is the director of the Austrian Science Forum for Logistics (WissLog) and a board member and vice-president of the Austrian Logistics Association (BVL). In February 2006 Prof. Dr. Sihn was invited to join the Paris-based International Academy for Production Engineering (CIRP). In November 2008 he was appointed managing director of the newly founded Fraunhofer Austria Research GmbH and is responsible for managing the project group for production and logistics management in Vienna. In March 2009 he became the head of the new transnational M.B.A. automotive industry programme of the Vienna University of Technology in cooperation with the Slovak University of Technology in Bratislava. He became a member of the board of the German Chamber of Commerce (DHK) in Vienna in October 2010 and a member of the American Chamber of Commerce in Austria (AMCHAM) in March 2011. He was appointed vice-president of the 'Europäische Lernfabrik-Initiative' in May 2011.

René Schmidpeter
Dr. René Schmidpeter is the academic head of the Centre for Humane Market Economy ("Zentrum für humane Marktwirtschaft") in Salzburg, member of the Club of Rome (Austrian Chapter) and member of the Jury for the Sustainable Entrepreneurship Award (SEA). He teaches CSR and Sustainability at several business schools and universities in Europe. He studied business administration, applied European studies, social ethics and social politics in Germany, Great Britain and the USA. For more than 10 years he has worked and done research in the field of corporate social responsibility and sustainability. He is Section Editor of the CSR Encyclopaedia and Editor of several publications on Corporate Social Responsibility with Springer: e.g. "Social Innovation" (2013), "Corporate Social Responsibility – Verantwortungsvolle Unternehmensführung in Theorie und Praxis" (2012); "Handbuch Corporate Citizenship" (2008) and "CSR across Europe" (2005).

Martin Schulz
Martin Schulz was born on 20 December 1955 and grew up in Hehlrath Germany, close to the German-Dutch-Belgian borders. After high school he decided to try to make a living out of his passion for books and he did an apprenticeship as a bookseller. In 1982 he opened his own bookstore in Würselen, which he successfully ran for 12 years. Joining the Social Democratic Party of Germany at the age of 19, he started out his political career. Aged 31, he was elected as the youngest mayor of North Rhine-Westphalia, a post he held for 11 years. Since 1994, Martin Schulz is a Member of the European Parliament and has served in a number of committees, first serving on the sub-committee on Human Rights and then on the Committee on Civil Liberties and Home Affairs. He led the SPD MEPs from 2000; 2004 he was elected group leader of the Socialists and Democrats. Martin Schulz campaigned for social justice, promoting jobs and growth, reforming financial markets, fighting climate change, championing equality and creating a stronger and more democratic Europe. Martin Schulz was elected President of the European Parliament on 17 January 2012 for a mandate of two and half years.

Antonio Tajani
European Commission Vice-President and Commissioner in charge of Industry and Entrepreneurship since February 2010. European Commission's Vice-President and Commissioner in charge of Transport from May 2008 to February 2010. Vice-Chair of the European People's Party, elected at the EPP Congress in Estoril in 2002, re-elected at the EPP Congress in Rome in 2006 and re-elected again at the EPP Congress in Bonn in 2009 and in Bucharest in 2012. Member of the Bureau of the Group of the European People's Party. He has taken part in all EPP summits in preparation of the European Councils. Member of the Convention on the Future of Europe, which drew up the text of the European Constitution. Elected as member of the European Parliament in 1994, 1999 and 2004 with over 120 000 preference votes. During his 15 years of parliamentary activity he took part in many committees (Foreign Affairs, Constitutional Affairs, at the time chaired by the President of the Republic Giorgio Napolitano, Transport and Tourism, Fisheries, Security and Defence).

Jakob von Uexkull

Jakob von Uexkull is the founder of the World Future Council (2007) and the Right Livelihood Award (1980), often referred to as the 'Alternative Nobel Prize', as well as co-founder (1984) of The Other Economic Summit. As a past Member of the European Parliament (1987–1989) he served on the Political Affairs Committee and later on the UNESCO Commission on Human Duties and Responsibilities (1998–2000). Jakob von Uexkull has received the Binding-Prize (Liechtenstein) for the protection of nature and the environment (2006) and the Order of Merit First Class of the Federal Republic of Germany (2009). In 2005, he was honoured by Time Magazine as a European Hero and in 2008 he received the Erich-Fromm-Prize in Stuttgart, Germany. Born in Uppsala, Sweden, Jakob von Uexkull is the son of the author and journalist Gösta von Uexkull. He graduated with an M.A. (Honours) in Politics, Philosophy and Economics from Christ Church, Oxford.

Daniel Velásquez Norrman

Daniel Velásquez Norrman, DDip.-Ing. M.Sc., is a researcher at the Division of Production and Logistics Management at Fraunhofer Austria and at the Institute of Management Science, Industrial Engineering of the Vienna University of Technology. He is primarily specialized in efficiency enhancement and cost-savings programs within the domains of Order Processing and Lean Management as well as Business Process Reengineering. In his work he has widespread experience from international research- and industry projects in amongst others the electronic, automotive and the machinery and plant engineering industry and the fields Supply Chain Management and Process Management.

Peter Vogel

Peter Vogel is an internationally renowned serial entrepreneur, entrepreneurship trainer and scholar. Peter is founding CEO of HR Matching AG, a VC-backed Red Herring Europe winner revolutionizing the recruiting market through technological innovation. He is also founder of The Entrepreneurs' Ship, an organization fostering entrepreneurship as a viable career option in low- and middle-income countries. He also pursues a Ph.D. at the EPFL investigating governmental programs to help unemployed individuals transition to self-employment. As the youngest partner of the Future work Forum, Peter consults on international labor-market issues. Peter is think-tank member of the World Entrepreneurship Forum, member of the WEF Global Shapers. He spoke at TEDxLausanne on Youth Unemployment and Entrepreneurship.

Christina Weidinger

In September 2009 Christina Weidinger founded diabla media Verlag, a Vienna-based publishing house where she is a managing partner. Diabla media Verlag is an international business publisher with a portfolio that includes business media for specific target groups, e.g. the European business magazine SUCCEED, and Unternehmer, a business magazine for Austria's self-employed elite. Further, diabla media Verlag specialises in corporate publishing and produces high-quality magazines such as Skylines, Vienna City Guide, Qatar and special issues of SUCCEED. In December 2011 Weidinger founded the Sustainable Entrepreneurship Award (SEA). The SEA aims to achieve a synergy of politics, industry and society on the topic of sustainability. It showcases companies who are making important contributions to sustainable entrepreneurship. These are companies that put three kinds of responsibility – namely, social, ecological and economic responsibility – into practice.

Ernst Ulrich von Weizsäcker

Ernst Ulrich von Weizsäcker, Ph.D. born in 1939 is Co-Chair, International Resource Panel and Co-President of the Club of Rome. Earlier stations: Professor of Biology, University President, Director UN Centre for Science and Technology for Development, New York, Director, Institute for European Environmental Policy, President, Wuppertal Institute for Climate, Environment, Energy. 1998–2005 MP, Germany, Chair of the Bundestag Environment Committee. Dean, Bren School for Environmental Science and Management, UC Santa Barbara, California. Publications: 1994 Earth Politics. 1997 Factor Four (w/A & H Lovins). 2010, Factor Five (w/K. Hargroves et al.).

Liangrong Zu

Dr. Zu is a senior specialist in corporate social responsibility, social entrepreneurship, business ethics and sustainability. He has been working for the United Nations and national government for over 20 years. Currently he is a Senior Program Officer at International Training Centre of the International Labor Organization (ILO). Dr. Zu holds a master degree in economics of education in Beijing Normal University, and Ph.D. in business and management at Nottingham University Business School, specializing in corporate social responsibility, business ethics, corporate governance and restructuring. He is the author of the book: "Corporate Social Responsibility, Corporate Restructuring and Firm's Performance: Empirical Evidence from China", co-editor of the Encyclopaedia of Corporate Social Responsibility, and co-editor of Dictionary of Corporate Social Responsibility. He has also authored many articles published in Journal of Business Ethics and other national and international journals.

Printed by Printforce, the Netherlands